Design of Reinforced Concrete

Design of Reinforced Concrete

Editor

Marek Lagunov

Design of Reinforced Concrete
Edited by **Marek Lagunov**

Printed in 2017

ISBN: 978-1-68117-115-9
Library of Congress Control Number: 2015951141

© 2016 by
SCITUS Academics LLC,
616, Corporate Way, Suite 2, 4766,
Valley Cottage, NY 10989

www.scitusacademics.com

This book contains information obtained from highly regarded resources. Copyright for individual articles remains with the authors as indicated. All chapters are distributed under the terms of the Creative Commons Attribution License, which permits unrestricted use, distribution, and reproduction in any medium, provided the original author and source are credited.

Notice

Reasonable efforts have been made to publish reliable data and views articulated in the chapters are those of the individual contributors, and not necessarily those of the editors or publishers. Editors or publishers are not responsible for the accuracy of the information in the published chapters or consequences of their use. The publisher believes no responsibility for any damage or grievance to the persons or property arising out of the use of any materials, instructions, methods or thoughts in the book. The editors and the publisher have attempted to trace the copyright holders of all material reproduced in this publication and apologize to copyright holders if permission has not been obtained. If any copyright holder has not been acknowledged, please write to us so we may rectify.

Preface

Concrete is one of the most popular materials for buildings because it has high compressive strength, flexibility in its form and it is widely available. The history of concrete usage dates back for over a thousand years. Contemporary cement concrete has been used since the early nineteenth century with the development of Portland cement. Despite the high compressive strength, concrete has limited tensile strength, only about ten percent of its compressive strength and zero strength after cracks develop. In the late nineteenth century, reinforcing materials, such as iron or steel rods, began to be used to increase the tensile strength of concrete. Today steel bars are used as common reinforcing material.Concrete is a mixture of coarse and fine aggregates with a paste of binder material and water. Reinforced concrete is a composite material in which concrete's relatively low tensile strength and ductility are counteracted by the inclusion of reinforcement having higher tensile strength and ductility. The reinforcement is usually steel reinforcing bars and is usually embedded passively in the concrete before the concrete sets. Reinforcing schemes are generally designed to resist tensile stresses in particular regions of the concrete that might cause unacceptable cracking and structural failure.

Modern reinforced concrete can contain varied reinforcing materials made of steel, polymers or alternate composite material in conjunction with rebar or not. Reinforced concrete may also be permanently stressed (in compression), so as to improve the behaviour of the final structure under working loads. In the United States, the most common methods of doing this are known as pre-tensioning and post-tensioning.Without reinforcement, constructing modern structures with concrete material would not be possible.

The aim of this book is to provide reinforced concrete design tools to help architecture students, researchers or working professionals to understand the design process.

Table of Contents

CHAPTER 1   A Unified Approach for Concrete Impact 1

ABSTRACT 1
INTRODUCTION 2
AN OVERVIEW AND ASSESSMENT OF THE UMIST FORMULAE 2
    An Overview of the UMIST Formulae 4
    An Assessment of the UMIST Formulae 6
FORMULATION OF A UNIFIED APPROACH 13
    Penetration 13
    Cone Cracking, Cabbing and Perforation 14
    Comparison with Available Experimental Data and Discussion 15
    The Modified UMIST Formulae 29
CONCLUSIONS 30
ACKNOWLEDGEMENT 31
APPENDIX 31
REFERENCES 32
GLOSSARY 34
CITATION 35

CHAPTER 2   Microstructure and Properties of Concrete Using Bottom Ash and Waste Foundry Sand as Partial Replacement of Fine Aggregates 37

ABSTRACT 37
INTRODUCTION 38
EXPERIMENTAL PROGRAM 40
    Materials and Mix Proportions 40
    Testing Procedure 42

RESULTS AND DISCUSSIONS 44
  Fresh Concrete Properties 44
  Compressive Strength 45
  Splitting Tensile Strength 47
  Flexural Strength 48
  Relationship of Flexural Strength to Compressive and Splitting Tensile Strength 49
DURABILITY PROPERTIES 53
  Resistance to Rapid Chloride Penetration 53
  Relation Between Compressive Strength and Resistance to Chloride Ion Penetration 55
  Deicing Salt Surface Scaling 56
  X-ray Diffraction (XRD) 56
  Scanning Electron Microscope (SEM) Analysis 60
CONCLUSIONS 65
REFERENCES 66
CITATION 70

CHAPTER 3   A Unified Method for Calculating Fire Resistance of Solid and Hollow Concrete-Filled Steel Tube Columns Based on Average Temperature 71

ABSTRACT 71
INTRODUCTION 72
  General Plastic Limit Analysis for CFST Columns under Fire 74
  Equivalent Reduction Factor of Strength and Elastic Modulus under Fire 74
FORMULA FOR CALCULATING AVERAGE TEMPERATURE OF STEEL AND CONCRETE 82
  Average Temperature of Steel Tube $\overline{T}_s$ 82
  Average Temperature of Concrete Core 82
UNIFIED FORMULATION OF FIRE RESISTANCE FOR CFST COLUMNS 83
  Calculation Model Based On Eurocode 4 (Method 1) 83
  Unified Formulation for CFST Columns under Normal and High Temperature (Method 2) 84
  Validation of the Fire Resistance Formulas for CFST Columns Subjected to Axial Compression 86

CALCULATION PROCEDURE AND DISCUSSION 101
   Restrictions on the Application of the Formulas 102
CONCLUDING REMARKS 102
ACKNOWLEDGEMENT 103
REFERENCES 103
CITATION 106

CHAPTER 4   Development of UHPC Mixtures from an Ecological Point of View 107

ABSTRACT 107
RESEARCH SIGNIFICANCE: SUSTAINABILITY IN CONCRETE CONSTRUCTION 108
   Substitution of cement in UHPC mixtures by SCM 109
   Degree of Substitution 109
   Mixture Proportions 110
   Characterization of Supplementary Cementitious Materials Used 111
   Material Properties of UHPC with Supplementary Cementitious Materials 113
      Fresh Concrete Properties of UHPC Mix Design with Reduced Cement Content 113
      HARDENED CONCRETE PROPERTIES 114
   COMPARISON OF THE ECOLOGICAL PROPERTIES OF DIFFERENT UHPC MIXTURES 115
   COMPARISON OF UHPC WITH NSC 115
   COMPARISON OF BUILDING MEMBERS MADE OF UHPC WITH NSC 118
CONCLUSIONS 120
ACKNOWLEDGEMENTS 121
REFERENCES 121
CITATION 123

CHAPTER 5   Numerical Study of FRP Reinforced Concrete Slabs at Elevated Temperature 125

ABSTRACT 125
INTRODUCTION 125
MATERIAL BEHAVIOUR AT HIGH TEMPERATURES 127

HEAT CONDUCTION SIMULATION IN REINFORCED CONCRETE
MEMBERS 128
LOAD CAPACITY MODEL 129
SLABS WITH ONE LAYER OF FRP 133
STRENGTH-DOMAIN AND TEMPERATURE-DOMAIN FAILURE 137
SLABS WITH TWO LAYERS OF FRP 138
CONCLUSIONS 140
ACKNOWLEDGEMENTS 141
REFERENCES 141
CITATION 143

CHAPTER 6 Achievements of Truss Models for Reinforced Concrete Structures 145

KEYWORDS 145
ABSTRACT 145
INTRODUCTION 146
TRUSS CONSTITUTIVE MODELS 148
TRUSS FINITE ELEMENT FOR PLANE RC FRAME 149
DETERMINATION OF BAR SECTIONS 149
NONLINEAR STATIC ANALYSIS 152
NONLINEAR DYNAMIC ANALYSIS 152
APPLICATIONS TO ANALYSIS OF SIMPLE PLANE RC FRAMES 153
APPLICATION TO CONFINEMENT OF A RC COLUMN 154
CONCLUSIONS 156
REFERENCES 157
CITATION 160

CHAPTER 7 Exploring Mechanical and Durability Properties of Ultra-High Performance Concrete Incorporating Various Steel Fiber Lengths and Dosages 161

ABSTRACT 161
INTRODUCTION 162
RESEARCH SIGNIFICANCE 163
EXPERIMENTAL PROGRAM 164
    MATERIALS COMPOSITION AND PROPORTIONS.........................164

MIXING OF UHPC CONSTITUENTS	165
SPECIMEN PREPARATION AND ENVIRONMENTAL CONDITIONS	166
MECHANICAL TESTING	166
DURABILITY TESTING	169
MICRO-STRUCTURAL ANALYSIS	170
RESULTS AND DISCUSSION	171
FRESH UHPC PROPERTIES	171
COMPRESSIVE STRENGTH AND MODULUS OF ELASTICITY	172
SPLITTING TENSILE STRENGTH	173
FLEXURAL STRENGTH	174
PERMEABLE VOIDS	181
SORPTIVITY VOIDS	182
ELECTRICAL RESISTANCE	183
CHLORIDE IONS PENETRATION	185
EFFECTS OF CHLORIDE IONS EXPOSURE ON MECHANICAL PROPERTIES	187
CONCLUSIONS	188
REFERENCES	189
CITATION	194

CHAPTER 8 Seismic Performance Evaluation of Corroded Reinforced Concrete Structures by Using Default and User-Defined Plastic Hinge Properties 195

INTRODUCTION	195
NONLINEAR MATERIAL MODELLING	196
DESCRIPTION OF STRUCTURES	198
MOMENT-CURVATURE RELATIONSHIPS	205
NONLINEAR STATIC ANALYSIS	207
SEISMIC PERFORMANCE ANALYSES	213
CONCLUSION	218
REFERENCE	218
CITATION	220

CHAPTER 9	Strain Rate Dependent Properties of Ultra High Performance Fiber Reinforced Concrete (Uhp-Frc) Under Tension 221

ABSTRACT 221
INTRODUCTION 222
STRAIN RATE EFFECT ON FRC UNDER TENSION 223
EXPERIMENTAL PROGRAM 225
    MATERIALS AND FABRICATION 227
    TEST SETUP AND PROCEDURE 227
    TEST RESULTS 228
    SEPARATE EFFECTS OF FIBER VOLUME FRACTION, ASPECT RATIO, LENGTH, AND TYPE 239
    EQUIVALENT BOND 240
    ENERGY ABSORPTION CAPACITY 242
DIF 244
    GENERAL TRENDS IN THE TEST DATA 246
CONCLUSIONS 247
ACKNOWLEDGMENTS 248
REFERENCES 248
CITATION 251

CHAPTER 10	Development of an Eco-Friendly Ultra-High Performance Concrete (Uhpc) with Efficient Cement and Mineral Admixtures Uses 253

ABSTRACT 253
INTRODUCTION 254
MATERIALS AND EXPERIMENTAL METHODOLOGY 256
    MATERIALS 256
EXPERIMENTAL METHODOLOGY 260
    MIX DESIGN OF UHPC 260
    DETERMINATION OF WATER DEMAND 262
    MIXING PROCEDURE 263
    FLOWABILITY OF UHPC 263
    MECHANICAL PROPERTIES OF UHPC 264
    WATER-PERMEABLE POROSITY OF UHPC 264

CALORIMETRY ANALYSIS OF UHPC 264
THERMAL TEST AND ANALYSIS OF UHPC..265
EXPERIMENTAL RESULTS AND DISCUSSION 265
 FRESH BEHAVIOR OF THE DESIGNED UHPC..265
 MECHANICAL PROPERTIES OF THE DESIGNED UHPC 266
 WATER-PERMEABLE POROSITY OF THE DESIGNED UHPC 268
 HYDRATION KINETICS OF THE DESIGNED UHPC.....................................272
 THERMAL ANALYSIS OF THE HARDENED UHPC 275
 ECOLOGICAL EVALUATION OF THE DESIGNED UHPC 278
CONCLUSIONS 279
ACKNOWLEDGEMENTS...280
REFERENCES 280
CITATION 287

CHAPTER 11 Concrete Mix Design for Service Life of RC Structures under Carbonation Using Genetic Algorithm............ 289

ABSTRACT 289
INTRODUCTION 290
BACKGROUND OF GA AND INFLUENCING PARAMETERS ON CARBONATION 291
 STUDY OF CARBONATION PARAMETERS AND PREDICTION TECHNIQUES.....293
CONCRETE MIX OPTIMIZATION USING GA 295
 EVALUATION OF GA APPLICABILITY TO GENERATING MIX PROPORTION........................303
DESIGN OF CONCRETE MIX PROPORTIONS FOR CARBONATION..308
 SCENARIO FOR MIX DESIGN CONSIDERING CARBONATION 308
CONCLUDING REMARK...312
ACKNOWLEDGMENTS 313
REFERENCES 313
CITATION 315

CHAPTER 12   Achievements of Truss Models for Reinforced Concrete Structures 317

ABSTRACT 317
INTRODUCTION 318
TRUSS CONSTITUTIVE MODELS 320
TRUSS FINITE ELEMENT FOR PLANE RC FRAME 321
DETERMINATION OF BAR SECTIONS 321
NONLINEAR STATIC ANALYSIS 324
NONLINEAR DYNAMIC ANALYSIS 325
APPLICATIONS TO ANALYSIS OF SIMPLE PLANE RC FRAMES 326
CONCLUSIONS 328
REFERENCES 329
CITATION 332

Index 333

CHAPTER 1

A Unified Approach for Concrete Impact

H.M. Wen and Y.X. Xian

CAS Key Laboratory for Mechanical Behavior and Design of Materials, University of Science and Technology of China, Hefei, Anhui Province 230027, PR China

ABSTRACT

A unified approach is presented herein for concrete impact which represents an extension and further development of the UMIST formulae. The paper consists of two parts, i.e. a critical overview and assessment of the accuracy of the UMIST formulae in the wake of the new test results and the data made available recently; derivation of a new set of empirical equations by introducing a previously suggested mean resistive pressure to predict the depth of penetration, through-thickness cone cracking, scabbing and perforation of reinforced concrete targets subjected to projectile impact. It is shown that the UMIST formulae can be used with reasonable confidence for large mass low velocity projectile impact onto reinforced concrete targets and that they considerably underestimate the scabbing of reinforced concrete targets struck by non-flat-ended missiles and significantly over predict the perforation of very thick concrete targets as compared to the experimental data. It is also shown that the present empirical formulae are in good agreement with available experimental data over a wider range of impact conditions, being applicable not only to flat-faced projectiles but also to non-flat-ended (i.e., conical-, spherical-, ogival-nosed) missiles; not only to low strength concrete but also to high strength concrete; not only to low impact velocity but also to high impact velocity and that the modified UMIST formulae are in reasonable agreement with available test data.

INTRODUCTION

Concrete is a common material which has been widely used to construct protective structures to resist impact by missiles and explosive loads. Hence, the problem of missile impact on reinforced concrete targets has long been of interest to military as well as civil engineers. In nuclear industry, potential missiles include crashing aircrafts, fragments generated by accidental explosions or resulting from pressured vessel failure, pipe break, turbine blade failure and high speed rotating equipment. These missiles may vary in shapes and sizes, impact velocities, deforming rigidities, hardiness and produce a wide damage spectrum in reinforced concrete targets [1] and [2]. Generally speaking, missiles can be categorized as either "soft" or "hard" depending on whether the missile deformability is large or small relative to the target deformability. "Soft" missile such as aircraft, which impact a large area, affect the whole reinforced concrete structure and are studied in connection with the overall survivability of the structure. "Hard" missile impact on concrete is a difficult problem to solve and major efforts have been invested in the development and validation of numerical modeling techniques but with very limited success. On the other hand, various empirical formulae have been developed to assess the local impact damages [2]. Though different failure modes have been identified with a hard/rigid flat-nosed projectile as it represents the most dangerous scenario in regard to nuclear power industry empirical equations for four damage modes such as penetration, through-thickness cone cracking, scabbing and perforation have been proposed [2],[3] and [4].

The objective of this paper is to formulate a unified approach for concrete impact which represents an extension and further development of the UMIST formulae. An overview and assessment of the UMIST formulae is first conducted in the wake of the new test data and the results made available recently and then various empirical equations are derived by introducing a previously proposed mean resistive pressure within a unified framework. Comparisons are made and discussed.

AN OVERVIEW AND ASSESSMENT OF THE UMIST FORMULAE

The UMIST formulae [2] and [4] have been proposed to predict the response and failure of reinforced concrete targets struck by rigid

AN OVERVIEW AND ASSESSMENT OF THE UMIST FORMULAE

projectiles with different nose shapes as shown schematically in Fig. 1. Various equations were derived which can be used to predict penetration, through-thickness cone cracking, scabbing and perforation of the targets as shown schematically in Fig. 2. In the following an overview of the UMIST formulae is first conducted and then the accuracy of the UMIST formulae is assessed in the wake of newly acquired test data.

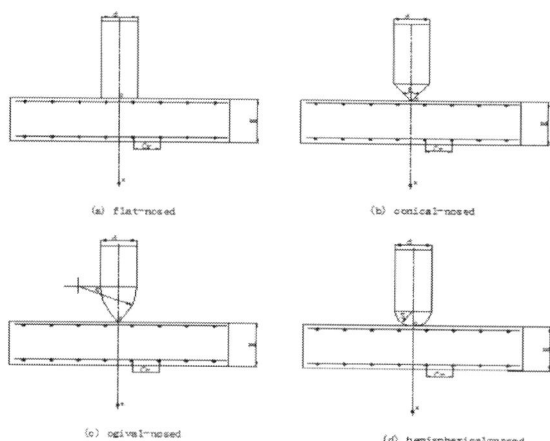

Figure 1: Schematic diagram showing concrete targets struck by projectiles with different nose shapes.

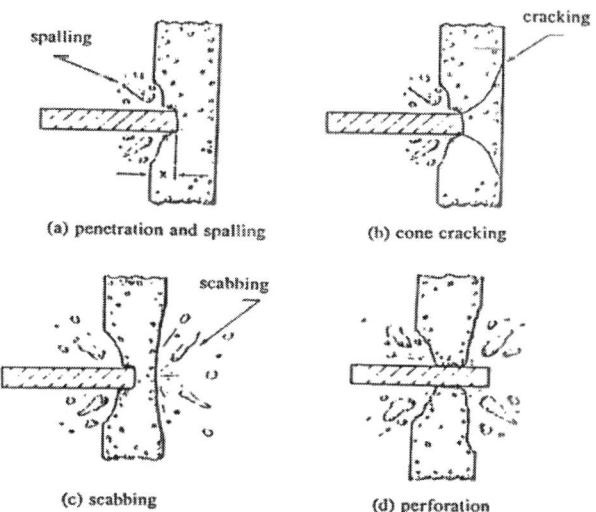

Figure 2: Schematic diagram showing penetration, cone cracking, scabbing and perforation of concrete slabs struck by "hard" missiles.

An Overview of the UMIST Formulae

The UMIST penetration formula is given by the following equation [2] and [4]

$$\frac{x}{d} = \left(\frac{2}{\pi}\right) \frac{N^* M V_i^2}{0.72 \sigma_t d^3} \tag{1}$$

Where x is penetration depth, the nose shape factor N^* is 0.72 for flat nose, 0.84 for a hemispherical nose, 1.0 for a blunt nose and 1.13 for a sharp nose. M is projectile mass, d projectile diameter and V_i projectile impact velocity. σ_t is defined by the following expression:

$$\sigma_t(\text{MPa}) = 4.2 f_c'(\text{MPa}) + 135.0 + \left(0.014 f_c'(\text{MPa}) + 0.45\right) V_i(\text{m/s}) \tag{2}$$

In which f_c' is the unconfined compressive strength of concrete. V_i is the impact velocity of the projectile. As can be seen from the Eq. (1), the penetration depth of projectile is relevant to the compressive strength f_c' of the concrete, the impact velocity V_i and the diameter d of the projectile. The effect of the projectile's diameter is equivalent to the slendness ratio of the projectile.

Eq. (1) has been verified for 50 < d < 600 (mm), 35 < M < 2500 (kg), 0 < x/d < 2.5 and 3 < V_i < 66.2 (m/s).

The critical impact energies of a projectile to cause through-thickness cone cracking (E_c), scabbing (E_s) and perforation (E_p) are given as a function of the ratio of target thickness to missile diameter.

(i) H/d < 5

The UMIST formula for cone cracking is expressed as follows [2], [3] and [4]:

$$\frac{E_c}{\eta \sigma_t d^3} = -0.00031 \left(\frac{H}{d}\right) + 0.00113 \left(\frac{H}{d}\right)^2 \quad 0 < \frac{H}{d} \leq 2 \tag{3a}$$

$$\frac{E_c}{\eta \sigma_t d^3} = -0.00325\left(\frac{H}{d}\right) + 0.00130\left(\frac{H}{d}\right)^3 \quad 2 < \frac{H}{d} < 5$$

(3b)

Where H is target thickness and the influence of the nose shape can be neglected.

The UMIST formula for scabbing is written as [2] and [4]

$$\frac{E_s}{\eta \sigma_t d^3} \frac{N^*}{0.72} = -0.005441\left(\frac{H}{d}\right) + 0.01386\left(\frac{H}{d}\right)^2 \quad 0 < \frac{H}{d} < 5$$

(4)

Where the nose shape factor N* is 0.72 for flat nose, 0.84 for a hemispherical nose, 1.0 for a blunt nose and 1.13 for a sharp nose.

The UMIST formula for perforation is written in the following forms [2], [3] and [4]

$$\frac{E_p}{\eta \sigma_t d^3} = -0.00506\left(\frac{H}{d}\right) + 0.01506\left(\frac{H}{d}\right)^2 \quad 0 < \frac{H}{d} \leq 1$$

(5a)

$$\frac{E_p}{\eta \sigma_t d^3} = -0.01\left(\frac{H}{d}\right) + 0.02\left(\frac{H}{d}\right)^3 \quad 1 < \frac{H}{d} \leq 5$$

(5b)

Where in above equations, η is determined by the following expressions [2], [3] and [4]

$$\eta = \begin{cases} 0.5 + \frac{3}{8}\left(\frac{d}{C_r}\right) r_t & \frac{d}{C_r} < \sqrt{\frac{d}{d_r}} \\ 0.5 + \frac{3}{8}\left(\sqrt{\frac{d}{d_r}}\right) r_t & \frac{d}{C_r} \geq \sqrt{\frac{d}{d_r}} \end{cases}$$

(6)

Where d_r is the rebar diameter, C_r is the rebar spacing and r_t is the total bending reinforcement (r_t = 4r with r being % EWEF, defined as $r = \pi d_r^2 / 4HC_r$).

(ii) H/d ≥ 5

For H/d ≥ 5, the effect of the reinforcement on the response and failure of the targets may be neglected. The majority of the impact energy of the missile will be dissipated in the tunneling process. The critical energies for cone cracking, scabbing and perforation are given by the following expressions [2], [3] and [4]:

$$\frac{E_c}{\sigma_t d^3} = \frac{\pi}{4}\left[\frac{H}{d} - 4.7\right] \tag{7}$$

$$\frac{E_s}{\sigma_t d^3}\frac{N^*}{0.72} = \frac{\pi}{4}\left[\frac{H}{d} - 4.3\right] \tag{8}$$

$$\frac{E_p}{\sigma_t d^3} = \frac{\pi}{4}\left[\frac{H}{d} - 3.0\right] \tag{9}$$

An Assessment of the UMIST Formulae

In the wake of the new test data and the data made available recently, the UMIST formulae are assessed and discussed in the following section.

Depth of Penetration

Comparisons are made between the UMIST penetration formula (Eq. (1)) and the experimental data available in the literature for flat-nosed projectiles [5], [6], [7] and [8] in Fig. 3 and for non-flat-nosed (i.e. conical-, ogival-nosed) missiles [9], [10], [11], [12], [13], [14], [15], [16], [17], [18], [19] and [20] in Fig. 4. In the legends of Fig. 3 and Fig. 4, (f) (c) and (o) indicate flat conical and ogival nosed projectiles respectively whilst (N) (H) denote normal and high strength concrete. It is seen from Fig. 3 that the predictions of the UMIST penetration formula for flat-nosed projectiles are in good agreement with the experimental data for $E_k/\sigma_t d^3 <$ 1.5, and are lower than the experimental data for $E_k/\sigma_t d^3 > 1.5$. The difference between the predictions of the UMIST penetration formula and the test results increases with increasing $E_k/\sigma_t d^3$. It is also seen from Fig. 4 that the predictions of the UMIST penetration formula for non-flat-nosed projectiles are in good agreement with the experimental data for $E_k/\sigma_t d^3 <$ 7.5, and are lower than experimental data for $E_k/\sigma_t d^3 \geq 7.5$. The difference between the predictions of the UMIST penetration formula and the experiment data also increases with increasing $E_k/\sigma_t d^3$. In other words, the

UMIST penetration formula (Eq. (1)) is most likely applicable to relatively low velocity impact. This is not surprising since the original UMIST penetration formula was derived on the basis of the test results acquired by nuclear power industry for large mass low velocity impact [3].

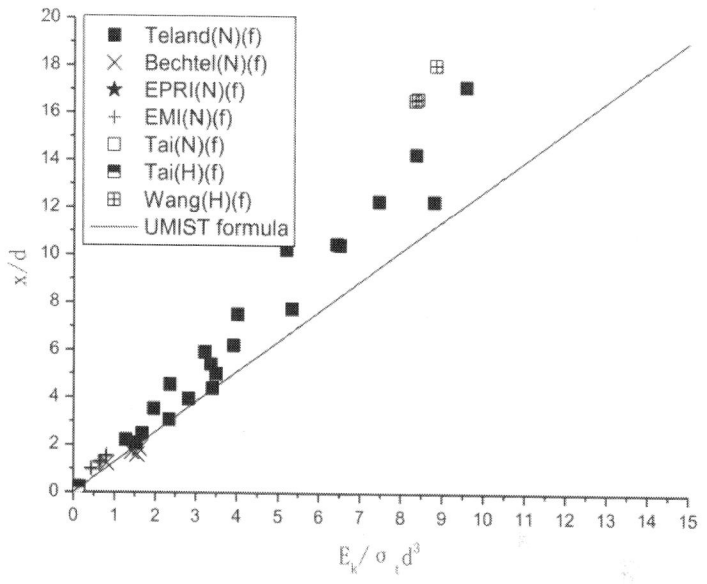

Figure 3: Comparison of the UMIST penetration formula (Eq. (1)) with the experimental data for flat–nosed projectiles [5], [6], [7] and [8].

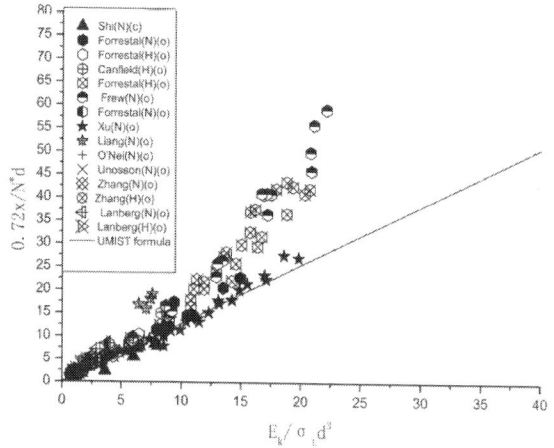

Figure 4: Comparison of the UMIST penetration formula (Eq. (1)) with the experimental data for non-flat –nosed projectiles [9], [10], [11], [12], [13], [14], [15], [16], [17], [18], [19] and [20].

Cone Cracking

Comparisons are made between the UMIST formulae for cone cracking and the experimental data for H/d < 5 [21] and [22] in Fig. 5 and for H/d ≥ 5 [21] in Fig. 6. It is seen from Fig. 5 and Fig. 6 that the predictions of the UMIST formulae for cone cracking are in good agreement with the available test data for flat-faced missiles.

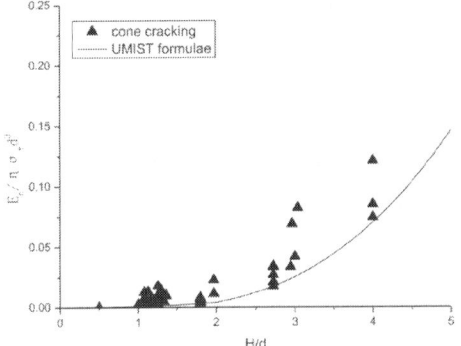

Figure 5: Comparison of the UMIST formulae for cone cracking (Eq. (3a) and (3b)) with the experimental data for flat–nosed projectiles [21] and [22].

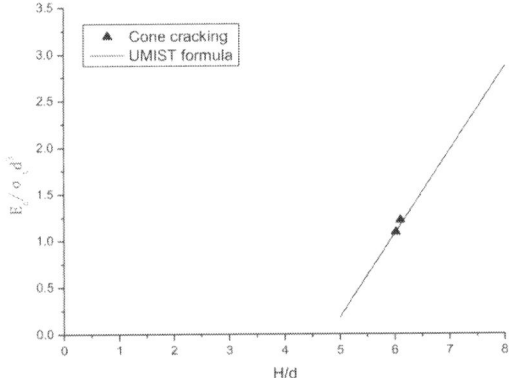

Figure 6: Comparison of the UMIST formula for cone cracking (Eq. (7)) with the experimental data for flat–nosed projectiles [21].

Scabbing

Comparisons are made between the UMIST scabbing formulae and the experimental results for flat-faced missiles for H/d < 5 [6] in Fig. 7a and for H/d ≥ 5 [6] in Fig. 8. Comparison is also made between the UMIST scabbing formula (Eq. (4)) and the test data for non-flat-ended projectiles for H/d < 5 [6], [23] and [24] in Fig. 9. In the legends of Fig. 7, Fig. 8 and Fig. 9, (f) (c) and (s) indicate flat conical and spherical nosed projectiles respectively whilst (N) (H) denote normal and high strength concrete. It is seen from Fig. 7 and Fig. 8 that the predictions of the UMIST scabbing formulae are generally in good agreement with the test results for flat-nosed projectiles. However, it should be noted that the UMIST scabbing formula Eq.(4) over predicts the test data a little bit for $H/d < 2.5$ as can be seen from the enlarged detail shown in Fig. 7b. It implies that for $H/d < 2.5$ the UMIST formula (Eq. (4)) may not be that safe to be used for design of structures to resist scabbing due to impact by flat-ended projectiles. The thick reinforced concrete targets damage failure mode with scabbing are rare. It can be seen from Fig. 8 that the predictions of the UMIST formula are agreement with experiment data. However, the thick reinforced concrete targets damage failure mode with scabbing are rare to demonstrate that its protective ability for scabbing is higher than thin reinforced concrete targets.

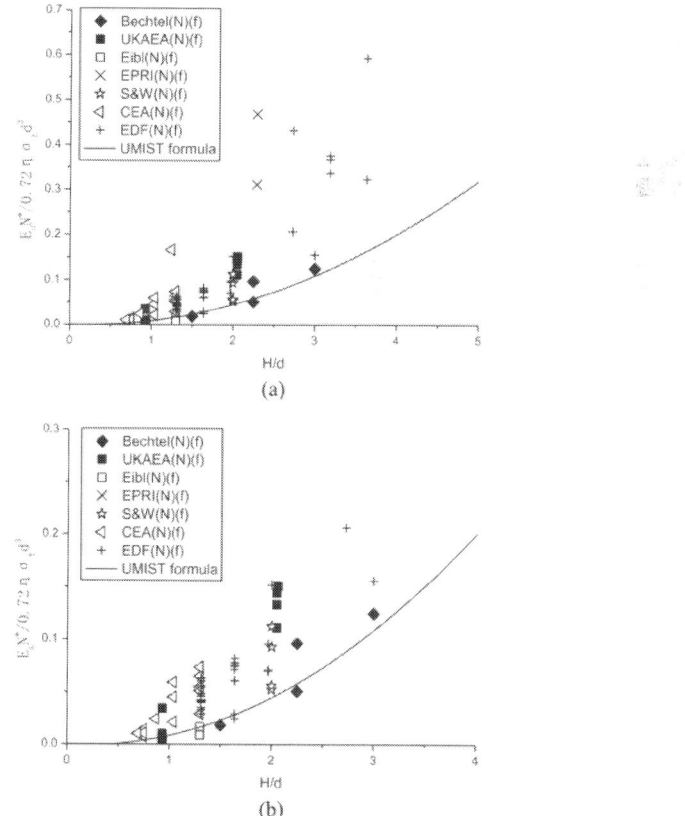

Figure 7: (a) Comparison of the UMIST scabbing formula (Eq. (4)) with the experimental data for flat–nosed projectiles [6]. (b) Enlarged detail of the partial Fig. 7a.

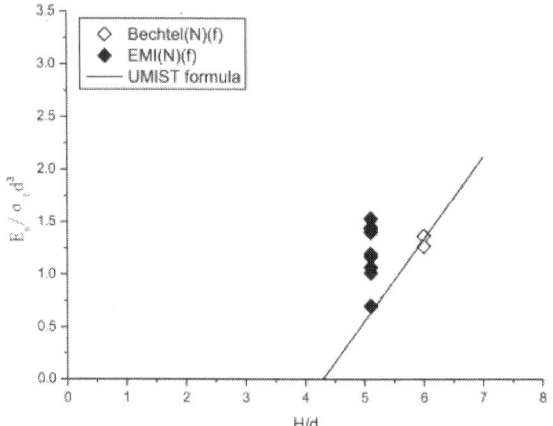

Figure 8: Comparison of the UMIST scabbing formula (Eq. (8)) with the test data for flat–nosed projectiles [6].

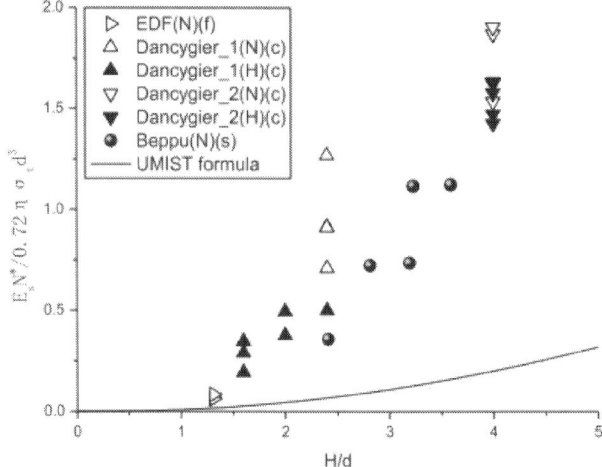

Figure 9: Comparison of the UMIST scabbing Formula (4) with the experimental data for non-flat–nosed projectiles [6], [23] and [24].

It is also seen from Fig. 9 that the UMIST scabbing formula (Eq. (4)) significantly underestimates the experimental data for non-flat-nosed missiles. In other words, the UMIST scabbing formula is too conservative for design of structures to resist scabbing resulting from impact by non-flat-faced (i.e. conical, spherical-nosed) missiles.

AN OVERVIEW AND ASSESSMENT OF THE UMIST FORMULAE

Perforation

Fig. 10 and Fig. 11 show comparisons between the UMIST perforation formulae (Eq. (5a) and (5b)) and the test data for flat-faced missiles [6] and non-faced-missiles [6], [23], [25] and [26] respectively for H/d < 5. Comparison is also made between the UMIST perforation formula (Eq. (9)) and the test results for flat-faced as well as non-flat-ended projectiles for H/d > 5 [6], [18], [24], [25], [27], [28], [29] and [30] in Fig. 12. In the legends of Fig. 10, Fig. 11 and Fig. 12, (f) (c) (s) and (o) indicate flat, conical, spherical and ogival nosed projectiles respectively whilst (N) (H) denote normal and high strength concrete. The predictions of the UMIST perforation formulae (Eq. (5a) and (5b)) are generally in reasonable agreement with the test results for flat-nosed projectiles as well as non-flat-ended missiles as can be seen from Fig. 10, Fig. 11 and Fig. 12 except for very large H/d where the UMIST perforation formula (Eq. (9)) significantly overestimate the experimental result as can be seen from Fig. 11. In other words, the UMIST perforation formulae fail to produce consistent results as compared to the experimental data especially for very thick concrete targets.

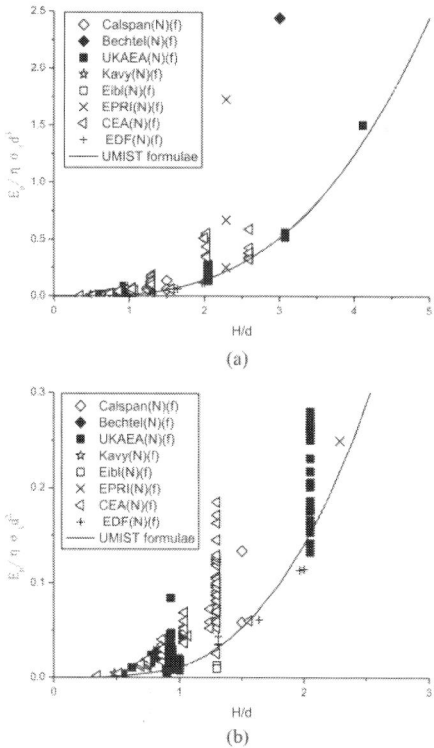

Figure 10: (a) Comparison of the UMIST perforation formulae (Eq. (5a) and (5b)) with the experimental data for flat–nosed projectiles [6]. (b) Enlarged detail of the partial Fig. 10a.

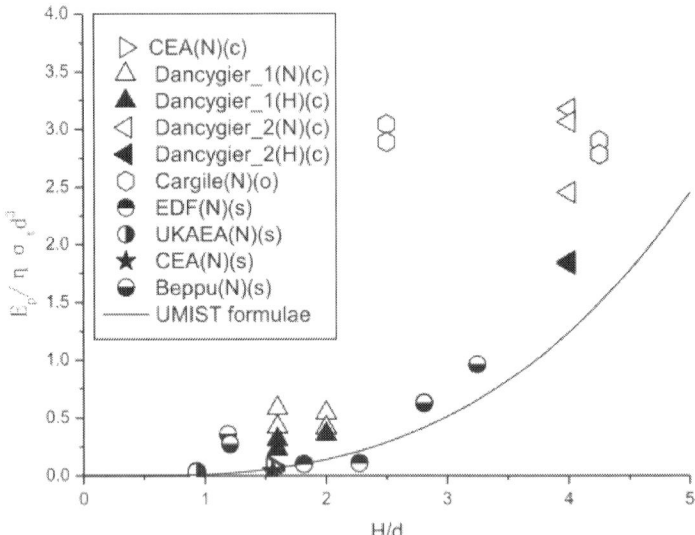

Figure 11: Comparison of the UMIST perforation formulae (Eq. (5a) and (5b)) with the experimental data [6], [23], [25] and [26] for non-flat–nosed projectiles.

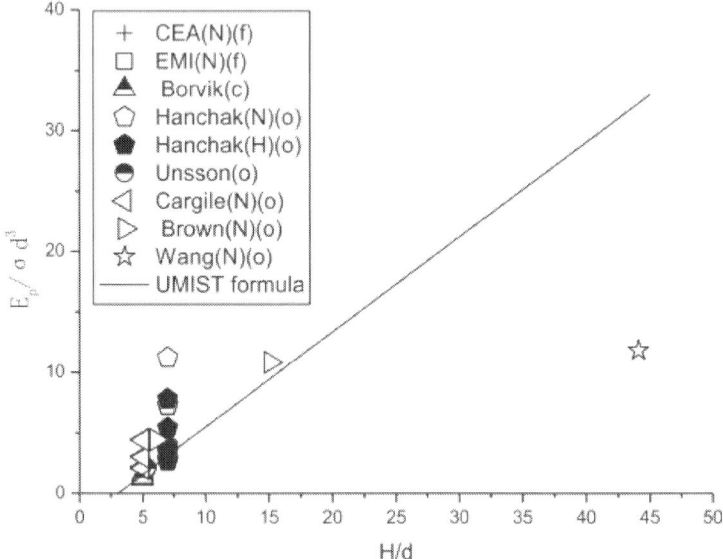

Figure 12: Comparison of the UMIST perforation formula (Eq. (9)) with the test results [6], [18], [24], [25], [27], [28], [29] and [30].

Furthermore, closer examination reveals that the UMIST perforation formulae (Eq. (5a) and (5b)) over predict the test data for thin concrete targets with H/d < 2.5 a little bit for flat-faced projectiles as well as non-flat-ended missiles, as can be seen from the enlarged detail shown in Fig. 10 and Fig. 11, respectively. Closer examination also reveals that for thick concrete targets with H/d ≥ 5 the UMIST perforation formula (Eq. (9)) overestimates the test results for H/d < 7.5 as well. It implies that the UMIST formulae (Eq. (4) and Eq. (9)) may not be safe to be used for design of structures to resist perforation due to impact by flat-ended projectiles as well as non-flat-ended missiles for thin concrete targets with H/d < 2.5, intermediate thick concrete targets with 5 ≤ H/d < 7.5 and very thick concrete targets with H/d > 15 approximately. It has to be mentioned here that care should be exercised where assessments are required for H/d ≤ 0.5, as also noted in Refs. [2] and [4].

FORMULATION OF A UNIFIED APPROACH

In the following sections a unified approach for concrete impact is formulated by closely following the study in Ref. [3] and by introducing a mean resistive pressure which caters for the effects of various parameters such as nose shape, impact velocity and unconfined compressive strength as proposed in Refs. [31] and [33]. Various empirical equations are obtained which can be used to predict the depth of penetration, through-thickness cone cracking, scabbing and perforation of reinforced concrete targets subjected to projectile impact at normal incidence.

Penetration

The depth of penetration of concrete targets struck transversely by projectiles with different nose shapes has been examined in Ref. [33]. For completeness the equation for depth of penetration is quoted as follows

$$\frac{x}{d} = \frac{4}{\pi} \frac{E_k}{\sigma d^3}$$

(10)

where x is penetration depth, $E_k = (1/2)MV_i^2$ is projectile impact energy with M and V_i being projectile mass and impact velocity, respectively, d projectile diameter and σ is mean resistive pressure which can be expressed by the following equation

$$\sigma = \left(\alpha + \beta\sqrt{\frac{\rho_t}{Y}V_i}\right)Y \tag{11}$$

Where α can be determined theoretically by spherical or cylindrical cavity expansion analysis [32] and β evaluated from experiments; ρ_t is the density of the concrete target material; Y is a measure of shear strength for concrete which can be expressed as [33]

$$Y = \begin{cases} 1.4f'_c + 45, & f'_c \leq 75 \text{ MPa} \\ 150, & 75 \text{ MPa} < f'_c < 150 \text{ MPa} \\ f'_c, & f'_c \geq 150 \text{ MPa} \end{cases} \tag{12}$$

Where f'_c is the unconfined compressive strength of concrete.

The form of the UMIST penetration formula for a flat-ended projectile is the same as that of the newly suggested penetration formula except for the mean resistive pressure which has a different expression. Close examination of Eqs. (2) And (11) reveal that for relatively low impact velocities the values of σ_t (Eq. (2)) and σ (Eq. (11)) are almost the same and that for higher impact velocities the difference between these two parameters becomes quite large. In fact, the difference between the values of σ_t (Eq. (2)) and σ (Eq. (11)) increases with increasing impact velocity.

Cone Cracking, Cabbing and Perforation

According to the experimental observations, reinforced concrete slabs struck normally by projectiles can be categorized into two regimes, namely, thin target and thick target. The response of a thin concrete slab is global bending together with local failure whereas the response of a thick concrete slab is localized [2].

For a thin reinforced concrete target subjected to impact by a projectile the critical energies E_c, E_s and E_p for cone cracking, scabbing and perforation can generally be written as [3] and [21]

$$\frac{E_c}{\phi\eta\sigma d^3} = C_1\left(\frac{H}{d}\right) + C_2\left(\frac{H}{d}\right)^2 + C_3\left(\frac{H}{d}\right)^3 + C_4\left(\frac{H}{d}\right)^4 \tag{13}$$

$$\frac{E_s}{\phi\eta\sigma d^3} = C_5\left(\frac{H}{d}\right) + C_6\left(\frac{H}{d}\right)^2 + C_7\left(\frac{H}{d}\right)^3 + C_8\left(\frac{H}{d}\right)^4 \tag{14}$$

$$\frac{E_p}{\phi\eta\sigma d^3} = C_9\left(\frac{H}{d}\right) + C_{10}\left(\frac{H}{d}\right)^2 + C_{11}\left(\frac{H}{d}\right)^3 + C_{12}\left(\frac{H}{d}\right)^4 \tag{15}$$

Where η is coefficient of reinforcement, ϕ is projectile nose shape effect and C_1–C_{12} are empirical constants.

For a thick reinforced concrete target struck by a projectile the critical energies E_c, E_s and E_p for cone cracking, scabbing and perforation can generally be written in the following forms [3] and [21]

$$\frac{E_c}{\sigma d^3} = \frac{\pi}{4}\left(\frac{H}{d} - \chi_c\right) \tag{16}$$

$$\frac{E_s}{\sigma d^3} = \frac{\pi}{4}\left(\frac{H}{d} - \chi_s\right) \tag{17}$$

$$\frac{E_p}{\sigma d^3} = \frac{\pi}{4}\left(\frac{H}{d} - \chi_p\right) \tag{18}$$

Where χ_c, χ_s and χ_p are constants determined by experiments.

Comparison with Available Experimental Data and Discussion

The values of various parameters in the equations given in Sections 3.1 and 3.2 can be determined experimentally. For concrete targets, the values of α, β and ϕ are listed in Table 1. α and β are constants in Eq. (11), ϕ is projectile nose shape effect, θ is the cone angle of the missile, ψ is the caliber-radius-head of the ogival nose projectile.

Table 1: Values of various parameters for concrete targets [33]

	α	β	ϕ
Conical-nosed ($\theta < 90°$)	$\frac{1}{2}\left[1 + \ln\frac{2E}{(5-4\nu)Y}\right]$	$2\sin\frac{\theta}{2}$	$1 + \frac{2}{\tan\left(\frac{\theta}{2}\right)}$
Conical-nosed ($90° \leq \theta < 180°$)	$\frac{1}{2}\left[1 + \ln\frac{2E}{(5-4\nu)Y}\right]$	$\sqrt{2}$	$1 + \frac{2}{\tan\left(\frac{\theta}{2}\right)}$
Flat-nosed	$\frac{1}{2}\left[1 + \ln\frac{2E}{(5-4\nu)Y}\right]$	$\sqrt{2}$	1
Ogival-nosed	$\frac{2}{3}\left[1 + \ln\frac{E}{3(1-\nu)Y}\right]$	$3/4\psi$	$1+\psi$
Hemispherical-nosed	$\frac{2}{3}\left[1 + \ln\frac{E}{3(1-\nu)Y}\right]$	$3/2$	$3/2$

For conical-nosed projectiles η can be determined by following expression.

$$\eta = \begin{cases} 0.5 + \frac{3}{8}\sin^4\left(\frac{\theta}{2}\right)\left(\frac{d}{C_r}\right)r_t & \frac{d}{C_r} < \sqrt{\frac{d}{d_r}} \\ 0.5 + \frac{3}{8}\sin^4\left(\frac{\theta}{2}\right)\left(\sqrt{\frac{d}{d_r}}\right)r_t & \frac{d}{C_r} \geq \sqrt{\frac{d}{d_r}} \end{cases} \quad (19)$$

Where θ is the cone angle of the missile, d_r is the rebar diameter, C_r is the rebar spacing and r_t is the total bending reinforcement ($r_t = 4r$ with r being % EWEF, defined as $r = \pi d_r^2/4HC_r)$.

For ogival-nosed projectiles

$$\eta = \begin{cases} 0.5 + \frac{3}{128\psi^2}\left(\frac{d}{C_r}\right)r_t & \frac{d}{C_r} < \sqrt{\frac{d}{d_r}} \\ 0.5 + \frac{3}{128\psi^2}\left(\sqrt{\frac{d}{d_r}}\right)r_t & \frac{d}{C_r} \geq \sqrt{\frac{d}{d_r}} \end{cases} \quad (20)$$

Where $\psi = S/d$ is the caliber-radius-head of the ogive nose projectile. S is the radius of the ogive nose.

FORMULATION OF A UNIFIED APPROACH

Based on the available experimental data and using polynomial curve-fitting technique, the various empirical constants in Eqs. (13), (14), (15), (16), (17) and (18) have been obtained. Hence, Eqs. (13)– (18) can be recast into the following forms:

For thin concrete targets

$$\frac{E_c}{\phi \eta \sigma d^3} = 0.0003 \left(\frac{H}{d}\right) + 0.0003 \left(\frac{H}{d}\right)^4 \quad 0 < \frac{H}{d} < 5 \tag{21}$$

$$\frac{E_s}{\phi \eta \sigma d^3} = -0.0066 \left(\frac{H}{d}\right) + 0.0120 \left(\frac{H}{d}\right)^2 \quad 0 < \frac{H}{d} < 5 \tag{22}$$

And

$$\frac{E_p}{\phi \eta \sigma d^3} = 0.0048 \left(\frac{H}{d}\right) + 0.0018 \left(\frac{H}{d}\right)^2 \quad 0 < \frac{H}{d} \le 1.3 \tag{23a}$$

$$\frac{E_p}{\phi \eta \sigma d^3} = -0.0300 \left(\frac{H}{d}\right) + 0.0220 \left(\frac{H}{d}\right)^3 \quad 1.3 < \frac{H}{d} < 4 \tag{23b}$$

For thick concrete targets

$$\frac{E_c}{\sigma d^3} = \frac{\pi}{4}\left(\frac{H}{d} - 4.7\right) \quad \frac{H}{d} \ge 5 \tag{24}$$

$$\frac{E_s}{\sigma d^3} = \frac{\pi}{4}\left(\frac{H}{d} - 4.3\right) \quad \frac{H}{d} \ge 5 \tag{25}$$

And

$$\frac{E_p}{\sigma d^3} = \begin{cases} \frac{\pi}{4}\left(\frac{H}{d} - 3.0\right) & \frac{H}{d} \ge 4, V_i < 300 \text{ (m/s)} \\ \frac{\pi}{4}\left(\frac{H}{d} - 2.0\right) & \frac{H}{d} \ge 4, V_i \ge 300 \text{ (m/s)} \end{cases} \tag{26}$$

The response (energy absorbing) mechanisms are different for thin and thick concrete slabs. Hence, the function of E_c is not necessarily continuous for $H/d = 5$. The same argument can be applied to the functions of E_s and E_p.

Penetration

Comparisons are made of the present formula (Eq. (10)) with test data [5], [6], [7], [8], [9], [10], [11], [12],[13], [14], [15], [16], [17], [18], [19] and [20] for the depth of penetration of concrete targets struck normally by flat-nosed projectiles in Fig. 13a and by non-flat-nosed projectiles in Fig. 14a. In the legends of Fig. 13, (f) (c) and (o) indicate flat, conical and ogival nosed projectiles respectively whilst (N) (H) denote normal and high strength concrete. It is seen from Fig. 13 that the present penetration formula is in good agreement with the experimental results available in the literature.

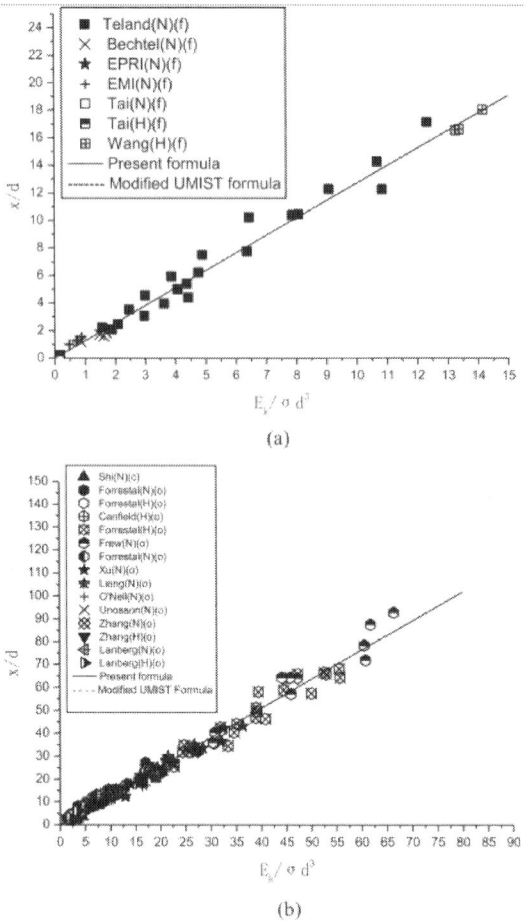

Figure 13: Comparisons of the present penetration formula (Eq. (10)) with the test data for the depth of penetration: (a) flat-nosed projectiles; (b) non-flat-nosed missiles [5], [6], [7], [8], [9], [10], [11], [12], [13], [14], [15], [16], [17], [18], [19] and [20].

FORMULATION OF A UNIFIED APPROACH

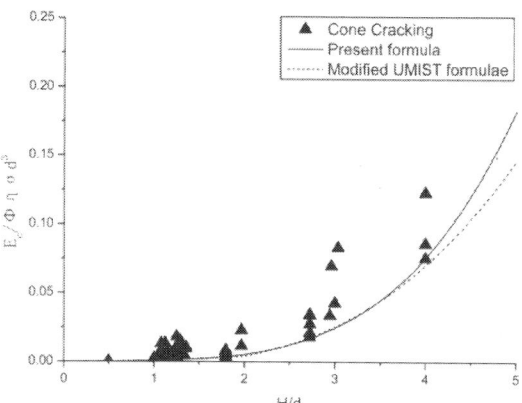

Figure 14: Comparison of the present formula (Eq. (21)) with the test data [21] and [22] for cone cracking of reinforced concrete targets with H/d < 5.

Cone Cracking

Comparisons are made between the test data for cone cracking of reinforced concrete targets with the present formulae for H/d < 5(Eq. (21)) in Fig. 14 [21] and [22] and for H/d ≥ 5(Eq. (24)) in Fig. 15 [21]. It can be seen from Fig. 14 and Fig. 15 that the present formulae for cone cracking are in good agreement with the available experimental data. It should be mentioned here that the experimental data presented in Fig. 14 and Fig. 15 were acquired for flat-nosed missiles only and, at the present time, there is paucity of the test results for non-flat-ended projectiles.

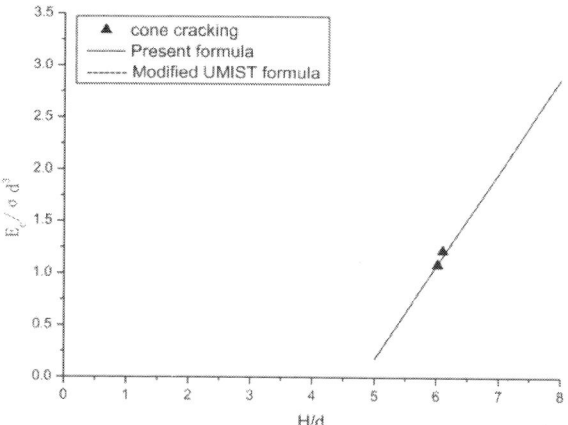

Figure 15: Comparison of the present formula (Eq. (24)) with the test data [21] for cone cracking of reinforced concrete targets with H/d ≥ 5.

Scabbing

Comparisons are made between the experimental results for scabbing of reinforced concrete targets with the present formulae for H/d < 5(Eq. (22)) in Fig. 16a for flat-nosed projectiles [6] and in Fig. 17 for non-flat-nosed missiles [23] and [24]. Comparison is also made between the experimental data for scabbing of reinforced concrete targets with the present formulae and for H/d ≥ 5(Eq. (25)) in Fig. 18 [6]. In the legends of Fig. 16, Fig. 17 and Fig. 18, (f) (c) and (s) indicate flat, conical and spherical nosed projectiles respectively whilst (N) (H) denote normal and high strength concrete. It is clear from Fig. 16, Fig. 17 and Fig. 18 that the present formulae for scabbing are in good agreement with the available test data.

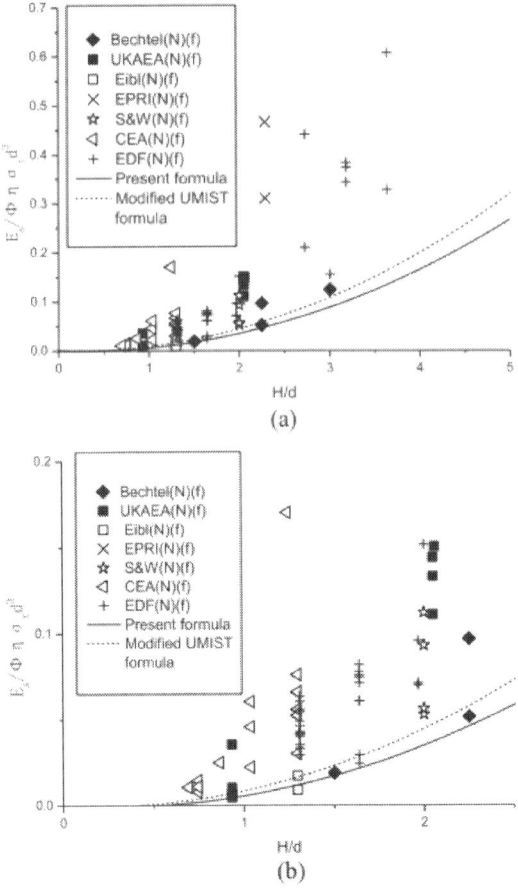

Figure 16: (a) Comparison of the present formula (Eq. (22)) with the test data [6] for scabbing of reinforced concrete targets with H/d < 5 struck by flat-nosed projectile. (b) Enlarged detail of the partial Fig. 16a.

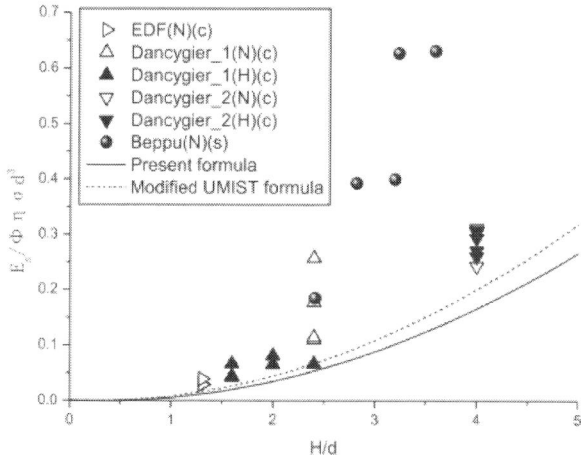

Figure 17: Comparison of the present formula (Eq. (25)) with the test data [6], [23] and [24] for scabbing of reinforced concrete targets with H/d < 5 struck by non-flat-nosed projectile.

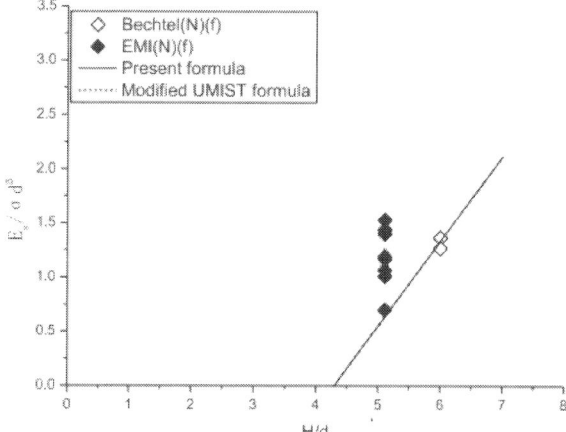

Figure 18: Comparison of the predictions of the present model with the test data [6] for scabbing of reinforced concrete targets (H/d ≥ 5) struck by flat-nosed projectile.

Perforation

Comparisons are made between the tests data for perforation of reinforced concrete targets with the present formulae for H/d < 4(Eq. (23a) and (23b)) in Fig. 19a for flat-nosed projectiles [6] and in Fig. 20a for non-flat-nosed missiles [6], [23], [25] and [26]. Comparisons are also made between the experimental results [6],[18], [24], [25], [27], [28], [29] and [30] for perforation of reinforced concrete targets with the present formulae for H/d ≥ 4(Eq. (26)) in Fig. 21a for impact velocities less than 300 m/s and in Fig. 21b for impact velocities greater than 300 m/s. In the legends of Fig. 19, Fig. 20 and Fig. 21, (f) (c), (s) and (o) indicate flat, conical, spherical and ogival nosed projectiles respectively whilst (N) (H) denote normal and high strength concrete. It is clear from Fig. 19, Fig. 20 and Fig. 21 that the present perforation formulae are in good agreement with the test data available in the literature.

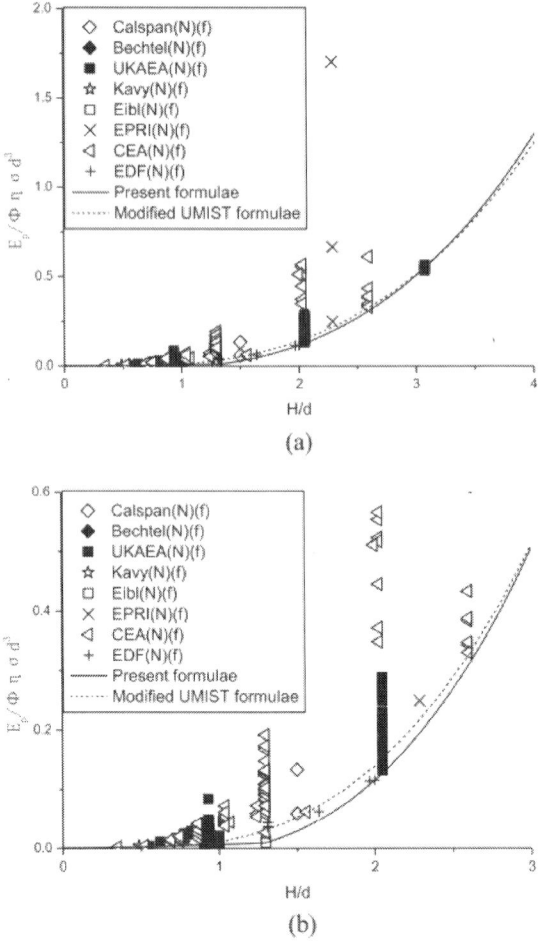

Figure 19: (a) Comparison of the predictions of the present model with the test data [6] for perforation of reinforced concrete targets (H/d < 4) struck by flat-nosed projectile. (b) Enlarged detail of the partial Fig. 19a.

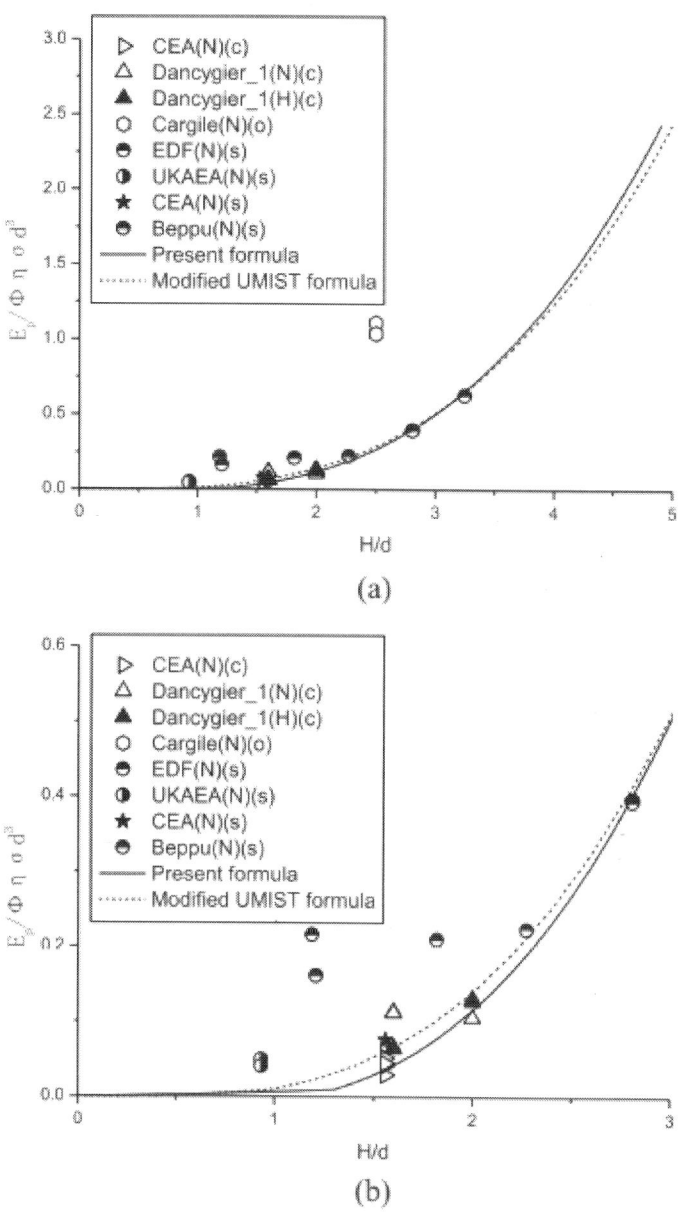

Figure 20: (a) Comparison of the predictions of the present model with the test data [6], [23], [25] and [26] for perforation of reinforced concrete targets (H/d < 4) struck by non-flat-nosed projectile. (b) Enlarged detail of the partial Fig. 20a.

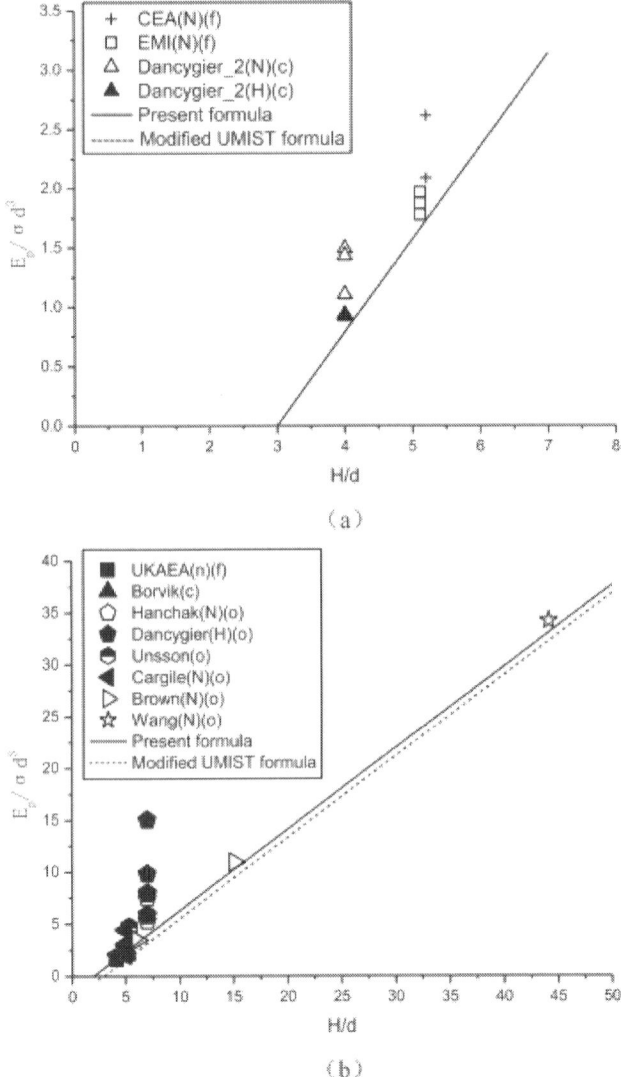

Figure 21: Comparison of the present formula (Eq. (26)) with the test data [6], [18], [24], [25], [27], [28] and [29] for perforation of reinforced concrete targets with H/d ≥ 4 (a) $V_i < 300$ (m/s); (b) $V_i \geq 300$(m/s).

Further Discussion

Unfortunately, the UMIST equations and the newly proposed formulae cannot be presented in one graph due to the use of different expressions for the mean resistive pressure (see Eqs (2) and (11) for more details). In order to make direct comparison between the UMIST equations and the newly proposed formulas, Table 2, Table 3 and Table 4 are given which contain the test data of the critical conditions respectively for through-thickness cone cracking (cone cracking limits), scabbing (scabbing limits) and perforation (ballistic limits) available in the literature and the predicted values from the UMIST formulae and the newly suggested equations are also listed in the tables. $V_{cracking}$, $V_{scabbing}$ and V_{50} are the experimentally obtained cone cracking, scabbing and ballistic limits respectively whilst V_c, V_s and V_b are the predicted cone cracking, scabbing and ballistic limits respectively. It can be seen from Table 2, Table 3 and Table 4 that the newly suggested empirical equations produce more consistent results than the UMIST formulae as compared to the experimental observations. It also can be seen from Fig. 16, Fig. 19, Fig. 20 and Fig. 21 that the present equations are better than the modified UMIST formulae.

Table 2: Comparison of the predictions of the present model and UMIST formulae with experimentally obtained cracking limits

Reference	Tests no	Nose shape	Mass (kg)	d (mm)	$V_{cracking}$ (m/s)	H (mm)	f'_c (MPa)	d_r (mm)	C_r (mm)	UMIST V_c (m/s)	Present V_c (m/s)	UMIST Terror ($V_c/V_{cracking}$)	Present error ($V_c/V_{cracking}$)
Reid and Wen [21]	8	Flat	2.622	60	3.70	300	33.5	16	150	3.28	3.75	0.89	1.01
	36	Flat	2.622	75	2.07	300	31.3	16	150	2.01	2.106	0.97	1.02
	72	Flat	4.044	167	4.27	300	32.5	16	150	4.14	4.47	0.97	1.05
	66	Flat	3.9660	110	5.07	300	36.3	16	150	5.06	5	1	0.99
	5	F	1	5	8.	30	29	16	15	7.9	7.	0.9	0.9

	1	lat	066	0	2	0	.2		0	6	89	7	6
	72	Flat	404	167	4.27	300	32.5	16	150	4.14	4.47	0.97	1.05
	77	Flat	399	50	14.35	300	32	16	150	13.42	13.26	0.94	0.93
Sinclare [22]	E5	Flat	48	150	4.94	161	27.5	16	175	5.26	4.51	1.06	0.91

Table 3: Comparison of the predictions of the present model and UMIST formulae with experimentally obtained scabbing limits

Ref	Test No	Nose-shape	Mass (kg)	d (mm)	V_{sc} (abbing) (m/s)	H (mm)	f_c' (MPa)	d_r (mm)	c_r (mm)	UMIST V_s (m/s)	Present V_s (m/s)	UMIST error (V_s/V_{sc} abbing)	Present error (V_s/V_{sc} abbing)
Bechtel [6]	4F	Flat	96.94	203	37.21	30 4.8	39.13	0	0	41.99	34.92	1.13	0.94
Bechtel [6]	18F	Flat	96.49	203	63.14	2 45 7.2	35.19	0	0	67.78	57.68	1.07	0.92
UKAEA [6]	di 30	Flat	310	89	5.3	83	39.97	2.5	10	6.61	5.11	1.25	0.96
EIBL [6]	vii	Flat	10 19	200	8.2	260	43	12	80	11.3	9.39	1.38	1.15
EDF [6]	ii i-19	Flat	160	305	77.5	500	35.5	16	125	84.98	69.91	1.1	0.9
Bechtel [6]	4h	Flat	3.65	25	67.1	15	40.9	9.5	15	67.14	-64.39	1	0.96

	ard			.4		2.4	4		2.4				
	5hard	Fl at	1.71	25.4	98.21	152.4	40.94	9.5	152.4	102.6	96.95	1.04	0.99
EMI [6]	380	Flat	2.25	45	155	230	35	6	40	147.14	137.8	0.95	0.89
Dancygier and Yankelevisky [23]	6PC-H	Conical	0.16	25	144	60	104	0	0	46.47	128.07	0.32	0.89

Table 4: Comparison of the predictions of the present model and UMIST formulae with experimentally obtained ballistic limits

Reference	Test No	Nose shape	Mass (kg)	d (mm)	V_{50} (m/s)	H (mm)	f_c (MPa)	dr (mm)	Cr (mm)	UMIST V_b (m/s)	Present V_b (m/s)	UMIST error (V_b/V_{50})	Present error (V_b/V_{50})
UKAEA [6]	di 13	flat	159.4	150	6.67	83	42.65	2.5	32	5.31	6.99	0.8	1.05
	di 42	flat	310	89	3.58	80.3	38.27	2.5	15	4.87	4.18	1.36	1.17
	di 60	flat	54.48	83	3.73	83	38.8	0	0	5.64	4.53	1.51	1.21
EIBL [6]	vi 1	flat	1019	200	8.2	260	37	12	80	15	8.03	1.83	0.98
CEA [6]	M 553	flat	310	100	15.16	260	44	0	0	18.89	18.01	1.25	1.19
EDF [6]	II I-31	flat	240	200	75.8	40	36	16	125	96.39	82.81	1.27	1.09
CEA [6]	M 507A	conical	133	100	18.7	156	45	6	50	18.63	21.06	1	1.13

Design of Reinforced Concrete

Reference	ID	Shape											
	M515	conical	252	100	16.68	156	48	6	50	10.89	17.17	0.65	1.03
Dancygier and Yankelevisky[23]	5SX-R	conical	0.16	25	158	50	35	0.5	7	65.1	147.9	0.41	0.94
	5SX-H	conical	0.16	25	206	50	110	0.5	7	89.77	195.3	0.44	0.95
Dancygier et al.[24]	7	conical	1.5	50	289	200	113	8	200/1000	359.98	264.7	1.25	0.92
Borvick et al.[27]	C75-4	conical	0.2	20	318.4	100	80.1	0	0	355.29	326.8	1.12	1.03
	C75	conical	0.2	20	302.6	100	80.1	0	0	355.29	326.8	1.17	1.08
	C150-6	conical	0.2	20	317.5	100	129.2	0	0	355.29	331.6	1.12	1.04
	C200-3	conical	0.2	20	366.2	100	180.5	0	0	355.42	335.2	0.97	0.92
Beppu et al.[26]	3	spherical	0.05	25	307.43	70.2	25	0	0	235.67	310	0.77	1.01
Cargile et al.[25]	6	ogive	2.27	50.8	307.91	215.9	39.9	0	0	251.19	293.4	0.82	0.95
Brown[29]	7	ogive	2.32	50.8	288.5	215.9	39.9	0	0	247.39	289.8	0.86	1

	1	ogive	0.77	37	6475	224	39.3	0	0	503.18	453.6	1.06	0.96
	2	ogive	0.77	37	9399	559	39.3	0	0	1499.8	900.1	1.6	0.96
Wang et al.[30]		ogive	52	144	1278.8	60000	35	0	0	321 1.3	1277	2.51	1

It should be mentioned here that all the test results presented in Fig. 5, Fig. 6, Fig. 7, Fig. 8, Fig. 9, Fig. 10, Fig. 11, Fig. 12, Fig. 13, Fig. 14, Fig. 15, Fig. 16, Fig. 17, Fig. 18, Fig. 19, Fig. 20 and Fig. 21 were for concrete slabs which failed either in cone cracking or in scabbing or in perforation. For example, the test data presented in Fig. 5 and Fig. 6, Fig. 14 and Fig. 15 were for concrete slabs which failed in through-thickness cone cracking which include just cone cracking (cracking limit condition), moderate cone cracking and severe cone cracking. The same argument can be applied to Fig. 7, Fig. 8 and Fig. 9, Fig. 16, Fig. 17 and Fig. 18 for scabbing and Fig. 10, Fig. 11 and Fig. 12, Fig. 19, Fig. 20 and Fig. 21 for perforation. In other words, the lines predicted from various empirical equations are demarcation lines or failure boundaries above which failure occurs and below which no failure happens.

It has to be mentioned here that care should be exercised where assessments are required for H/d ≤ 0.5, as also noted in Refs. [2] and [4].

The Modified UMIST Formulae

The modified UMIST formulae (see Appendix for more details) are obtained directly by substituting σ for σ_t of the Eq. (1) with $N^* = 0.72$ for depth of penetration; φσ for σ_t of the Eqs. (3a), (3b), (4), (5a) and (5b) with $N^* = 0.72$ for H/d < 5 and σ for σ_t of the Eqs. (7), (8) and (9) with the coefficients and the functional forms being kept unchanged. It should be noted here that the modified UMIST formulae for penetration (Eq. (A1)) and cone cracking (Eq. (A5)), scabbing (Eq. (A6)), perforation (Eq. (A7)) for thick concrete targets are the same as those of the present empirical formulae given in Section 3 except for perforation where the newly suggested empirical formula for $V_i \geq 300$ m/s is slightly different from that of the modified UMIST formula. The results predicted from the modified UMIST formulae are also shown in Fig. 13, Fig. 14, Fig. 15, Fig. 16, Fig.

17, Fig. 18, Fig. 19, Fig. 20 and Fig. 21. As can be seen from Fig. 13, Fig. 14, Fig. 15, Fig. 16, Fig. 17, Fig. 18, Fig. 19, Fig. 20 and Fig. 21 the modified UMIST formulae are generally in reasonable agreement with the experimental data available in the literature.

CONCLUSIONS

A new set of empirical equations have been derived to predict the depth of penetration, through-thickness cone cracking, scabbing and perforation of reinforced concrete targets struck by projectiles at normal incidence within a unified framework. The present work represents an extension and further development of the UMIST formulae. The main conclusions are as follows:

1. The UMIST formulae can be used with reasonable confidence to predict the depth of penetration, cone cracking, scabbing and perforation of reinforced normal strength concrete targets subjected to impact by large mass low velocity missiles. On the other hand, for scabbing of reinforced concrete targets struck by non-flat-ended missiles the UMIST formulae are too conservative whilst they significantly over predict the perforation of very thick concrete targets as compared to the experimental data. This is not surprising since the UMIST formulae were originally developed for flat-faced missiles and nuclear power industry where the threat by large mass low velocity projectile impact is a major concern.
2. The predictions of the new set of empirical equations have been shown to be in good agreement with the experimental data available in the literature for reinforced concrete targets subjected to impact by projectiles with a wide range of impact conditions. It has been demonstrated that these new equations are not only applicable for flat-nosed projectiles, but also for non-flat-nosed projectiles; that not only applicable for low velocity impact, but also for high velocity impact and that not only applicable for normal strength concrete, but also for high strength concrete.
3. The new empirical formulae have been verified against a wider range of parameters, i.e., 0.059 kg < M < 2622 kg, 7.6 mm < d < 600 mm, 0 < V_i < 1300 m/s, $13.5\ \text{MPa} < f_c < 210\ \text{MPa}$.
4. The modified UMIST formulae have also been found to be in reasonable agreement with the experimental data available in the literature.

ACKNOWLEDGEMENT

This work was supported by National Natural Science Foundation of China through grant no. 11172298.

APPENDIX

The UMIST formulae originally developed for flat-faced projectiles [2] and [3] can be directly extended to obtain the modified UMIST formulae by substituting σ for σ_t of the Eq. (1) with $N^* = 0.72$ for depth of penetration; $\phi\sigma$ for σ_t of the Eqs. (3a), (3b), (4), (5a) and (5b) and σ for σ_t of the Eqs. (7), (8) and (9) with $N^* = 0.72$ whilst the coefficients and the functional forms are kept unchanged.
Depth of penetration

$$\frac{x}{d} = \left(\frac{2}{\pi}\right)\frac{MV_i^2}{\sigma d^3} \tag{A1}$$

Cone cracking, scabbing and perforation for H/d < 5

$$\frac{E_c}{\eta\phi\sigma d^3} = -0.00031\left(\frac{H}{d}\right) + 0.00113\left(\frac{H}{d}\right)^2 \quad \left(0 < \frac{H}{d} \leq 2\right) \tag{A2a}$$

$$\frac{E_c}{\eta\phi\sigma d^3} = -0.00325\left(\frac{H}{d}\right) + 0.00130\left(\frac{H}{d}\right)^3 \quad 2 < \frac{H}{d} < 5 \tag{A2b}$$

$$\frac{E_s}{\eta\phi\sigma d^3} = -0.005441\left(\frac{H}{d}\right) + 0.01386\left(\frac{H}{d}\right)^2 \quad 0 < \frac{H}{d} < 5 \tag{A3}$$

$$\frac{E_p}{\eta\phi\sigma d^3} = -0.00506\left(\frac{H}{d}\right) + 0.01506\left(\frac{H}{d}\right)^2 \quad 0 < \frac{H}{d} \leq 1 \tag{A4a}$$

$$\frac{E_p}{\eta\phi\sigma d^3} = -0.01\left(\frac{H}{d}\right) + 0.02\left(\frac{H}{d}\right)^3 \quad 1 < \frac{H}{d} < 5$$

(A4b)

Cone cracking, scabbing and perforation for H/d≥5

$$\frac{E_c}{\sigma d^3} = \frac{\pi}{4}\left[\frac{H}{d} - 4.7\right]$$

(A5)

$$\frac{E_s}{\sigma d^3} = \frac{\pi}{4}\left[\frac{H}{d} - 4.3\right]$$

(A6)

$$\frac{E_p}{\sigma d^3} = \frac{\pi}{4}\left[\frac{H}{d} - 3.0\right]$$

(A7)

η is estimated by Eqs. (19) and (20).

REFERENCES

1. Kennedy RP. A review of procedures for the analysis and design of concrete structures to resist missile impact effects Nucl Eng Des 1976;37(2):183e203.
2. Li QM, Reid SR, Wen HM, Telford AR. Local impact effects of hard missiles on concrete targets. Int J Impact Eng 2005; 32(1e4):224e84.
3. Reid SR, Wen HM. Predicting penetration, cone cracking, scabbing and perforation of reinforced concrete targets struck by flat-faced projectiles. 2001. UMIST Report ME/AM/02.01/TE/G/018507/Z.
4. BNFL. Appendix H. Reinforced concrete slab local damage assessment. R3 impact assessment procedure, vol. 3. Magnox Electric Plc & Nuclear Electric Limited; 2003.
5. Teland JA, Sjol H. Penetration onto concrete by truncated projectiles. Int J Impact Eng 2003; 30:447e64.
6. Bainbridge P. World impact datadSRD impact database version Pre 3i. July 1988. CCSD/CIWP (88)107(P).
7. [7] Tai YS. Flat ended projectile penetrating ultra-high strength concrete plate target. Theor Appl Fract Mec 2009; 51(2):117e28.

REFERENCES

8. Wang DR, Ge T, Zhou ZP, Wang MY. Investigation of calculation method for anti-penetration of reactive power steel fiber concrete (RPC). Explo Shock Wave 2006; 26(4):367e72.
9. Shi ZY, Tang WH, Zhao GM, Zhang RQ. Similarity study of the penetration depth for the concrete targets. J Ballist 2005; 17(1):62e6.
10. Forrestal MJ, Altman BS, Cargile JD, Hanchak SJ. An empirical equation for penetration depth of ogive-nose projectile into concrete targets. Int J Impact Eng 1994;15(4):395e405.
11. Canfield J, Clator I. Development of a scaling law and techniques to investigate penetration in concrete. Dahlgren, VA: Naval Weapons Labortory; 1966. NWL Report No.2057 U.S.
12. Forrestal MJ, Frew DJ, Hanchak SJ, Brar NS. Penetration of grout and concrete targets with ogive-nose steel projectiles. Int J Impact Eng 1996; 18(5):465e76.
13. Frew DJ, Hanchak SJ, Green ML, Forrestal MJ. Penetration of concrete targets with ogive-nose steel rods. Int J Impact Eng 1998; 21(6):489e97.
14. Forrestal MJ, Frew DJ, Hickerson JP, Rohwer TA. Penetration of concrete targets with deceleration-time measurements. Int J Impact Eng 2003; 28(5): 479e97.
15. Xu JB, LIN JD. The penetration of steel bar projectiles into concrete targets. Explo Shock Wave 2002;22(2):174e8.
16. Liang B, Chen XW, Ji YQ, Huang HJ, et al. Experimental study on deep penetration of reduced-scale advanced earth penetrating weapon. Explo Shock Wave 2008;28 (1):1e9.
17. O'Neil EF, Neeley BD, Cargile JD. Tensile properties of very-high-strength concrete for penetration-resistant structures. Shock Vib 1999;6(5e6):237e46.
18. Unosson M, Nilsson L. Projectile penetration and perforation of high performance concrete: experimental results and macroscopic modelling. Int J Impact Eng 2006; 32 (7):1068e85.
19. Zhang MH, Shim VPW, Lu G, Chew CW. Resistance of high-strength concrete to projectile impact. Int J Impact Eng 2005; 31(7):825e41.
20. Langberg H, Markeset G. High performance concrete-penetration resistance and material development. Norwegain Defense Construction Service; 1999.
21. Reid SR, Wen HM. Review and feasibility study of research into concrete impact and pipe-on-pipe impact. 1996. UMIST Report ME/AM 10.96/BL/G/46833/S.
22. [Sinclair ACE. Drop tests on a reinforced concrete floor at Rogerstone Power Station. 1993. TD/SEB/REP/41555/93, May 1993.
23. Dancygier AN, Yankelevisky DZ. High strength concrete response to hard projectile impact. Int J Impact Eng 1996; 18 (6):583e99.

24. Dancygier AN, Yankelevsky DZ, Jaegermann C. Response of high performance concrete plates to impact of non-deforming projectiles. Int J Impact Eng 2007; 34:1768e79.
25. Cargile JD, Giltrud ME, Luk VK. Perforation of thin unreinforced concrete slabs. Albuquerque, NM (United States): Sandia National Labs; 1993.
26. Beppu M, Miwa K, Itoh M, Katayama M, Ohno T. Damage evaluation of concrete plates by high-velocity impact. Int J Impact Eng 2008; 35:1419e26.
27. Borvik T, Langseth M, Hopperstad OS, Polanco-Loria MA. Ballistic perforation resistance of high performance concrete slabs with different unconfined compressive strengths. In: Brebbia CA, DeWilde WP, editors. High performance structures and composites. Southampton: WIT Press; 2002. p. 273e82.
28. Hanchak SJ, Forrestal MJ, Young ER, Ehrgott JQ. Perforation of concrete slabs with 48 MPa (7 ksi) and 140 MPa (20 ksi) unconfined compressive strengths. Int J Impact Eng 1992; 12 (1):1e7.
29. Brown SJ. Energy release protection for pressurized systems: part II review of studies into impact/terminal ballistics. Appl Mech Rev 1986;39(2):177e201.
30. Wang B, Cao YR, Tan DW. Experimental study on penetration of reinforced concrete by a high-speed penetrator with large mass. Explo Shock Wave 2013;33(1):98e102.
31. Wen HM. Predicting the penetration and perforation of targets struck by projectiles at normal incidence. Mech Struct Mach 2002; 30 (4):543e77.
32. Hill R. The mathematical theory of plasticity. Oxford University Press; 1950.
33. Wen HM, Yang Y. A note on the deep penetration of projectiles into concrete. Int J Impact Eng 2014; 66:1e4.

GLOSSARY

X	Penetration depth into semi-infinite target
D	Projectile diameter
N^*	Nose shape factor
H	Target thickness
M	Projectile mass
V_i	Projectile impact velocity
f'_c	Unconfined compressive strength of concrete
E_k	Initial kinetic energy of missile

E_c	Critical impact energy for through-thickness cone cracking
E_s	Critical impact energy for scabbing
E_p	Critical impact energy for perforation
χ_i	Constants defined in Eqs. (16), (17) and (18), i = c,s,p
σ_t	Mean dynamic resistive pressure, defined by Eq. (2)
σ	Mean dynamic resistive pressure, defined by Eq. (11)
η	Effect of rebar quantity and spacing, defined in Eq. (6)
d_r	Rebar diameter
C_r	Rebar spacing
r_t	Total bending reinforcement
r	Each way each face (% EWEF)
α	Parameter defined in Eq. (11)
β	Parameter defined in Eq. (11)
ρ_t	Concrete density
Y	A measure of shear strength for concrete
Φ	Projectile nose shape factor
Θ	Cone angle of a conical-nosed missile
Ψ	Caliber-radius-head of an ogival-nosed projectile
S	Radius of an ogive nose
$V_{cracking}$	Limit velocity for cone cracking obtained by experiment
$V_{scabbing}$	Limit velocity for scabbing obtained by experiment
V_{50}	Limit velocity for perforation obtained by experiment
V_c	Limit velocity for cone cracking obtained by model
V_s	Limit velocity for cone scabbing obtained by model
V_b	Limit velocity for perforation obtained by model

CITATION

H.M. Wen, Y.X. Xian, A unified approach for concrete impact, International Journal of Impact Engineering, Volume 77, March 2015, Pages 84-96, ISSN 0734-743X, http://dx.doi.org/10.1016/j.ijimpeng.2014.11.015.

CHAPTER 2

Microstructure and Properties of Concrete Using Bottom Ash and Waste Foundry Sand as Partial Replacement of Fine Aggregates

Yogesh Aggarwal[a]*, Rafat Siddique[b]

[a]Civil Engineering Department, National Institute of Technology, Kurukshetra, India
[b]Civil Engineering Department, Thapar University, Patiala 147004, India

ABSTRACT

The possibility of substituting natural fine aggregate with industrial by-products such as waste foundry sand and bottom ash offers technical, economic and environmental advantages which are of great importance in the present context of sustainability in the construction sector. The study investigated the effect of waste foundry sand and bottom ash in equal quantities as partial replacement of fine aggregates in various percentages (0–60%), on concrete properties such as mechanical (compressive strength, splitting tensile strength and flexural strength) and durability characteristics (rapid chloride penetration and deicing salt surface scaling) of the concrete along with microstructural analysis with XRD and SEM. The results showed that the water content increased gradually from 175 kg/m^3 in control mix (CM) to 238.63 kg/m^3 in FB60 mix to maintain the workability and the mechanical behavior of the concrete with fine aggregate replacements was comparable to that of conventional concrete except for FB60 mix. The compressive strength was observed to be in the range of 29–32 MPa, splitting tensile strength in the range of 1.8–2.46 MPa, and flexural strength in the range of 3.95–4.10 MPa on the replacement of fine aggregates from 10% to 50% at the interval of 10%.

Furthermore, it was observed that the greatest increase in compressive, splitting tensile strength, and flexural strength compared to that of the conventional concrete was achieved by substituting 30% of the natural fine aggregates with industrial by-product aggregates. The inclusion of waste foundry sand and bottom ash as fine aggregate does not affect the strength properties negatively as the strength remains within limits except for 60% replacement. The morphology of the formations arising as a result of the hydration process was not observed to change in the concrete with varying percentages of waste foundry sand and bottom ash.

INTRODUCTION

High consumption of natural sources, high amount production of industrial wastes and environmental pollution are some of the factors which are responsible for obtaining new solutions for a sustainable development. A sustainable development can be achieved only if the resource efficiency increases. The resource efficiency increment is possible by the reduction in use of energy and materials. Thus, solution is utilization of industrial by-products or solid wastes such as fly ash (FA), bottom ash (BA), waste foundry sand (FS), slag, silica fume, and waste glass in producing concrete. These concrete technologies reduce the negative effects on economical and environmental problems of concrete industry by having low costs, high durability properties and environmental friendliness.

When coal is burned in a coal fired boiler, it leaves behind ash, some of which is removed from the bottom of the furnace known as bottom ash, and some of which is carried upward by the hot combustion gases of the furnace, and removed by collection devices (fly ash). Worldwide, coal-fired power generation presently accounts for roughly 38% of total electricity production. Coal use in some of the more developed countries is static or is in decline. Significant increases in coal-fired generation capacity are taking place in many of the developing nations and large capacity increases are planned. During coal-fired electric power generation three types of coal combustion products (CCPs) are obtained. These by-products; fly ash, bottom ash and boiler slag are the largest sources of industrial waste. Utilization of CCPs in construction industry is an important issue involving reduction in technical and economical problems of plants, besides reducing the amount of solid wastes, greenhouse gas emissions and conserving existing natural resources. Some authors have reported the use of bottom ash in concrete as partial replacement of portland cement [1],[2] and [3] or as a partial replacement of fine aggregates [4], [5], [6], [7], [8] and [9].

INTRODUCTION

A foundry produces metal castings by pouring molten metal into a preformed mold to yield the resulting hardened cast. The metal casts include iron and steel from the ferrous family and aluminum, copper, brass and bronze from non-ferrous family. Waste foundry sand is high quality silica sand with uniform physical characteristics. It is a by-product of ferrous and non-ferrous metal casting industries, where sand has been used for centuries as a molding material because of its thermal conductivity. Foundries successfully recycle and reuse the sand many times. When the sand can no longer be reused in the foundry, it is removed from the foundry and is termed as waste foundry sand. Several authors have reported the use of used-waste foundry sand in various civil engineering applications such as highway applications [10], [11], [12], [13], [14],[15], [16] and [17], leaching aspect of usage of foundry sand [18], [19], [20] and [21], controlled low strength materials [22], [23] and [24], concrete and concrete related products like bricks, blocks and paving stones [25], [26], [27], [28] and [29], asphalt concrete [30].

Coal-combustion bottom ash and used foundry sand are abundant by-products which appear to possess the potential, to partially replace regular sand as a fine aggregate in concrete mixtures, providing a recycling opportunity for them. If these recycled materials can be substituted for part of the cementitious and virgin aggregate materials in concrete mixtures without sacrificing, or even improving strength and durability, there are clear economic and environmental gains. One of the primary impediments to beneficial reuse of industrial by-products such as waste foundry sands and bottom ash is a lack of engineering data that designers can use to evaluate the efficacy and economy of using the by-product in place of the natural sand. The engineering properties and behavior of sands can be readily estimated from the literature for use in preliminary design. In contrast, there is a dearth of similar information for industrial by-products and there are insufficient data to confirm that industrial by-products, which appear similar to sands, also have comparable engineering properties. With emphasis now being placed on engineering for sustainable development, there is a pressing need to provide this practical information to designers. Fulfilling this need is the primary purpose of this study. The objective is to provide practical information, regarding the strength, durability and micro-structural properties of bottom ash and waste foundry sand as replacement of fine aggregates in concrete. Both waste foundry sand and bottom ash have been studied as aggregate replacement, separately. The value of the current research is the use of both together. The present experimental study was conceived following the general purpose of testing new sustainable building processes and modern production systems, aimed not only at saving natural raw materials and reducing energy consumption, but also to recycle industrial by-products. The objectives of this study are to investigate the effect of use of bottom ash and waste foundry sand in equal quantities as partial replacement of fine aggregates in various percentages (0–60%), on concrete properties such as mechanical and durability characteristics of the concrete along with micro-structural analysis with XRD and SEM.

EXPERIMENTAL PROGRAM

The effect of using various percentages of bottom ash and waste foundry sand as partial replacement of the fine aggregate in concrete was investigated. Also, the effect of incorporating waste foundry sand and bottom ash, in concrete on the mechanical, durability properties and microstructure were evaluated.

Materials and Mix Proportions
Portland Pozzolana Cement (53 MPa) conforming to Indian standard specifications IS: 1489-1991 [31] with consistency as 27%, specific gravity as 3.56 and fineness as 5%, was used. Locally available natural sand with 4.75 mm maximum size was used as fine aggregate, fulfilling the requirements of ASTM C 33-02a [32]and IS:383-1970 [33] along with crushed stone of 20 mm maximum size used as coarse aggregate. Locally available waste foundry sand was used as partial replacement of fine aggregates (regular sand). The waste foundry sand showed lower fineness modulus and bulk density than the regular sand. As per the particle size distribution of the waste foundry sand, the size corresponding to 50% of passing (d_{50}) was around 33 μm and average diameter of waste foundry sand particle was observed to be 28.8 μm. Coal bottom ash obtained from Panipat Thermal Power Station, Panipat, Haryana, was also used as partial replacement of fine aggregates. The properties of coal bottom ash conformed to IS:3812-2003. The particle size distribution of bottom ash was measured, which showed that, of the particles 100% were smaller than 56 μm and 38% were smaller than 31.3 μm with average diameter of the particle size distribution was 33.4 μm with standard mean deviation of 8.1 μm for bottom ash. Table 1 gives the chemical composition of waste foundry sand and bottom ash while Table 2 gives the physical properties of the aggregates used. A polycarboxylic ether based superplasticizer of CICO brand complying with ASTM C-494 type F [34], IS:9103-1999 [35] and IS:2645-2003[36] was used.

EXPERIMENTAL PROGRAM

Table 1: Chemical properties of coal bottom ash and waste foundry sand

Constituents	Coal bottom ash		Waste foundry sand	
	Percent by weight	Codal requirement	Percent by weight	Requirements (American foundry men's society, 1991)
Silica (SiO_2)	57.76	35%(min)	78.81	87.9%
Iron oxide (Fe_2O_3)	8.56	70%(min)SiO_2 + Fe_2O_3 + Al_2O_3	4.83	0.94%
Alumina (Al_2O_3)	21.58	70%(min)SiO_2 + Fe_2O_3 + Al_2O_3	6.32	4.70%
Calcium oxide (CaO)	1.58	-	1.88	0.14%(min)
Magnesium oxide (MgO)	1.19	5%(max)	1.95	0.3%
Total sulphur (SO_3)	0.02	3%(max)	0.05	0.09%
Alkalies (a) sodium oxide (Na_2O)	0.14	1.5%(max)	-	-
(b) Potassium oxide (K_2O)	1.08			
Chloride	0.01	0.05%(max)	0.04	-
Loss on ignition	5.80	5%(max)	2.15	5.15%(max)

Table 2: Physical properties of aggregates

Aggregates	Specific gravity	Unit weight (kg/m³)	Fineness modulus
Sand	2.63	1890	3.03
Waste foundry sand	2.61	1638	1.78
Bottom ash	1.93	948	1.60
Coarse aggregates	2.77	1650	6.74

Seven mix proportions were prepared. First was control mix (without bottom ash and waste foundry sand), and the other six mixes contained bottom ash and waste foundry sand in equal proportions. Fine aggregate (sand) was replaced with bottom ash and waste foundry sand by weight. The proportions of fine aggregate replaced ranged from 10% to 60% at the increment of 10%. Mix proportions are as given in Table 3. The control mix without waste foundry sand and bottom ash was proportioned as per Indian standard specifications IS:10262-1982 [37], to obtain a 28-day cube compressive strength of 36 MPa.

Table 3: Mix proportions of concrete mixes containing bottom ash & waste foundry sand

Mix no.	CM	FB10	FB20	FB30	FB40	FB50	FB60
Cement (kg/m³)	350	350	350	350	350	350	350
Foundry Sand (%)	0	5	10	15	20	25	30
Foundry Sand (kg/m³)	0	30.25	60.50	90.75	121.00	151.25	181.50
Bottom ash (%)	0	5	10	15	20	25	30
Bottom ash (kg/m³)	0	30.25	60.50	90.75	121.00	151.25	181.50
Water (kg/m³)	175	180.30	185.60	190.90	201.50	212.12	238.63
W/C	0.5	0.52	0.53	0.55	0.58	0.61	0.68
Sand SSD (kg/m³)	605	544.5	484.0	423.5	363.0	302.5	242.0
Fine aggregate (kg/m³)	605	605	605	605	605	605	605
Coarse aggregate (kg/m³)	1260	1260	1260	1260	1260	1260	1260
Superplasticizer (kg/m³)	1.75	1.75	1.75	1.75	1.75	1.75	1.75
Slump (mm)	30	30	30	30	30	30	30
Compaction factor	0.83	0.81	0.78	0.81	0.78	0.78	0.81
Vee-bee consistometer (sec)	5.98	5.20	6.42	5.54	6.44	6.68	5.26
Air temperature (°C)	23	25	24	26	25	25	34
Concrete temperature (°C)	25	25	25	26	25	27	28
Air content (%)	2.1	2.6	2.6	2.7	2.7	2.9	3.4
Fresh concrete density (kg/m³)	2392	2397	2402	2408	2418	2428.87	2455.38

For these mix proportions, required quantities of materials were weighed. The mixing procedure adopted was as follows: First, the cement, waste foundry sand, and coal bottom ash were dry mixed till a uniform color was obtained without any clusters of cement, waste foundry sand and bottom ash particles. Weighed quantities of coarse aggregates and sand were then mixed in dry state, thoroughly until a homogeneous mix was obtained. Water was then added in three stages as 50% of total water to the dry mix of concrete in first stage; 40% of water and superplasticizer to the wet mix; Remaining 10% of water was sprinkled on the above mix and it was thoroughly mixed. All the moulds were properly oiled before casting the specimens. The casting immediately followed mixing, after carrying out the tests for fresh properties. The top surface of the specimens was scraped to remove excess material and achieve smooth finish. The specimens were removed from moulds after 24 h and cured in water till testing or as per requirement of the test.

Testing Procedure

Fresh concrete properties such as slump flow, compaction factor, vee-bee consistometer were determined according to an Indian Standard specification IS:1199-1959 [38]. The results are presented in Table 3. The 150 mm concrete cubes were cast for compressive strength, cylinders of size 150 mm × 300 mm for splitting tensile strength and beams of size 100 × 100 × 500 mm for flexural strength. After required period of curing, the specimens were taken out of the curing tank and their surfaces were wiped off. The various tests performed were compressive strength test of cubes (150 mm side), splitting tensile strength of cylinders (150 mm × 300 mm), at 7, 28, 90, and 365 days and flexural strength of beams (100 × 100 × 500 mm) at 28, 90, and 365 days, as per IS:516-1959 [39].

The cylinders (100 mm × 200 mm) were cast for rapid chloride penetration resistance test and were sliced 51-mm (2-in.) thick of 102-mm (4-in.) nominal diameter. Rapid chloride penetration resistance test (according to ASTM C 1202-97 [40] covered the determination of the electrical conductance of concrete to provide a rapid indication of its resistance to the penetration of chloride ions. The test method consisted of monitoring the amount of electrical current passed through 51-mm (2-in.) thick slices of 102-mm (4-in.) nominal diameter cores or cylinders for a 6-h period. A potential difference of 60 V dc was maintained across the ends of the specimen, one of which was immersed in a sodium chloride solution, the other in a sodium hydroxide solution. The total charge passed, in coulombs, was related to the resistance of the specimen to chloride ion penetration.

The test method (according to ASTM C 672 [41]) covers the determination of the resistance to scaling of a horizontal concrete surface exposed to freezing and thawing cycles in the presence of deicing chemicals. It is intended for the use in evaluating surface resistance qualitatively by visual examination. Specimens of size 225 × 225 × 25 mm were prepared for all mixes. A dike of 25 mm wide and 20 mm high was placed along the perimeter of the top surface of the specimens. The specimens were removed from the moulds after 24 h and then cured. Since, the concretes with strength at different ages are to be compared, the specimen were cured till that age. At the desired age, the specimens were removed from moist storage and stored in air for 14 days. After completion of moist and air curing, the flat surface of the specimen was covered with 6 mm thick layer solution of calcium chloride solution and water. 100 mL of solution contains 4 g of anhydrous calcium chloride. The specimens were placed in freezing environment for 16–18 h. The specimens were removed from freezer and placed in air for 6–8 h. Water was added at each cycle, as necessary, to maintain proper depth of solution. Cycle was repeated after every 24 h, flushing the surface at the end of each 5 cycles. After making the visual examination, the solution was replaced and the test continued for 50 cycles. Visual rating of the surface on the basis of the scale given in Table 4 was carried out.

Table 4: Rating for deicing salt surface scaling (ASTM C 672)

Rating	Condition of surface
0	No scaling
1	Very slight scaling (3 mm depth, max., no coarse aggregates visible)
2	Slight to moderate scaling
3	Moderate scaling (some coarse aggregate visible)
4	Moderate to severe scaling
5	Severe scaling (coarse aggregate visible over entire surface)

X-ray diffraction analysis (XRD) was done on Philips PW 1140/09. Diffractometer operated at 35 KV, using Cu Kα radiation and Ni filler ($\lambda = 1.5418$ Å). The samples for X-ray diffraction analysis were prepared in powdered form. The concrete sample was taken from the inner core of the matrix. X-ray diffraction is based on the fact that, in a mixture, the measured intensity of a diffraction peak is directly proportional to the content of the substance producing it (Soroka [42]). Since 2d is a known constant, the 2θ setting of each peak corresponds to a certain wave length.

The type, amount, size, shape, and distribution of phases present in a solid constitute its microstructure. It is the application of transmission and scanning electron microscopy techniques which has made it possible to resolve the microstructure of the materials to a fraction of one micrometer. Although, concrete is the most widely used structural material, its microstructure is heterogeneous and highly complex. Also, the microstructure-property relationships in concrete are not fully developed.

At the macroscopic level, concrete may be considered as a two-phase material, consisting of aggregate particles dispersed in a matrix of cement paste. At the microscopic level, complexities of the concrete microstructure are evident that two phases are neither homogeneously distributed with respect to each other, nor are they themselves homogeneous (Lea's [43]).

Original microstructure and morphology of the hydrate mixes were observed on fractured surfaces. Fractured small samples were mounted on the SEM stubs with gold coating. The scanning electron microscopic studies of various concrete samples and constituent materials were carried out using Philips XL20 Scanning Electron Microscope. The concrete specimens were first cured in water for 365 days and then oven dried at 105 °C for 24 h.

RESULTS AND DISCUSSIONS

Fresh Concrete Properties
The workability of fresh concrete is a composite property which includes the diverse requirements of stability, mobility, compactibility, placeability, and finishability. Slump is a measure indicating the consistency or workability of concrete. Slump for control mix CM and FB mixes was observed to be 30 mm. The compaction factor values for control mix, and FB mixes corresponded to the slump flow values as perTable 3. The presence of finer waste foundry sand particles and bottom ash, in concrete lead to the increase in the water demand, as compared to the regular sand particles. Thus, to maintain the workability within specified range (slump has been kept constant at 30 mm), the water content was gradually increased with increase in replacement of sand with waste foundry sand and bottom ash, which gave an idea about the increase in water demand due to increase in replacement of sand with waste foundry sand and bottom ash. The water content increased gradually from 175 kg/m^3 in control mix (CM) to 238.63 kg/m^3 in FB60 mix. It was observed that for initial replacements of 10%, 20% and 30%, the increase in water content was constant and thereafter for 40% and 50%, again it remained constant but almost twice the value of initial replacements. For FB60 mix, the water content increased drastically which reflected on various strengths.

Compressive Strength

Compressive strength results of FB mixes made with waste foundry sand and bottom ash in equal percentages are as given in Table 5. There is a decrease in the compressive strength of concrete mixes with the inclusion of waste foundry sand and bottom ash as replacement of regular sand. The percentage decrease in comparison to reference mix at various ages is as shown in Table 6. It was observed that the mixes with replaced fine aggregates had less difference from the reference mix as age increased to 365 days; the difference between CM and FB mixes left between 3% and 9% for all mixes (except FB60 mix which has showed higher decrease at all ages, but that also has decreased with increase in age).

Table 5: Various strengths of CM and FB mixes

Age	7-day (MPa)		28-day (MPa)			90-day (MPa)			365-day (MPa)		
Mix	Compressive strength	Splitting Tensile strength	Compressive strength	Splitting Tensile strength	Flexural strength	Compressive strength	Splitting Tensile strength	Flexural strength	Compressive strength	Splitting Tensile strength	Flexural strength
CM	25.58	1.30	36.27	2.08	4.44	43.91	2.66	5.03	44.42	2.97	5.37
FB10	17.67	1.34	29.02	1.80	4.10	33.47	2.04	4.90	40.59	2.12	5.24
FB20	18.25	1.39	29.63	2.05	4.00	34.69	2.45	4.80	41.44	2.78	5.05
FB30	19.36	1.54	31.81	2.46	4.34	37.37	2.83	4.97	42.69	3.22	5.30
FB40	18.42	1.46	29.95	2.35	3.87	35.85	2.72	4.61	42.03	2.98	4.83
FB50	18.91	1.41	30.53	2.25	3.95	36.86	2.58	4.71	42.27	2.86	4.93
FB60	13.45	0.84	21.08	1.45	3.60	24.03	1.52	4.46	28.22	1.53	4.55

Table 6: Decrease in compressive strength at various ages in comparison to reference mix(CM)

Mix	7-day (%)	28-day(%)	90-day(%)	365-day(%)
FB10	30.92	19.99	23.78	8.62
FB20	28.66	18.31	21.00	6.71
FB30	24.32	12.30	14.89	3.89
FB40	27.99	17.42	18.36	5.38
FB50	26.08	15.83	16.06	4.84
FB60	47.42	41.88	45.27	36.47

Original microstructure and morphology of the hydrate mixes were observed on fractured surfaces. Fractured small samples were mounted on the SEM stubs with gold coating. The scanning electron microscopic studies of various concrete samples and constituent materials were carried out using Philips XL20 Scanning Electron Microscope. The concrete specimens were first cured in water for 365 days and then oven dried at 105 _C for 24 h.

The strength variation was observed to be marginal in the replaced mixes i.e. FB mixes, as the waste foundry sand tends to increase the strength as observed in the waste foundry sand mixes given elsewhere [44] and inclusion of same proportion of bottom ash tends to decrease the strength [4]. Thus, not much difference in strength was observed from 10% to 50% replacement of sand with equal percentages of waste foundry sand and bottom ash. Also, the strength observed at FB30 was highest as compared to the other FB mixes, but less than that of reference mix. The maximum strength was obtained at replacement of 30% (15% waste foundry sand and 15% bottom ash) in the replaced mixes, which can be adjudged as optimum mix.

Compressive strength of all FB mixes increased with age. At 7 days, all FB mixes also showed the strength lower than CM mix but as the age increases to 365 days gradually attained strengths marginally lower than the reference mix CM as shown in Fig. 1. The difference in the strengths of various mixes from reference mix was observed to decrease for various FB mixes as given in Table 6 with increase in age from 7 to 365 days. The reference mix gained increase in strength from 7 to 28 days of 41.79%, from 28 to 90 days of 21.06% and from 28 to 365 days of 1.16% which is characteristic of normal concrete. The FB mixes attained a relatively constant strength of 56–64% from 7 to 28 days, 14–19% from 28 to 90 days, and 14–21% from 90 to 365 days

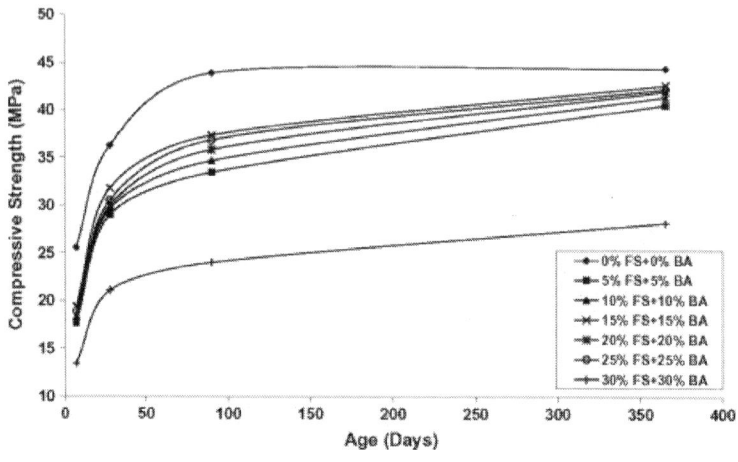

Figure 1: Variation of cube compressive strength with age for CM & FB mixes.
The FB mixes showed 7-day strength between 60% and 63% of 28-day strength; 90-day strength between 114% and 120%; and 365-day strength 133–140% of the 28-day strength. It was observed that at 7, and 90 days the gain of strength was less than that of the CM mix but at 365 days the strength gain for FB mixes, exceeded the strength gain of CM mix.

Splitting Tensile Strength

The results of splitting tensile strength of FB mixes are indicated in Table 5. The increase of 3.1%, 6.92%, 18.46%, 12.31% and 8.46% was observed for the mixes FB10–FB50 and decrease of 35.38% was observed for the mix FB60 at 7 days with regards to reference mix. The decrease of 13.46%, 1.44% and 30.29% for mixes FB10, FB20 and FB60 was observed at 28 days, with increase of 18.27%, 12.98%, and 8.17% for the mixes FB30, FB40 and FB50 in comparison to the reference mix CM. Similarly, at 90 days, decrease of 23.31%, 7.89%, 3.01% and 42.86% for mixes FB10, FB20, FB50, and FB60 was observed, with increase of 6.39% and 2.26% for the mixes FB30 and FB40 in comparison to the reference mix CM. The trend observed was same at 365 days as indicated in Fig. 2.

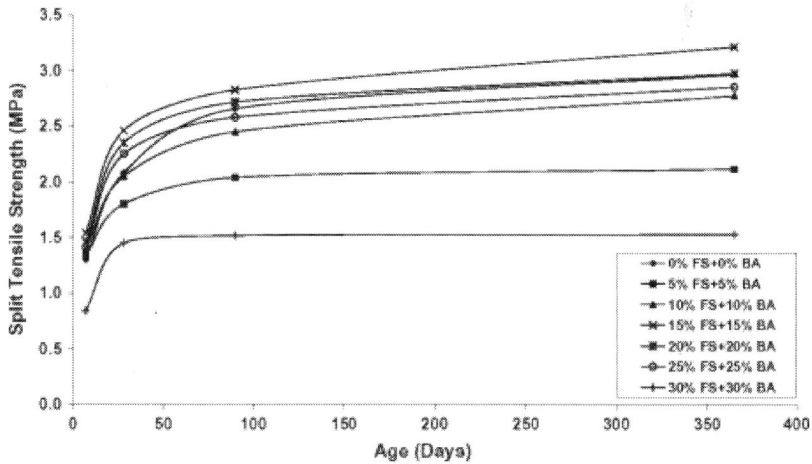

Figure 2: Variation of splitting tensile strength with age for CM & FB mixes.

At 7 days, all FB mixes showed the strength higher than CM mix except FB60, but as the age increases to 365 days all mixes showed almost comparable strengths to that of reference mix CM.

Table 7 gives the ratio of splitting tensile strength to the cube compressive strength of FB mixes. The results indicate that splitting tensile vary from 0.062 to 0.080; 0.062 to 0.078; 0.061 to 0.076 and 0.052 to 0.075 times the compressive strength, at ages of 7, 28, 90 and 365 days, respectively.

Table 7: Ratio of splitting tensile strength and cube compressive strength for FB mixes

Mix	7-day	28-day	90-day	365-day
FB10	0.076	0.062	0.061	0.052
FB20	0.076	0.069	0.071	0.067
FB30	0.080	0.077	0.076	0.075
FB40	0.079	0.078	0.076	0.071
FB50	0.075	0.074	0.070	0.068
FB60	0.062	0.069	0.063	0.054

For FB mixes – The ratio of splitting tensile strength to cube compressive strength of concrete varies from 5% to 8%.

For Normal strength mixes – The ratio of splitting tensile strength to cube compressive strength of concrete is in the range of 7–9% [45].

Flexural Strength

The flexural strength of all FB mixes was observed to be less than the strength of reference mix CM and FB30 attained maximum strength i.e. more than all other FB mixes at all the ages. The flexural strength results of concrete mixes are shown in Table 5. Like compressive and splitting tensile strength, flexural strength of concrete mixes varied marginally with the increase in waste foundry sand and bottom ash content. The 28-day flexural strength of CM mix was observed as 4.44 MPa, whereas mixes FB10, FB20, FB30, FB40, FB50, and FB60 showed a decrease of 7.66%, 9.91%, 2.25%, 12.84%, 11.04%, and 18.92%. At 90 days, a decrease of 2.58%, 4.57%, 1.19%, 8.35%, 6.36% and 11.33% was observed for mixes FB10, FB20, FB30, FB40, FB50 and FB60, in comparison to the reference mixture CM. The same trend was observed at the age of 365 days with FB10, FB20, FB30, FB40, FB50, and FB60 showing decrease of 2.42%, 5.96%, 1.3%, 10.06%, 8.19%, and 15.27%.

From the results given in Fig. 3, it is also evident that flexural strength of FB mixes increased with the age. The FB mixes showed decrease in the strengths in comparison to reference mix as 7.66–2.42%, 9.91–5.96%, 2.25–1.30%, 12.84–10.06%, 11.04–8.19%, and 18.92–15.27% for mixes FB10, FB20, FB30, FB40, FB50, and FB60 with increase in age from 28 to 365 days. Also, an increase in strength from 28 to 90 days was observed to be 13.29% for CM mix whereas FB mixes showed increase in strength from 14.52% to 23.89%, from 28 to 90 days. Between 90 and 365 days, an increase in strength for CM mix was 6.76% and the FB mixes showed an increase of 2.02–6.94%.

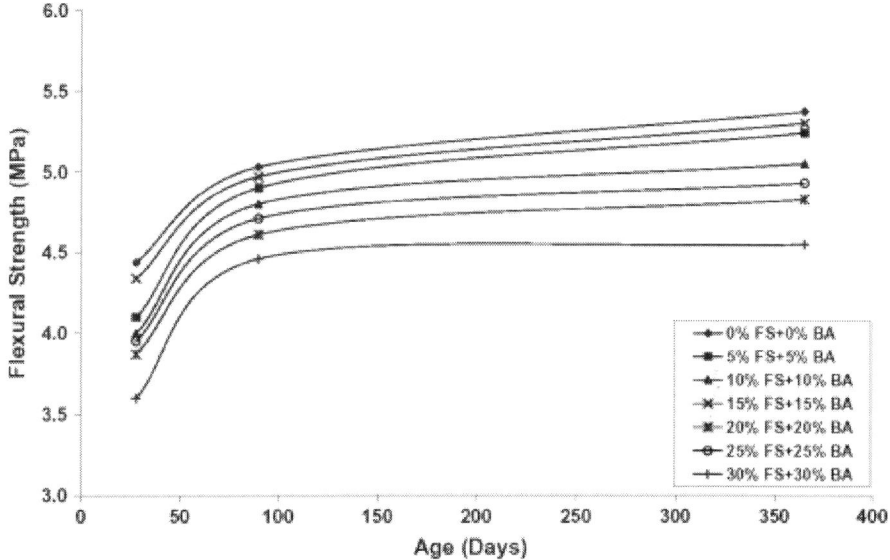

Figure 3: Variation of flexural strength with age for CM & FB mixes.

Relationship of Flexural Strength to Compressive and Splitting Tensile Strength

The flexural strength as observed from Table 8 was 12.2%, 11.5% and 12.1% of the cube compressive strength at the age of 28 days, 90 days, and 365 days, respectively of the reference mix CM. The FB mixes showed the variation of 12.9–17.1% at 28 days, 12.8–18.6% at 90 days and 11.5–16.1% at 365 days of the cube compressive strength as has also been observed by Mehta and Monterio [45] and Price [46].

Table 8: Ratio of flexural strength and cube compressive strength for CM & FB mixes

Mix	28-day	90-day	365-day
CM	0.122	0.115	0.121
FB10	0.141	0.146	0.129
FB20	0.135	0.138	0.122
FB30	0.136	0.133	0.124
FB40	0.129	0.129	0.115
FB50	0.129	0.128	0.117
FB60	0.171	0.186	0.161

For normal concrete – The ratio of flexural strength to cube compressive strength of concrete is nearly 0.11 to 0.18 [45].

The flexural strength was observed to be 2.13, 1.89 and 1.81 times of the splitting tensile strength at the age of 28 days, 90 days and 365 days, respectively of the reference mix CM. The FB mixes showed the variation of 1.64–2.4 times at 28 days, 1.69–2.9 times at 90 days and 1.6–2.9 times at 365 days of the splitting tensile strength as given in Table 9.

Table 9: Ratio of flexural strength and splitting tensile strength for CM & FB mixes

Mix	28-day	90-day	365-day
CM	2.135	1.891	1.808
FB10	2.278	2.402	2.472
FB20	1.951	1.959	1.817
FB30	1.764	1.756	1.646
FB40	1.647	1.695	1.621
FB50	1.756	1.826	1.724
FB60	2.483	2.934	2.974

For normal concrete – Flexural strength is about 1.5–2 times the splitting tensile strength of concrete [45], [61] and [62].

Table 10 shows the computations of ratios of flexural strength (fr) to the square root of the cube compressive strength ($\sqrt{fck}$) of experimental values of present investigation and the theoretical values of flexural strength based on expressions proposed by earlier investigators (ACI Committee [47], IS:456-2000 [48]). An average value so obtained for different concrete mixes has been found to be 0.744. This, in the general form, would give an expression as

$$Fr = 0.744\sqrt{fck} \qquad (1)$$

Further, a comparison of the experimental results has been made with those of other authors and shown in Fig. 4. The results of the flexural strength of present study are less than the values of the flexural strength as reported by Siddique et al. [25] and higher than those reported by ACI Committee 363 and slightly higher than IS:456-2000 which gives the average value as 0.7. It was observed that the concrete mixes containing waste foundry sand and bottom ash behave in similar manner to that as plain concrete.

Table:10 Comparison of experimental values of flexural strength (fr) with the theoretical values predicted by other researchers

Mix	28 days cube compressive strength, fck (N/mm²)	Flexural Strength, Fr (N/mm²)			Ratios based on experimental values fr/√fck
		Exp.	Theoretical values as per references		
			ACI committee	IS:456-2000	
CM	36.27	4.44	4.19	4.22	0.7372
FB10	29.02	4.10	3.55	3.77	0.7611
FB20	29.63	4.00	3.60	3.81	0.7348
FB30	31.81	4.34	3.75	3.95	0.7695
FB40	29.95	3.87	3.64	3.83	0.7072
FB50	30.53	3.95	3.70	3.87	0.7149
FB60	21.08	3.60	3.01	3.21	0.7841
Average value					0.744

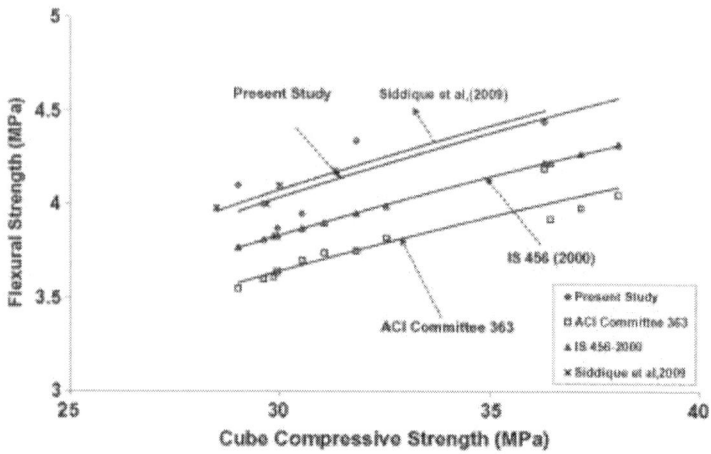

Figure 4: Comparison of experimental values of flexural strength (Fr) with the theoretical values predicted by other researchers.

Theoretical expressions for flexural strength:

ACI Committee 363, proposed the expression for flexural strength as:

$$Fr = 0.94\sqrt{fcc} \text{ in MPa} \quad \text{for } 21\text{ MPa} < fcc \leqslant 83\text{ MPa}$$

IS 456-2000, proposed the expression for flexural strength for NSC as:

$$Fr = 0.7\sqrt{fck} \text{ in MPa} \quad \text{for } fck \leqslant 60\text{ MPa}$$

Compressive strength is assumed as an adequate index for all types of strength, and therefore a direct relationship ought to exist between the compressive and tensile or flexural strength of a given concrete. It has been observed that relationship among various types of strength is influenced by factors like the methods by which the tensile strength is measured (i.e., direct tension test, splitting test, or flexure test), the quality of concrete (i.e., low, moderate or high-strength), the aggregate characteristics (e.g., surface texture and mineralogy), and admixtures (e.g., air-entraining and mineral admixtures) [45].

Bottom ash when used as aggregate replacement from 20% to 50% showed decrease in compressive strength, from control mix which was designed for almost comparable strength as in the present research[49]. Waste foundry sand when used as aggregate replacement from 10% to 60% showed increase in strength, from the control mix which was same as in the present research, with strength at 30% replacement of fine aggregates, being highest in the replaced mixes and even higher than control mix [44]. The use of waste foundry sand and bottom ash together compensate for the increase and decrease of strength due to replacement with waste foundry sand and bottom ash, respectively. It provides an opportunity to use two by-products together and achieve strength comparable to that of reference mix.

DURABILITY PROPERTIES

Durability and more specifically, resistance to chloride ion penetration and deicing salt surfacing are of major importance for reinforced concrete structures. Before using any industrial by-product such as waste foundry sand and bottom ash, these behavior needs to be investigated to study the effect of use of waste foundry sand and bottom ash on concrete.

Resistance to Rapid Chloride Penetration

The ability of concrete to resist the penetration of chloride ions is a critical parameter in determining the service life of steel-reinforced concrete structures exposed to deicing salts or marine environments. The effect of fly ash on the mass transfer properties of concrete has been well documented; however, no documentation of waste foundry sand and bottom ash together, as replacement of fine aggregates in concrete mixes is available. The measurement concerns the chloride ions that come into concrete and also those flowing through the samples.

The RCPT values of FB mixes, at the age of 90 and 365 days are given in Table 11. It can be observed that the RCPT value decreases with increase in age. For the FB mixes, RCPT values were found to be more than reference mix CM with maximum value observed for FB60 mix. Results reported [50], [51], [52] and [53] for the normal concrete and concrete with various additives also indicate decrease of RCPT values with increase in age.

Table 11: Charge passed and rating for FB mixes

Mix	Charge passed in coulombs (90-day)	Charge passed in coulombs (365-day)	Chloride ion penetrability
CM	578	323	Very low
FB10	628	357	Very low
FB20	616	306	Very low
FB30	600	321	Very low
FB40	664	383	Very low
FB50	652	377	Very low
FB60	741	486	Very low

The RCPT values in coulombs, from literatures available, were observed to be very high for normal concretes as shown in Table 12, mostly above 1500 coulombs for most of the mixes, such as 7890 coulombs [54]; 2766 coulombs [55], 1802 coulombs [56], 2869 coulombs [57], 5250 coulombs [58] at 28 days, 1725 coulombs [51] and 2971 coulombs [52] at 90 days; 3767 coulombs at 180 days [50]. The concrete with other additions like slag and rice husk ash also showed higher RCPT values at 28 days [53] and [59]. As compared to these concretes, it was observed that all FB mixes in this study showed very low RCPT values, less than 750 coulombs at 90 days and 500 coulombs at 365 days on the addition of waste foundry sand and bottom ash, as is evident from Table 11, which comes under very low category as per ASTM-1202C.

Table 12: Charge passed at various ages for various types of concretes.

Author	Type of concrete	Fly ash content (%)	RCPT Values (Coulombs)			
			28	56	90	180
Ramezanianpour and Malhotra [50]	Normal	0	4251			3767
Oh et al. [55]	Normal	0	2766			
Naik et al. [51]	Normal	0	3150		1725	
Mackechnie and Alexander [56]	Normal	0	1802			
Feng et al. [57]	Normal	0	2869			
Yang and Chiang [54]	Normal	0	7890			
Guneyisi [52]	Normal		4093		2971	
Gu et al. [58]	Normal		5250			
Gastaldini et al. [53]	Rice husk ash	0	3166	2136		
		20	1557	692		
Cho and Chiang [59]	Slag	0	9639			
		20	6355			
		40	2709			
		50	2148			
		70	1350			

It is observed that cement type, w/c ratio, curing condition, and testing age have effect on chloride permeability of concrete. The normal concretes or the concrete with various additives could vary in above parameters, thereby effecting RCPT values. The FB mixes in the present study showed very less RCPT value thereby indicating good resistance to permeability on addition of waste foundry sand and bottom ash in concrete.

Relation Between Compressive Strength and Resistance to Chloride Ion Penetration

The fundamental destructive effect of chlorides is their influence upon the reinforcement corrosion process, is primarily due to their capacity to negate the corrosion inhibiting properties of the alkaline cement paste pore solution. This risk increases with increasing concentration of free chlorides in the pore solution. It is generally believed that there is a threshold concentration of the chloride ions, which must be exceeded before corrosion occurs. The threshold concentration may depend on the concrete composition and on environmental parameters. The corrosion due to chloride ingress progresses at a much higher rate than that due to carbonation. In extreme cases, the corrosion rate in real structures can be 5 mm/year compared to 0.05 mm/year for carbonation-induced corrosion. The correlation as obtained from Fig. 5 shows a good relation between compressive strength and chloride permeability (R^2 as 0.80 and 0.78 for 90 days and 365 days). A decrease in RCPT values with increase in strength of FB mixes is also evident from Fig. 5.

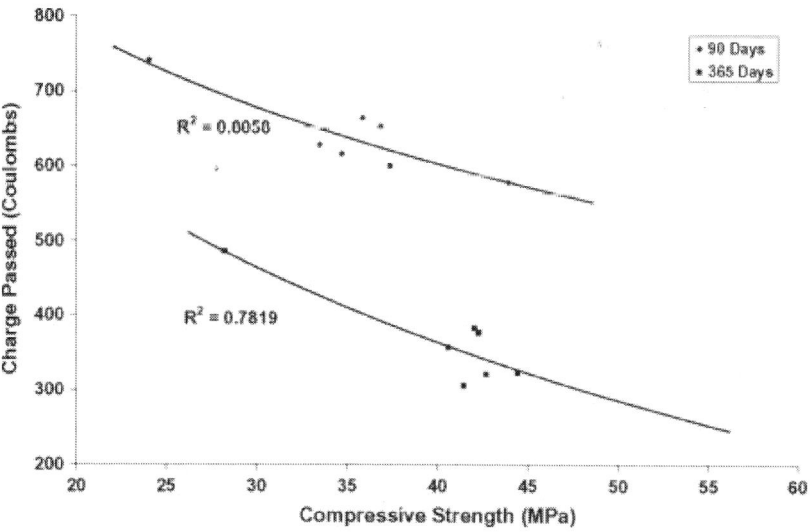

Figure 5: Relation between compressive strength and resistance to chloride ion penetration.

Deicing Salt Surface Scaling

For the average cumulative mass of scaled-off material obtained after 50 freezing-thawing cycles along with average visual surface ratings determined as per the ASTM C672, results are given in Table 13. For reference mix, weight loss was observed as 0.81 kg/m². All FB mixes showed weight loss lower than the reference mix which was observed as maximum of 0.79 kg/m² for FB60 mix at 90 days. Except for the FB mixes FB30 and FB60, all other mixes showed weight loss lower than reference mix at 365 days. The mass loss was observed to be between 0.31% and 0.66% at 90 days and between 0.92% and 1.87% at 365 days for the FB mixes. In the present study, the visual rating as per ASTM C 672, for most of the mixes was between 0 and 1, and never exceeded 2 as given in Table 13.

Table 13: Weight loss and visual rating ASTM C 672 for FB mixes

Mix no.	90-Day			365-Day		
	Weight loss (kg/m²)	Weight loss (%)	Visual rating	Weight loss (kg/m²)	Weight loss (%)	Visual rating
CM	0.8099	0.61	0	1.7975	1.36	1
FB10	0.3160	0.31	0	1.0074	0.92	0
FB20	0.4148	0.36	0	1.4222	1.17	1
FB30	0.6123	0.49	0	2.0741	1.67	1
FB40	0.4741	0.43	0	1.8765	1.62	1
FB50	0.4938	0.40	0	1.6198	1.30	1
FB60	0.7901	0.66	0	2.2716	1.87	1

Wang et al. [63] reported mass loss of the normal concrete specimens immersed in $CaCl_2$ deicing solution, as also used in the present study, about 2% of the total weight of the sample at 20 wetting dry cycles. For the present study, the mass loss was observed to be 0.61% and 1.36% at 90 and 365 days for the CM mix.

X-ray Diffraction (XRD)

XRD technique was conducted to analyze the components of concrete mixes and the results are shown inFig. 6, Fig. 7a, Fig. 7b, Fig. 7c, Fig. 7d, Fig. 7e and Fig. 7f. The X-ray diffraction pattern and analysis of the concrete mixes i.e. reference mix, and FB mixes was carried out at age of 365 days. One of the major problems encountered in the qualitative and quantitative analysis of cement is that there are strong overlapping of major diffraction peaks of all the main phases of cement components in the angular range of 2θ values from 30° to 35° making the identification of the individual components extremely difficult. In all the mixes, C_2S, C_3S, and C_4AF peaks are not visible indicating that they may be totally consumed or overlapping of the peaks of unhydrated cement by that of Si may have occurred as all analyzed mixes were concrete specimens with large number of aggregate particles containing quartz which resulted in intensive Si peaks. Hence, as shown, SiO_2 peak indicating free silica, in CM mix was observed at 1800. The X-ray diffraction pattern observed in FB10 mix was similar to CM mix as the overall replacement of the sand was only 10%, with waste foundry sand and bottom ash as 5% and 5%, respectively. The FB20 to FB50 mix showed SiO_2 peak between 4000 and 4500. The strength variation in all the FB mixes was comparatively less, thus the FB40 and FB50 mixes show almost same intensity of SiO_2 peak at 4200. FB60 gave the SiO_2 peak at 3100. Phase determination could not be carried out as the mixes are complex and XRD analysis is done for single crystalline and poly-crystalline (two) for phase determination. Using the software library, the analysis for various mixtures was carried out which showed that the main component consisted of highly crystalline quartz (compounds shown in the graphs were obtained at various 2θ values, from the standard library of the software itself). Since, in the material with crystalline structure, X-rays scattered by ordered features will be scattered coherently "in-phase" in certain directions meeting the criterion for constructive interference. The conditions required for constructive interference are determined by Bragg's Law.

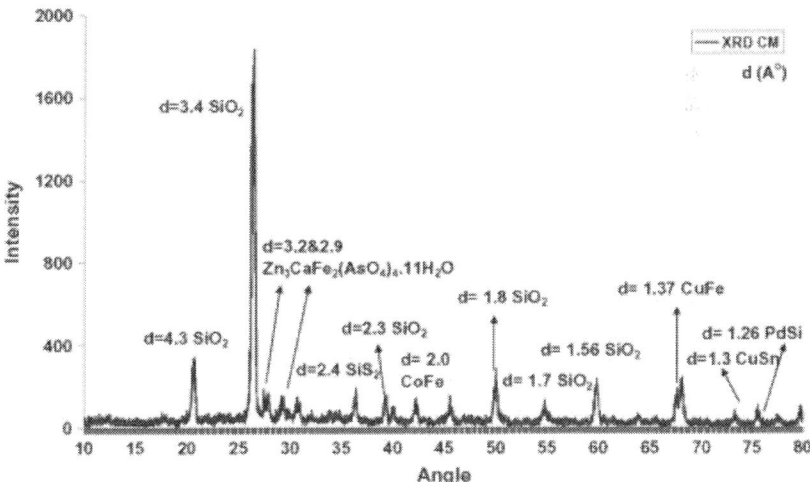

Figure 6: X-ray diffraction pattern of reference (CM) mix.

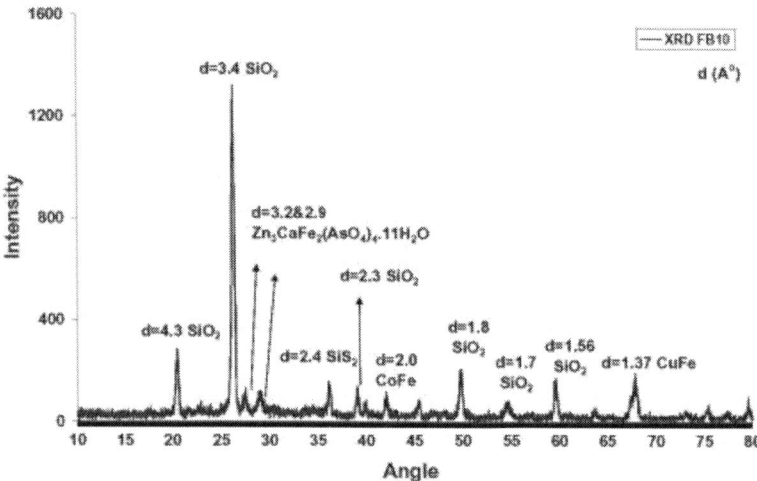

Figure 7a: X-ray diffraction pattern of FB10 mix

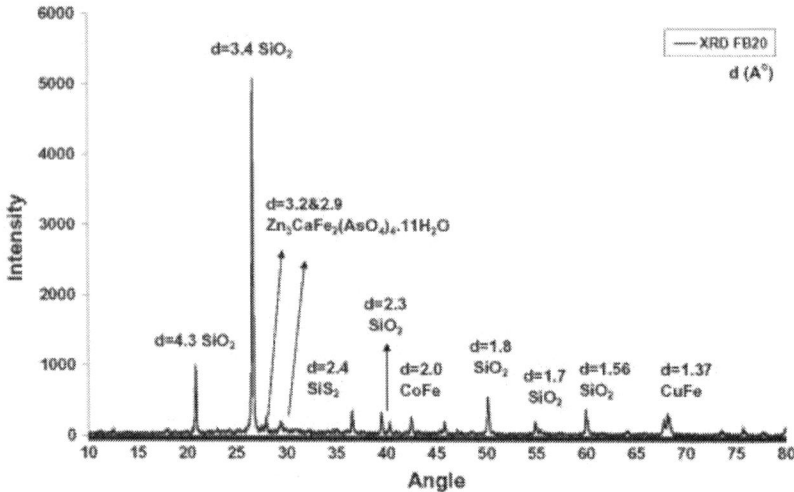

Figure 7b: X-ray diffraction pattern of FB20 mix

DURABILITY PROPERTIES

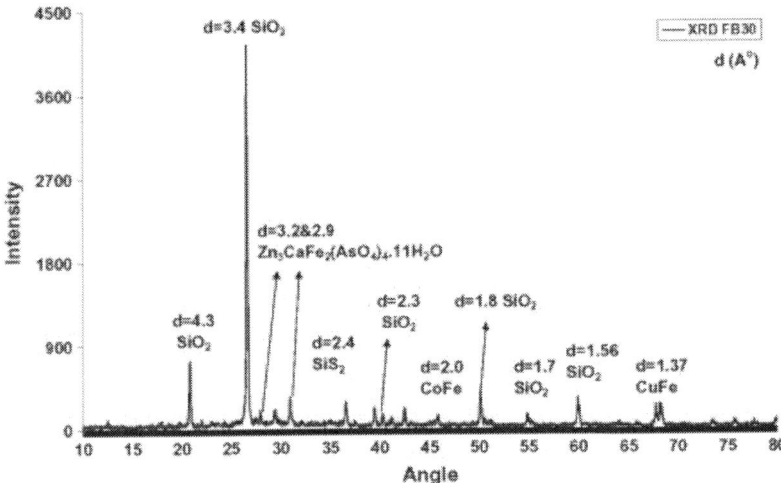

Figure 7c: X-ray diffraction pattern of FB30 mix.

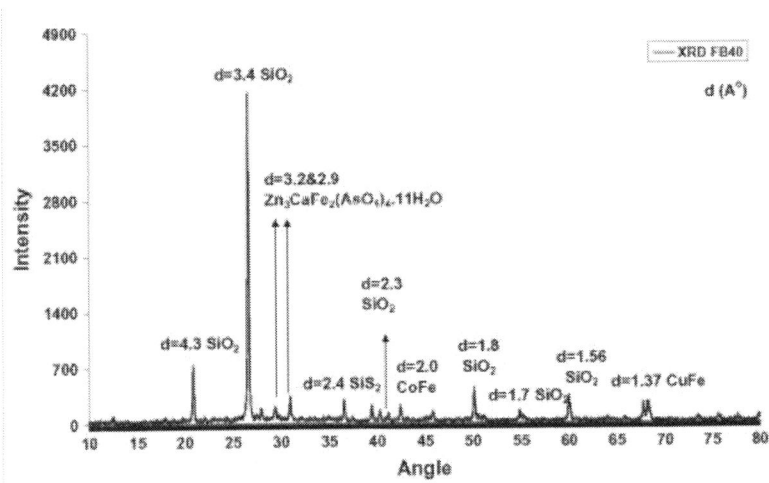

Figure 7d: X-ray diffraction pattern of FB40 mix.

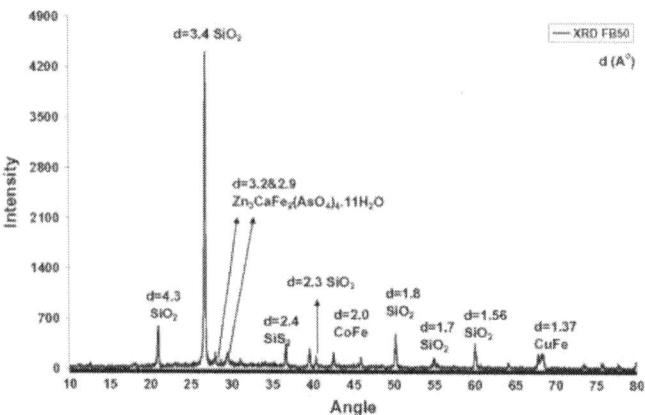

Figure 7e: X-ray diffraction pattern of FB50 mix.

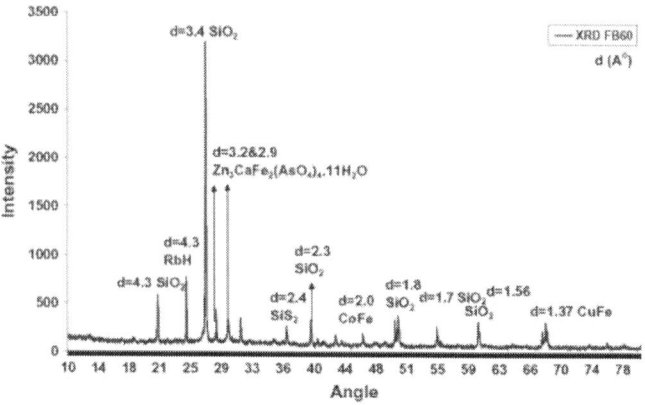

Figure 7f: X-ray diffraction pattern of FB60 mix.

Scanning Electron Microscope (SEM) Analysis

It is well known that, the calcium–silica–hydrate (C–S–H) is major phase present. The factors that influence the mechanical behavior of C–S–H phases are: size and shape of the particles, distribution of particles, particle concentration, particle orientation, topology of the mixture, composition of the dispersed/continuous phases and the pore structure. Considering various scanning electron microscope images, the phases were indicated studying the literature available [Lea's [41], Yazici [60]]. It was assumed that the bright and dark matter in the images stands for C–S–H gel/paste and inert aggregates respectively, after having some idea about the presence of C–S–H gel/paste and inert aggregates respectively, then further, referring to the above literatures, the differentiation in various particles of inert aggregates, was tried to be carried out as the medium dark particles considered as waste foundry sand particles while the spherical like particles considered as bottom-ash particles. The assumptions regarding presence of particles is based on the facts that these medium dark particles are seen in almost every sample indicating waste foundry sand particles except the reference mix CM (every sample except the reference mix CM contains waste foundry sand and bottom ash), while the spherical like particles are visible indicating bottom-ash [60]. These assumptions can be justified based on the fact that the basic structure of the concrete in all the samples is the same i.e. the mix designed for the reference mix has been kept constant in all the samples changing only the waste foundry sand and the bottom ash part in these mixes.

Fig. 8 is micrograph of reference mix i.e. the SEM image at 1.5 KX magnifications. It shows the formation of proper and clear C–S–H gel in various stages. The encircled portions represent the voids while rest of the picture consists of C–S–H gel and inert aggregates (both fine and coarse). The important point to be noted in the micrograph is that the C–S–H gel i.e. the bright masses with nodules and big chalky gel parts are spread over the entire micrograph, as it is evident from various literatures, the C–S–H gel gets spread over the aggregates thus acting as binders for the paste.

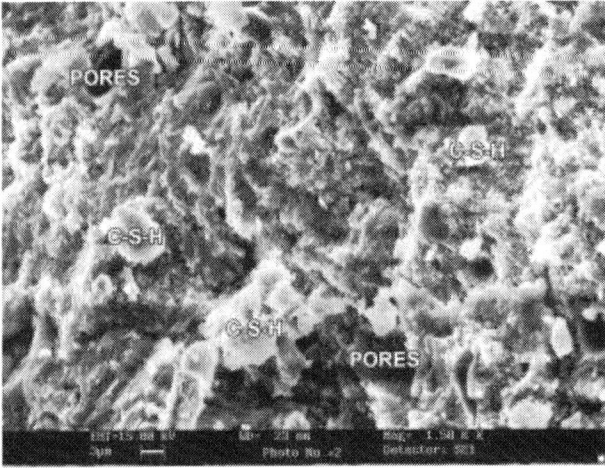

Figure 8: Micrograph of reference (CM) mix.

Fig. 9a, micrograph of FB10 mix shows two major features. Firstly, the number of voids in the mix has significantly reduced and secondly, the C–S–H gel paste is not as widely spread as it was in the reference mix, showing some aversion to the binder paste but more importantly, the effect of waste foundry sand and bottom ash has been negative on the strength because of lesser quantity of waste foundry sand and bottom ash, clearly evident as the strength of the mix has deteriorated significantly. The microstructure also shows the presence of waste foundry sand and bottom ash particles of various sizes at various places. The decrease in strength could be attributed to the non formation of proper C–S–H gel as compared to CM mix microstructure, although, at few places the formation of C–S–H gel could be detected as the percentage of the waste foundry sand and bottom ash, added was only 10%.

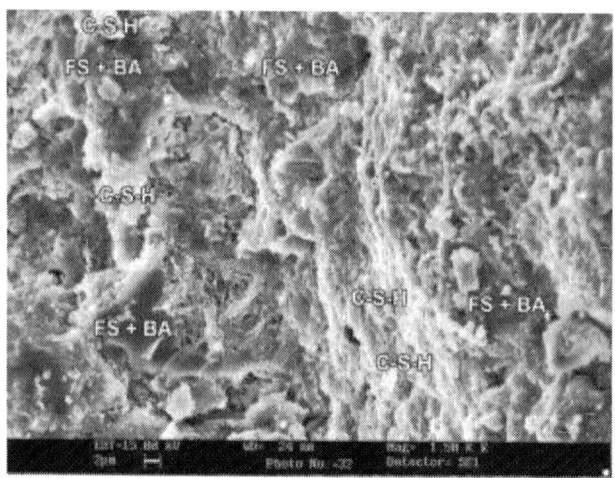

Figure 9a: Micrograph of FB10 mix.

Fig. 9b, Fig. 9c, Fig. 9d and Fig. 9e, the micrographs of FB20, FB30, FB40 and FB50 show the presence of needle like structures around waste foundry sand and bottom ash particles at various places. The reaction or the formation of C–S–H gel is better, thereby indicating comparative densification of the mixes till 50% replacement.

DURABILITY PROPERTIES 63

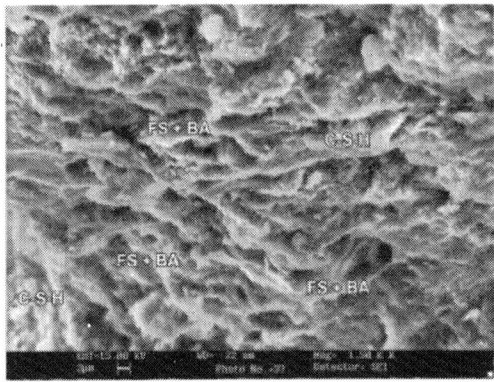

Figure 9b: Micrograph of FB20 mix.

Figure 9c: Micrograph of FB30 mix.

Figure 9d: Micrograph of FB40 mix.

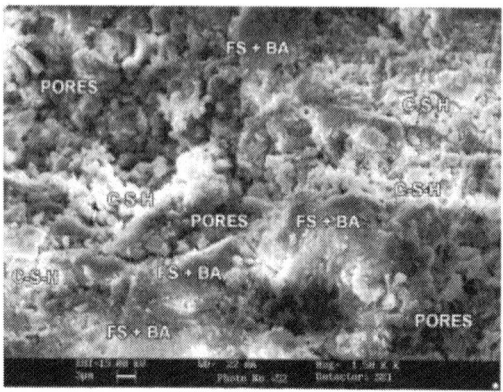

Figure 9e: Micrograph of FB50 mix.

Fig. 9f, micrograph of FB60 mix shows that the mix has crumbled with coming out of waste foundry sand and bottom ash particles from the mix. The C–S–H gel could not be seen at many places in the micrograph. The most important inference from the image is that the paste is crumbling, as the amount of replacement goes so high in this sample that the equilibrium falls and leads to lower strength.

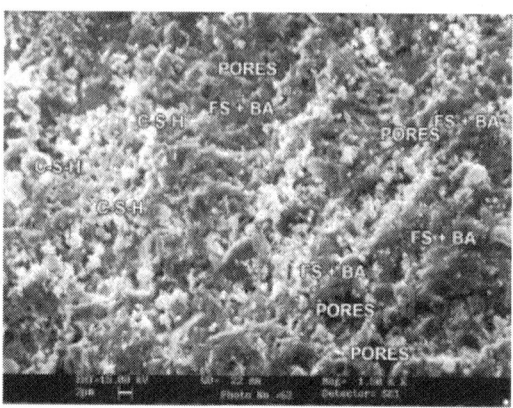

Figure 9f: Micrograph of FB60 mix.

The micrographs from Fig. 9b, Fig. 9c, Fig. 9d and Fig. 9e show similarity in the pattern formation of C–S–H gel in these mixes with all of them nearly having same strength except for the mix with 60% replacement.

Fig. 9f, micrograph of FB60 mix shows that the mix has crumbled with coming out of waste foundry sand and bottom ash particles from the mix. The C–S–H gel could not be seen at many places in the micrograph. The most important inference from the image is that the paste is crumbling, as the amount of replacement goes so high in this sample that the equilibrium falls and leads to lower strength.

In fact, in the present study the mixes with amount of replacement of sand more than 50% with waste foundry sand and bottom ash, also lead to crumbling at the time of curing done experimentally. These results simply imply that more than 50% replacement of sand with waste foundry sand and bottom ash leads to flaws in concrete, but the best mixture in any case is inarguably the 30% replacement mix. Further, FB30 mix showed large formation of C–S–H gel thus, development of dense microstructure. The fibrous C–S–H formation acts as a thick impermeable membrane for the ingress of chloride ions into concrete. This makes the concrete more resistant to aggressive environment as observed from RCPT values.

CONCLUSIONS

The following conclusions could be arrived at from the study:

1. The studies carried out indicate the viability of using waste from the foundry industry and bottom ash from electrostatic precipitators as recycled fine aggregates in the production of concrete for structural purposes.
2. As, it was observed that for initial replacements of 10%, 20% and 30%, the increase in water content was constant and thereafter for 40% and 50%, again it remained constant but almost double the value of initial replacements. The mixes can be developed by varying the water content at constant rate as specified in the study till 30% and thereafter till 50% replacement of fine aggregates. The mix FB60 is not recommended as the water content of this mix is high which also reflects on various strengths.
3. The mechanical behavior of the concrete with waste foundry sand and bottom ash showed strengths comparable to that of conventional concrete except for FB60 mix, at the age of 365 days. Furthermore, it was observed that the greatest increase in compressive, splitting tensile strength and flexural strength was achieved by substituting 30% of the natural fine aggregate with industrial by-product aggregate in replaced mixes. Also, the maximum replacement could be taken as 50%.

4. The splitting tensile strength for FB30 mix was observed to be more than the control mix at all ages.
5. An increase in strength from 28 to 90 days was observed to be 13.29% for CM mix whereas FB mixes showed increase in strength from 14.52% to 23.89%. Between 90 and 365 days, an increase in strength for CM mix was 6.76% and the FB mixes showed an increase of 2.02–6.94%.
6. The inclusion of waste foundry sand and bottom ash as fine aggregate does not affect the strength properties negatively as the strength remains within limits. The concrete was endowed with comparable mechanical properties and greater resistance to aggressive agents (chemical, physical and environmental).
7. The morphology of the formations arising as a result of the hydration process was not observed to change in the concrete with varying percentages of waste foundry sand and bottom ash except in FB60.
8. The possibility of substituting natural fine aggregate with industrial by-product aggregate such as waste foundry sand and bottom ash offers technical, economic and environmental advantages which are of great importance in the present context of sustainability in the construction sector.

REFERENCES

1. H. Kurama, M. KayaUsage of coal combustion bottom ash in concrete mixture, Constr Build Mater, 22 (9) (2008), pp. 1922–1928
   I. Kula, A. Olgun, V. Sevinc, Y. Erdogan, Effects of colemanite waste, coal bottom ash, and fly ash on the properties of cement, Cem Concr Res, 31 (2002), pp. 491–494
   I. Kula, A. Olgun, V. Sevinc, Y. Erdogan, An investigation on the use of tincal ore waste, fly ash, and coal bottom ash as Portland cement replacement materials, Cem Concr Res, 32 (2002), pp. 227–232
   I. Yuksel, T. Bilir, Usage of industrial by-products to produce plain concrete elements, Constr Build Mater, 21 (2007), pp. 686–694
   I. Yuksel, T. Bilir, O. Ozkan, Durability of concrete incorporating non-ground blast furnace slag and bottom ash as fine aggregate, Build Environ, 42 (2007), pp. 2651–2659
2. Y. Bai, F. Darcy, P.A.M. Basheer, Strength and drying shrinkage properties of concrete containing furnace bottom ash as fine aggregate, Constr Build Mater, 19 (2005), pp. 691–697

REFERENCES

3. Bai Y, Basheer PAM. Influence of furnace bottom ash on properties of concrete. In: Ice proceedings, structures and buildings, special issue new materials and new uses of for old materials 2003; 156(1): 85–92.
4. N. Ghafoori, Y. Cai, Laboratory-made roller compacted concretes containing dry bottom ash: Part 1–mechanical properties, ACI Mater J, 95 (1998), pp. 121–130
5. N. Ghafoori, J. Bucholc, Properties of high-calcium dry bottom ash concrete, ACI Mater J, 94 (2) (1997), pp. 90–101
6. Foundry Industry Recycling Starts Today (FIRST). Foundry sand facts for civil engineers. Washington, DC: Federal Highway Administration Environmental Protection Agency; 2004.
7. J.R. Kleven, T.B. Edil, C.H. Benson, Evaluation of excess foundry system sands for use as sub base material, Trans Res Rec Transportation Research Board, Washington DC (2000) p.40–8
8. T. Abichou, C.H. Benson, T.B. Edil, Database on beneficial reuse of foundry by-products, C. Vipulanandan, D. Elton (Eds.), Recycled materials in geotechnical applications, vol. 79Geotechnical Special Publication ASCE (1998), pp. 210–223
9. Javed S, Lovell CW. Use of Waste foundry sand in Highway construction. Report, JHRP/INDOT/FHWA-94/2J, Final Report. West Lafayette, IN: Purdue School of Engineering; 1994a.
10. S. Javed, C.W. Lovell, Use of waste foundry sand in civil engineering, Trans Res Rec 1486Transportation Research Board, Washington, DC (1994) p.109–13
11. MOEE. Spent foundry sand – alternative uses study. Report prepared by John Emery Geotechnical Engineering Limited for Ontario Ministry of the Environment and Energy and the Canadian Foundry Association. Queen's Printer for Ontario. 1993.
12. American Foundrymen's Society (AFS). Alternative utilization of foundry waste sand. Des Plaines, IL: Final Report (Phase I) prepared by American Foundrymen's Society Inc. for Illinois Department of Commerce and Community Affairs; 1991.
13. Traeger PA. Evaluation of the constructive use of foundry wastes in highway construction. MS thesis, Madison, WI: The University of Wisconsin-Madison; 1987.
14. R.K. Ham, W.C. Boyle, F.J. Blaha, Comparison of leachate quality in foundry waste landfills to leach test abstracts, J Haz Indl Solid Waste Testing Dispos, 6 (1990), pp. 29–44
15. E.C. Engroff, E.L. Fero, R.K. Ham, W.C. Boyle, Laboratory leachings of organic compounds in ferrous foundry process waste, Final Report to American Foundrymen's Society, Des Plaines, IL (1989)

16. R.L. Fero, R.K. Ham, W.C. Boyle, An investigation of ground water contamination by organic compounds leached from iron foundry solid wastes, Final Report to American Foundrymen's Society, Des Plaines, IL (1986)
17. R.K. Ham, W.C. Boyle, Leachability of foundry process solid wastes, J Env Div, ASCE, 107 (1) (1981), pp. 155–170
18. T.R. Naik, S.S. Singh, B.W. Ramme, Performance and leaching assessment of flow able slurry, J Environ Eng (2001), pp. 359–368
19. P.J. Tikalsky, M. Gaffney, R. Regan, Properties of controlled low-strength material containing foundry sand, ACI Mater J, 97 (6) (2003), pp. 698–702
20. P.J. Tikalsky, E. Smith, R. Regan, Proportioning spent casting sand in controlled low-strength materials, ACI Mater J, 95 (6) (1998), pp. 740–746
21. R. Siddique, G.D. Schutter, A. Noumowe, Effect of used-foundry sand on the mechanical properties of concrete, Constr Build Mater, 23 (2009), pp. 976–980
22. T.R. Naik, R.N. Kraus, Y.M. Chun, W.B. Ramme, R. Siddique, Precast concrete products using industrial by-products, ACI Mater J, 101 (3) (2004), pp. 199–206
23. T.R. Naik, R.N. Kraus, Y.M. Chun, W.B. Ramme, S.S. Singh, Properties of field manufactured cast-concrete products utilizing recycled materials, J Mater Civil Eng, ASCE, 15 (4) (2003), pp. 400–407
24. Khatib JM, Ellis DJ. Mechanical properties of concrete containing foundry sand. ACI Spl Pub SP-200. ACI. 2001; p.733–48.
25. S. Fiore, M.C. Zanetti, Foundry wastes reuse and recycling in concrete production, Am J Environ Sci, 3 (3) (2007), pp. 135–142
26. S. Javed, C.W. Lovell, L.E. Wood, Waste foundry sand in asphalt concrete, Trans Res Board, 1437 (1994), pp. 27–34
27. IS:1489-1991 (Part-I). Portland Pozzolana Cement Specification. New Delhi, India: Bureau of Indian Standard.
28. ASTM C33-86. Standard Specification for Concrete Aggregates.
29. IS:383-1970. Specification for coarse and fine aggregates from natural sources for concrete, New Delhi, India: Bureau of Indian Standard.
30. ASTM C494 type F, Standard specification for chemical admixtures for concrete.
31. IS: 9103-1999. Concrete admixtures specification. New Delhi, India: Bureau of Indian Standard.
32. IS: 2645-2003. Specification for integral cement waterproofing compounds, New Delhi, India: Bureau of Indian Standard.
33. IS:10262-1982. Recommended guidelines for concrete mix design. New Delhi, India: Bureau of Indian Standard.
34. IS:1199-1959. Methods of sampling and analysis of concrete, New Delhi, India: Bureau of Indian Standard.

REFERENCES

35. IS 516:1959. Methods of tests for strength of concrete, New Delhi, India: Bureau of Indian Standard.
36. ASTM C1202. Standard test method for electrical induction of concrete's ability to resist chloride ion penetration.
37. ASTM C672. Standard test method for scaling resistance of concrete surfaces exposed to deicing chemicals.
38. Soroka I. Portland cement paste and concrete. New York: The Macmillan Press Ltd.
39. Lea's Chemistry of Cement and Concrete. 4th ed., Hewlett Peter C, editor. Jordan Hill, Oxford OX2 8DP: Butterworth-Heinemann Lincare House; 2001.
40. R. Siddique, Y. Aggarwal, P. Aggarwal, E.l.-H. Kadri, R. Bennacer, Strength, durability, and micro-structural properties of concrete made with used-foundry sand (UFS), Constr Build Mater, 25 (2011), pp. 1916–1925
41. Mehta PK, Monterio PJM. Concrete microstructure, properties, and materials, Tata McGraw-Hill Edition. 2006.
42. ACI committee 363. State of the art report on high strength concrete. ACI J Proc 1984; 81(4): 364–411.
43. IS:456:2000. Plain and reinforced concrete code of Practice. New Delhi, India: Bureau of Indian Standard.
44. P. Aggarwal, Y. Aggarwal, S.M. Gupta, Effect of bottom ash as replacement of bottom ash as replacement of fine aggregates in concrete, Asian J Civ Eng (Build Housing), 8 (1) (2007), pp. 49–62
45. A.A. Ramezanianpour, V.M. Malhotra, Effect of curing on the compressive strength, resistance to chloride ion penetration porosity of concretes incorporating slag fly ash or silica fume, Cem Concr Compos, 17 (1995), pp. 125–133
46. R.N. Naik, S. Singh, B. Ramme, Mechanical properties and durability of concrete made with blended fly ash, ACI Mater J, 95 (1998), pp. 454–462
47. Guneyisi E. Effect of binding material composition on strength and chloride permeability of high strength concrete. MS thesis, Department of Civil Engineering, Bogazic University; 1999.
48. A.L.G. Gastaldini, G.C. Isaia, N.S. Gomes, J.E.K. Sperb, Chloride penetration and carbonation in concrete with rice husk ash and chemical activators, Cem Concr Compos, 29 (2007), pp. 176–180
49. C.C. Yang, C.T. Chiang, On the relationship between pore structure and charge passed from RCPT in mineral-free cement-based materials, Mater Chem Phys, 93 (2005), pp. 202–207
50. B.H. Oh, S.W. Cha, B.S. Jang, S.P. Jang, Development of high performance concrete has high resistance to chloride penetration, Nucl Eng Des, 212 (2001), pp. 221–231

51. Mackechnie JR, Alexander MG. Rapid chloride test comparisons, Concr Intl. 2000; p.40–5.
52. N. Feng, X. Feng, T. Hao, F. Xing, Effect of ultra fine mineral powder on the charge passed of the concrete, Cem Concr Res, 32 (2002), pp. 623–627
53. P. Gu, J.J. Beaudoin, M.H. Zhang, V.M. Malhotra, Performance of steel reinforcement in Portland cement and high-volume fly ash concretes exposed to chloride solution, ACI Mater J, 96 (1999), pp. 551–558
54. T. Cho, Prediction of cyclic freeze–thaw damage in concrete structures based on response surface method, Constr Build Mater, 21 (2007), pp. 2031–2040
55. H. Yazici, Utilization of coal combustion byproducts in building blocks, Fuel, 86 (2007), pp. 929–937
56. M.L. Gambhir, Concrete technology, (2nd ed.)Tata McGraw Hill Publishing Company Ltd., New Delhi (1988)
57. A.M. Neville, Properties of concrete, Longman Singapore Publication, Pvt. Ltd., Singapore (1995)
58. K. Wang, D.E. Nelsen, W.A. Nixon, Damaging effects of deicing chemicals on concrete materials, Cem Concr Compos, 28 (2006), pp. 173–188

CITATION

Yogesh Aggarwal, Rafat Siddique, Microstructure and properties of concrete using bottom ash and waste foundry sand as partial replacement of fine aggregates, doi.org/10.1016/j.conbuildmat.2013.12.051

CHAPTER 3

A Unified Method for Calculating Fire Resistance of Solid and Hollow Concrete-Filled Steel Tube Columns Based on Average Temperature

Min Yu [a,b], Xiaoxiong Zha [b], Jianqiao Ye [a,c], Baolin Wang [b]

[a]School of Civil Engineering, Wuhan University, Wuhan 430072, China
[b]Shenzhen Graduate School, Harbin Institute of Technology, Shenzhen 518055, China
[c]Department of Engineering, Lancaster University, Lancaster, LA1 4YR, UK

ABSTRACT

This paper presents a new method for calculating fire resistance of axially loaded Concrete-Filled Steel Tube (CFST) columns with different section profiles, including circular and polygonal sections that can be solid and hollow. The uniqueness of this new method is that the fire resistance is calculated on the basis of the average temperature of the columns' cross-sections. This is done by taking the bearing capacity of a CFST column at room temperature as a special case of the bearing capacity of the same column at the start of a fire. The equivalent strength and equivalent elastic modulus in relation to the average temperature of steel and concrete are investigated, and a unified method of calculation is proposed, by which the calculation of fire resistance of a CFST column can be divided into two steps, i.e. (a) calculation of the equivalent strength and elastic modulus of steel and concrete at elevated temperature based on the average temperature, and (b) calculation of fire resistance using the formulas at room temperature by replacing the equivalent material strength and elastic modulus at elevated temperature. The two sets of formulas for calculating fire resistance of CFST columns are given by combining the unified method, respectively, with Eurocode 4 and the authors' previous work. The proposed formulas and procedure are validated through comparisons with the experimental results of a number of solid and hollow CFST columns with circular and square sections.

INTRODUCTION

Fire resistance of Concrete-Filled Steel Tube (CFST) column is one of the most important factors that must be taken into account in the design process of modern building structures. For example, structural fire design is now required by Eurocode 4 [1], DBJ13-51 [2], ASCE/SFPE 29-99 [3] and ACI 216 [4]. There are many different forms of CFST columns characterized by their cross sectional profiles, as shown in Fig. 1[5]. The hollow ones are normally formed by pouring concrete into the steel tubes using the centrifugal method.

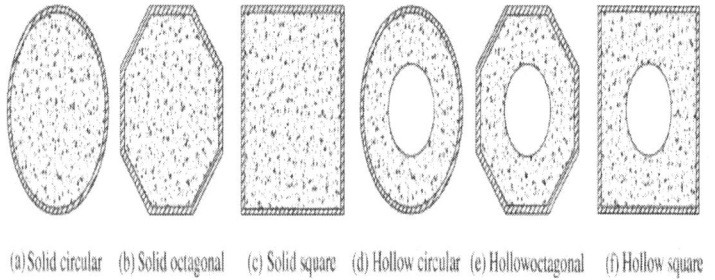

(a) Solid circular (b) Solid octagonal (c) Solid square (d) Hollow circular (e) Hollow octagonal (f) Hollow square

Figure 1: Common section profiles of CFSTs [5].

Extensive research has been published on experimental studies of solid CFST columns [6], [7], [8], [9],[10], [11] and [12], bar-reinforced CFST columns [13] and [14], concrete-filled double skin tube columns[15] and [16], and rectangular CFST columns exposed to fire [17] and [18] under axial or eccentrically compression. There are also limited published work on experimental study of hollow CFST columns subjected to fire, e.g., the work done by Yu [19] and Blaževičius and Kvedaras [20]. Since fire experiments are complex and expensive, numerical simulations have played an important role in studying fire resistance of the CFST columns [21], [22], [23], [24], [25], [26], [27] and [28]. There are also some analytical formulas proposed by, e.g., Kodur [29] who conducted parametric analysis through experiments and numerical calculations, and proposed formulas for calculating fire resistance time of solid circular and

INTRODUCTION

square CFST columns using a regression analysis. On the basis of Eurocode 4 [1], and Wang and Kodur[30] developed an approach for evaluating squash load and rigidity of solid CFST columns at elevated temperature. The formulas were developed for columns with "standard" steel and concrete strength; Wang[31] presented a method for both protected and unprotected circular CFST columns, which required interpolations to obtain their squash loads and rigidities; Li et al. [32] proposed a formula for bearing capacity of solid circular CFST columns under fire on the basis of a parametric analysis and regression; Han et al. [8] calculated strength index of circular and square solid CFST columns based on the results of parametric and experimental studies, and proposed also a formula for calculating the thickness of fireproof; Tan and Tang [33] applied the "Rankine method" to the analysis of reinforced and plain solid CFST columns at elevated temperature; Espinos et al. [34] presented a simple calculation method for evaluating fire resistance of circular solid CFST columns based on Eurocode 4, where the concept of equivalent temperature was adopted. The method is suitable for circular cross sections with a diameter of 139.7–508 mm, normal strength concrete C20/25–C40/C50, and a fire resistance time within 2 h. Detailed and more comprehensive review on current methods for calculating fire resistance of CFST columns can be found from Zhao et al. [35], Rush et al. [36] and Espinos et al. [34]. It can be seen from the above literature review that there has been extensive research on fire-resistance of solid CFST columns. In most, if not all, of the research, it has been always the case that different formulas were proposed for different section profiles to predict their axial load bearing capacity or fire resistance time. It was also noticed the research on hollow CFST columns is relatively rare.

This paper attempts to unify the solution procedure by taking (a) a solid CFST column as a special case of a hollow one, and (b) the bearing capacity of a column at room temperature as a special case of the bearing capacity of the same column under fire at time $t = 0$. Consequently, the analysis of the bearing capacity of a column at room and room temperatures for solid and hollow CFST columns becomes a continuous and integrated process of calculation. To this end, a new method is developed for computing fire resistance on the basis of introducing the concept of *average* temperature. Theoretical and finite element analysis are combined with Eurocode 4 and the unified formula proposed by Yu et al. [5], resulting in, respectively, two sets of formulas for calculating fire resistance of solid and hollow CFST columns. The formulas are validated through comparisons with existing experimental results of the columns with circular and square sections.

UNIFIED METHOD FOR CFST COLUMNS UNDER FIRE BASED ON AVERAGE TEMPERATURE

General Plastic Limit Analysis for CFST Columns under Fire

Eurocode 4 [1] uses the plastic limit method to calculate fire resistance of composite columns, without considering the confinement effect and any partial factors of the materials. The discrete form of plastic resistance and flexural stiffness of a cross-section are calculated by dividing the composite cross-section into a number of sub-areas, calculating the resistance and stiffness of each of the sub-areas and then taking the sum. This process can also be expressed by the following integrations:

$$N_{0,T} = \iint_{A_c} f_{c,x,y,T} dA + \iint_{A_c} f_{s,x,y,T} dA \tag{1}$$

$$(EI)_{sc,T} = \iint_{A_c} y^2 \cdot E_{c,x,y,T} dA + \iint_{A_c} y^2 \cdot E_{s,x,y,T} dA \tag{2}$$

where A_s and A_c denote the areas of steel and concrete; $f_{s,x,y,T}$ and $f_{c,x,y,T}$ are the respective strength of steel and concrete at location (x, y) of a cross-section at temperature T; $E_{s,x,y,T}$ and $E_{c,x,y,T}$ are the temperature, T, dependent elastic modulus of steel and concrete at location (x, y), respectively.

For fire resistance design of a CFST column, the use of column buckling curve "c" is recommended in Eurocode 4 [1]. The curve is a function of non-dimensional slenderness that can be determined by the plastic resistance of cross-section, $N_{0,T}$, flexural stiffness, $(EI)_{sc,T}$ and effective length L_0. Therefore, for a given $N_{0,T}$ and $(EI)_{sc,T}$, the stability bearing capacity of a CFST column under fire can be calculated in a straight forward manner.

Equivalent Reduction Factor of Strength and Elastic Modulus under Fire

In order to evaluate the temperature of a CFST cross section, the average temperature of the steel and concrete sections can be used, as follows:

$$\overline{T}_\alpha(t) = \frac{\iint_{A_\alpha} f(x, y, t) dA}{A_\alpha} \quad (3)$$

where $\overline{T}$ and T denote average and actual temperatures, respectively; α takes s for steel tube and c for concrete core; A_α is the area of material α. To calculate plastic resistance $N_{0,T}$ and flexural stiffness $(EI)_{sc,T}$ of a cross-section from Eqs. (1) and (2), the effects of the steel and the concrete are considered separately in the following calculations.

A: Steel Tube

A steel tube is normally thin and has good thermal conductivity. As a result, it can be treated as a lumped capacitance with uniform temperature distribution, i.e. $T_s(x, y, t) = \overline{T}_s$. The plastic resistance $N_{0,s,T}$ and flexural stiffness $(EI)_{s,T}$ of the steel part can be written as:

$$N_{0,s,T} = \iint_{A_s} f_{s,x,y,t} dA = k_{s,T}(\overline{T}_s) f_y A_s \quad (4)$$

$$(EI)_{s,T} = \iint_{A_s} y^2 \cdot E_{s,x,y,t} dA = k_{Es,T}(\overline{T}_s) E_s I_s \quad (5)$$

where $\overline{T}_s$ is the average temperature of steel; $k_{s,T}$ is the reduction factor of steel strength as a function of temperature; $k_{Es,T}$ is the reduction factor of steel elastic modulus, also, as a function of temperature; f_y is the strength of steel at room temperature; E_s is the elastic modulus of steel at room temperature; A_s and I_s are the area and second moment of area of steel tube, respectively.

Eurocode 4 [1] provides tables for the reduction factors of steel strength $k_{s,T}$ and elastic modulus $k_{Es,T}$ under elevated temperature. For the convenience of the following calculations, two formulas are proposed through curves fitting of the discrete points from Eurocode 4, as shown in Fig. 2a. The reduction factors of steel strength and elastic modulus can be calculated accurately by the following new formulas:

$$k_{s,T} = \frac{f_{y,T}}{f_y} = \begin{cases} 1 & 20\,°C \leqslant T \leqslant 400\,°C \\ e^{-\left(\frac{T-400}{240}\right)^{1.5}} & 400\,°C < T \leqslant 1200\,°C \end{cases} \quad (6)$$

$$k_{E_s,T} = \frac{E_{s,T}}{E_s} = e^{-\left(\frac{T-20}{560}\right)^{2.5}}, \quad 20\,°C \leqslant T \leqslant 1200\,°C \quad (7)$$

where $k_{s,T}$ and $k_{E_s,T}$ are the respective reduction factors of steel strength and elastic modulus; $f_{y,T}$ is strength of steel at elevated temperature; $E_{s,T}$ denotes elastic modulus of steel at elevated temperature; T is temperature of steel in °C.

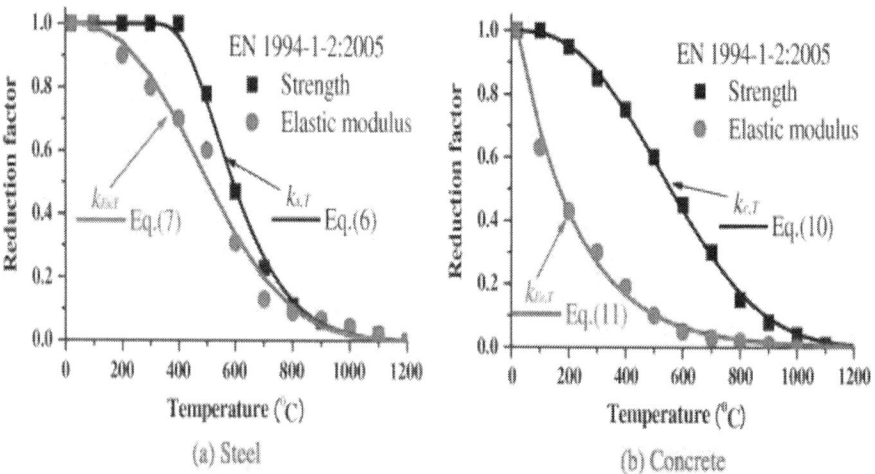

Figure 2: Reduction factors of steel strength and elastic modulus with temperature.

B: Concrete Core

Since the concrete part is much thicker than the steel tube and has lower thermal conductivity, the temperature gradient in the concrete core under fire has to be considered. In theory, the integrals of Eqs.(1) and (2) can be calculated by the mean value theorem for integration. Hence, the plastic resistance of $N_{0,c,T}$ and flexural stiffness $(EI)_{c,T}$ of concrete core can be expressed, respectively, as:

$$N_{0,c,T} = \iint_{A_c} f_{ck,x,y,t} dA = k_{c,T}(\tilde{T}_{c,N}) f_{ck} A_c \tag{8}$$

$$(EI)_{c,T} = \iint_{A_c} y^2 \cdot E_{ck,x,y,t} dA = k_{Ec,T}(\tilde{T}_{c,EI}) E_c I_c \tag{9}$$

where $\tilde{T}_{c,N}$ and $\tilde{T}_{c,EI}$ are the temperature at certain interior points of the concrete section; $k_{c,T}$ and $k_{Ec,T}$ are the respective temperature-dependent reduction factors of concrete strength and elastic modulus; f_{ck} is the strength of concrete and E_c the elastic modulus of concrete at room temperature; A_c and I_c are the area and second moment of area of concrete core, respectively.

Eurocode 4 [1] also provides tables for the reduction factors of concrete strength $k_{c,T}$ and the peak values of strain $\varepsilon_{cu,T}$ under elevated temperature. The reduction factor of concrete elastic modulus can be calculated using $k_{Ec,T} = k_{c,T} \varepsilon_{cu}/\varepsilon_{cu,T}$ [34]. Following exactly the same curves fitting procedure used above for the steel tube, as shown in Fig. 2b, the reduction factor of concrete strength $k_{c,T}$ and the reduction factor of concrete elastic modulus $k_{Ec,T}$ can be calculated accurately by the following two formulas:

$$k_{c,T} = \frac{f_{ck,T}}{f_{ck}} = e^{-\left(\frac{T-20}{622}\right)^{2.5}}, \quad 20\,°C \leqslant T \leqslant 1200\,°C \tag{10}$$

$$k_{Ec,T} = \frac{E_{c,T}}{E_c} = e^{-\frac{T-20}{211}}, \quad 20\,°C \leqslant T \leqslant 1200\,°C \tag{11}$$

where $f_{ck,T}$ is the strength of concrete at elevated temperature; $E_{c,T}$ is the secant elastic modulus of concrete ($0.3\,f'_c$) at elevated temperature; T is the temperature of concrete in °C.

The values of $\tilde{T}_{c,N}$ in Eq. (8) and $\tilde{T}_{c,EI}$ in Eq. (9) are generally different, and, normally, it is difficult to estimate them in a practical design. To overcome this difficult, the average temperature of concrete is adopted instead since it can be easily computed as long as the temperature distribution is known. In the following calculations, equivalent reduction factors, which are equal to the reduction factors in values, but now are functions of the average temperature, are introduced, i.e.:

$$k_{c,T}(\widetilde{T}_{c,N}) = \bar{k}_{c,T}(\overline{T}_c) \text{ where } \bar{k}_{c,T}(\overline{T}_c) = \frac{\iint_{A_c} k_{c,T}(T_{x,y,t})dA}{A_c} \quad (12)$$

$$k_{Ec,T}(\widetilde{T}_{c,El}) = \bar{k}_{Ec,T}(\overline{T}_c) \text{ where } \bar{k}_{Ec,T}(\overline{T}_c) = \frac{\iint_{A_c} y^2 \cdot k_{Ec,T}(T_{x,y,t})dA}{I_c} \quad (13)$$

In the above equations, $\overline{T}_c$ is the average temperature of concrete core; $\bar{k}_{c,T}$ is the equivalent reduction factor of concrete strength that is now a function of $\overline{T}_c$. $\bar{k}_{Ec,T}$ is the equivalent reduction factor of concrete elastic modulus that is also a function of $\overline{T}_c$.

In order to determine the equivalent reduction factor of concrete strength $\bar{k}_{c,T}$ and the equivalent reduction factor of concrete elastic modulus $\bar{k}_{Ec,T}$ for solid and hollow CFST columns with different section profiles, such as circular, octagonal and square sections, the finite element software of COMSOL Multiphysics was used to obtain the temperature field of the CFST sections. The average temperature of steel tube and the concrete core, and the equivalent reduction factor of concrete strength and elastic modulus can be obtained by Eqs. (12) and (13). The heat transfer model for CFST columns in fire is described in Yu et al.[25] and Wang and Tan [37].

In the parametric studies on heat transfer of circular CFST columns under fire, a series of tube diameters, tube thickness and hollow ratios were taken into account. The tube diameter was taken as 100 mm, 500 mm, 900 mm, 1300 mm, 1700 mm and 2100 mm; the tube thickness was 1 mm, 5 mm, 9 mm, 13 mm, 17 mm and 21 mm, and the hollow ratio was 0.0, 0.15, 0.30, 0.45, 0.60 and 0.75. In total, 216 cases were considered by taking combinations of the above geometric parameters. For octagonal and square CFST columns, 216 cases were also considered, where each of the cases was a counterpart of a circular column from the 216 circular cases with identical area of steel, area of concrete section and hollow ratio.

The total calculation time of fire was 4 h, and the values of $\overline{T}_c$ and $\bar{k}_{c,T}$ were calculated by Eqs. (3) and (12) for every 10 min. Thus there were 24 sets of ($\overline{T}_c, \bar{k}_{c,T}$) for each case, and a total of 5184 (216 × 24) sets of results for a section profile. Similarly, for the equivalent reduction factor of concrete elastic modulus, there were also 5184 sets of ($\overline{T}_c, \bar{k}_{Ec,T}$) for each section. The equivalent reduction factor of concrete strength $\bar{k}_{c,T}$ and the equivalent reduction factor of concrete elastic modulus $\bar{k}_{Ec,T}$ for circular, octagon and square sections are respectively shown in Fig. 3 and Fig. 4.

INTRODUCTION

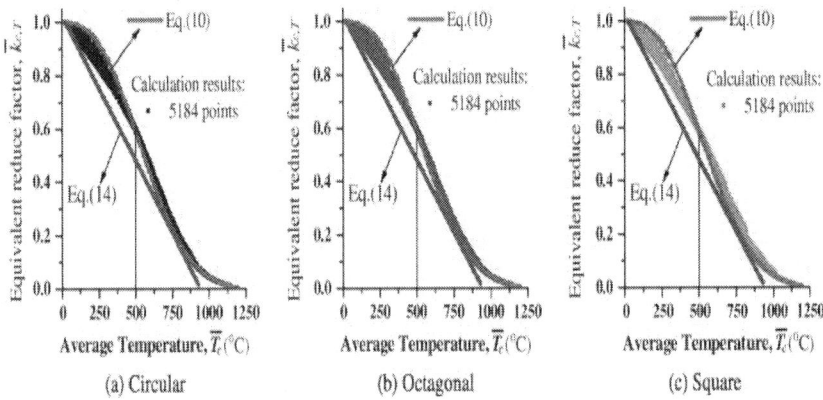

Figure 3: Equivalent reduction factor of concrete strength with average temperature.

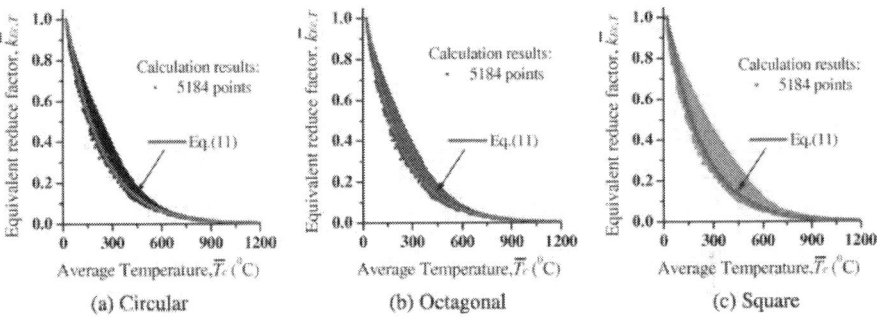

Figure 4: Equivalent reduction factor of concrete elastic modulus with average temperature.

It can be seen from Fig. 3 that the influence of section profiles on the equivalent reduction factors is not significant. It was found from the 5184 points that the average ratio between the reduction factors from Eq.(10) and the FEA results is 1.030, 1.029 and 1.027, with a variance of 0.004, 0.004 and 0.006, respectively, for the circular, octagonal and square CFST columns. It is clear from Fig. 3 that when the temperature is below 500 °C, the concrete equivalent strength reduction factor calculated from Eq. (10) is slightly over-estimated and otherwise slightly underestimated. To achieve a safer design, an approximate linear function of Eq. (14) is introduced, as shown in Fig. 3, to calculate the reduction factor. With this approximation, when the average temperature of concrete is below 800 °C, the average ratio of the reduction factors form Eq. (14) and the FEA results is 0.894, 0.894 and 0.889, with a variance of 0.004, 0.004 and 0.005, respectively, for the circular, octagonal and square CFST columns. For the elastic modulus of concrete, Fig. 4 shows that Eq. (11) is generally conservative. The average ratio between the reduction factors from Eq. (11) and the FEA results is 0.914, 0.911 and 0.879, with a variance of 0.012, 0.012 and 0.023, respectively, for the circular, octagonal and square CFST columns. Thus, the calculation of the equivalent reduction factor of concrete strength $\bar{k}_{c,T}$ and the equivalent reduction factor of concrete elastic modulus $\bar{k}_{Ec,T}$ can be further simplified as:

$$\bar{k}_{c,T}(\bar{T}_c) = 1 - \frac{\bar{T}_c - 20}{918}, \quad 0 \leqslant \bar{k}_{c,T} \leqslant 1 \tag{14}$$

$$\bar{k}_{Ec,T}(\bar{T}_c) = \bar{k}_{Ec,T}(\bar{T}_c) = e^{-\frac{\bar{T}_c - 20}{211}} \tag{15}$$

From the above analysis, it can be observed that the concrete strength and the elastic modulus from Eqs.(14) and (15) are about 10% different to the FEA results and the values of steel strength and elastic modulus obtained by Eqs. (6) and (7) are very close to the FEA results. Therefore, the relative error of the combined strength and stiffness of the CFST columns will be smaller than 10%, which are acceptable for practical design.

2.3. Unified calculation method for CFST columns under fire based on average temperature.

When a CFST column is subjected to fire, the temperature of the thin steel tube is assumed to be uniform. The strength and elastic modulus of the concrete are estimated by the average value of the non-uniform temperature within the concrete. Thus, the respective equivalent strength $f_{y,\bar{T}_s}$ and the equivalent elastic modulus $E_{s,\bar{T}_s}$ of steel in fire are:

$$f_{y,\bar{T}_s} = k_{s,T}(\bar{T}_s)f_y \tag{16}$$

$$E_{s,\bar{T}_s} = k_{Es,T}(\bar{T}_s)E_s \tag{17}$$

and the equivalent strength $\bar{f}_{ck,\bar{T}_c}$ and the equivalent elastic modulus $\bar{E}_{c,\bar{T}_c}$ of concrete core in fire are:

$$f_{ck,\bar{T}_c} = \bar{k}_{c,T}(\bar{T}_c) f_{ck} \tag{18}$$

$$\bar{E}_{c,\bar{T}_c} = \bar{k}_{Ec,T}(\bar{T}_c) E_c \tag{19}$$

where $k_{s,T}$ and $k_{Es,T}$ are the respective reduction factors of strength and elastic modulus of steel from Eqs. (6) and (7); $\bar{k}_{c,T}$ and $\bar{k}_{Ec,T}$ are the respective equivalent reduction factors of strength and elastic modulus of concrete from Eqs. (14) and (15).

The analysis of a CFST column under fire can now follow the solution procedure for a CFST column under normal temperature condition with reduced strength and stiffness that depend on the average temperature. Hence, the bearing capacity, stiffness and stability of a column under room and elevated temperature can be calculated by the unified formulations shown symbolically in Table 1.

Table 1: Formation of the unified theory of CFST column under fire

Item	Formulae at room temperature	Formulae at elevated temperature
Cross-sectional plastic resistance	$N_0 = f(n, A_s, A_c, \psi, f_y, f_{ck})$	$N_{0,T} = f(n, A_s, A_c, \psi, f_{y,\bar{T}_s}, \bar{f}_{ck,\bar{T}_c})$
Flexural stiffness	$(EI)_{sc} = E_c I_c + E_s I_s$	$(EI)_{sc,T} = \bar{E}_{c,\bar{T}_c} I_c + E_{s,\bar{T}_s} I_s$
Stability factor	$\varphi_{sc} = f(N_0, (EI)_{sc}, L_0, \ldots)$	$\varphi_{sc,T} = f(N_{0,T}, (EI)_{sc,T}, L_0, \ldots)$
Ultimate capacity	$N_u = \varphi_{sc} N_0$	$N_{u,T} = \varphi_{sc,T} N_{0,T}$

In summary, the unified method for fire resistance analysis requires only the average temperature, by which the calculation of fire resistance of a CFST column can be divided into two steps, i.e. (a) calculation of the equivalent strength and elastic modulus of steel and concrete at elevated temperature based on average temperature, and (b) calculation of fire resistance using the formulas at room temperature by using the reduced material strength and elastic modulus at elevated temperature. One of the obvious advantages of the method is that this procedure does not require calculation of the complex non-uniform temperature distribution.

FORMULA FOR CALCULATING AVERAGE TEMPERATURE OF STEEL AND CONCRETE

It is clear that one of the key steps is to calculate the time dependent average temperatures of the steel and the concrete of a CFST column. In Yu [19], it has been shown that the average temperatures of the steel and the concrete of a polygonal CFST can be approximated by those of a circular CFST with the same steel and concrete areas. Using the finite element method and a nonlinear regression analysis, a unified formula for calculating average temperatures of steel and concrete, which is valid for both solid and hollow CFST columns with circular and polygonal sections, can be obtained as follows [19]:

Average Temperature of Steel Tube $\overline{T}_s$

$$\overline{T}_s = A\left(1 - \frac{1}{1 + (t/B)^C}\right) + 20 \quad [°C] \tag{20}$$

where $A = 1200$, $B = 0.337 + 8.5\bar{d}$, $C = 0.996 + 14.0\bar{d}$; t is the fire exposure time in hours; $\bar{d}$ is the equivalent thickness of steel in meters, $\bar{d} = \sqrt{(A_c + A_k + A_s)/\pi} - \sqrt{(A_c + A_k)/\pi}$; A_c, A_s and A_k are the cross sectional areas of steel, concrete and the hollow part, respectively.

Average Temperature of Concrete Core

$$\overline{T}_c = \frac{2}{1 + \sqrt{\psi}} \times A\left(1 - \frac{1}{1 + (t/B)^C}\right) + 20 \quad [°C] \tag{21}$$

Where

$A = 120 + 1080e^{-4.47\bar{L}}$, $B = 0.337 + 8.5\bar{d} + 30\bar{L}(\bar{L}^2 - 1.46\bar{L} + 0.64)$, $C = 0.996 + 14.0\bar{d}$; $\bar{\mathrm{J}}$ is the equivalent thickness of concrete in meters, $\bar{L} = \sqrt{(A_c + A_k)/\pi} - \sqrt{A_k/\pi}$; ψ is the hollow ratio, $\psi = A_k/(A_c + A_k)$.

In order to assess the accuracy of above formulas, the average temperatures of both solid (equivalent diameter is 500 mm, tube thickness is 5 mm, and hollow ratio is 0.0) and hollow (equivalent diameter is 500 mm, tube thickness is 5 mm, and hollow ratio is 0.45) CFST columns with three different section profiles were computed using both the proposed formulas and the FEA. The results are shown and compared in Fig. 5. Evidently, they results for both steel and concrete agree with each other very well.

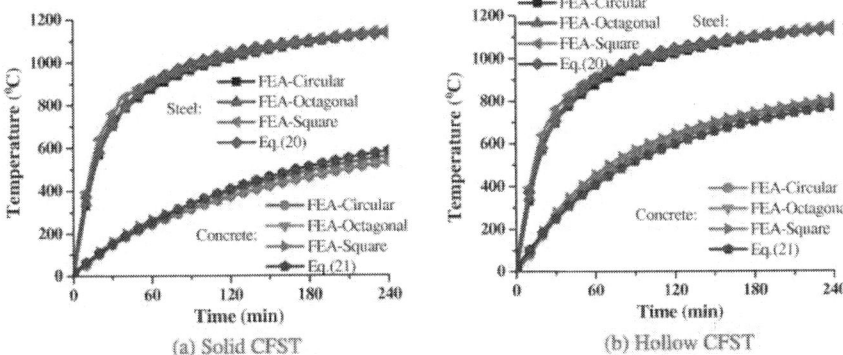

Figure 5: Average temperature obtained from the formula and the FEA results.

UNIFIED FORMULATION OF FIRE RESISTANCE FOR CFST COLUMNS

Calculation Model Based On Eurocode 4 (Method 1)

According to the unified calculation method described above and Eurocode 4, the plastic resistance of a cross-section in fire, without considering the confinement effect and partial factors of the materials, can be given by:

$$N_{0,T} = A_c \bar{f}_{c,\bar{T}_c} + A_s f_{y,\bar{T}_s}$$

(22)

where $f_{y,\bar{T}_s}$ is the strength of steel at average temperature $\bar{T}_s$ (Eq. (16)); $\bar{f}_{c,\bar{T}_c}$ is the equivalent compressive cylinder strength of concrete at average temperature $\bar{T}_c$, and $\bar{f}_{c,\bar{T}_c} = \bar{k}_{c,T}(\bar{T}_c) f'_c$ (see Eq.(18)).

The Euler buckling load is given by:

$$N_{cr,T} = \frac{\pi^2 (EI)_{sc,T}}{L_0^2} \qquad (23)$$

where $(EI)_{sc,T}$ is the effective flexural stiffness under fire, $(EI)_{sc,T} = \bar{E}_{c,\bar{T}_c} I_c + E_{s,\bar{T}_s} I_s$.

According to Eurocode 4, the reduction factor using the buckling curve 'c' is:

$$\varphi_{sc,T} = \frac{1}{\Phi + \sqrt{\Phi^2 - \bar{\lambda}_{sc,T}^2}} \leqslant 1.0 \qquad (24)$$

where $\Phi = 0.5[\bar{\lambda}_{sc,T}^2 + \alpha(\bar{\lambda}_{sc,T} - 0.2) + 1]$; $\bar{\lambda}_{sc,T}$ is the non-dimensional slenderness ratio under fire and is given by $\bar{\lambda}_{sc,T} = \sqrt{N_{0,T}/N_{cr,T}}$; α is the imperfection factor and, for buckling curve "c", α is 0.49.

Thus, the fire resistance of a CFST column subject to axial compression is obtained as:

$$N_{u,T} = \varphi_{sc,T} N_{0,T} \qquad (25)$$

where $\varphi_{sc,T}$ is the reduction factor of Eq. (24); $N_{0,T}$ is the plastic resistance of Eq. (22).

Unified Formulation for CFST Columns under Normal and High Temperature (Method 2)

The method described in Section 4.1 does not consider the confinement effect. To take this into account, the unified formulas proposed by the authors [5] for circular and polygon CFST columns under axial compression at room temperature are adopted and extended to include degradation of confinement effect under fire. In order to unify the formulas under room and higher temperature, it is assumed that a formula of a CFST column under fire takes the same structure as that of the same column at room temperature, by replacing the strength of steel and concrete at room temperature with the equivalent strength under fire. On the basis of the above assumption, the strength at room temperature from the authors' previous work [5] can be extended to include fire condition as:

$$N_{0,T} = (1 + \eta_T)[A_c \bar{f}_{ck,\bar{T}_c} + A_s \bar{f}_{y,\bar{T}_s}] \tag{26}$$

where η_T is the enhanced confining coefficient under fire, $\eta_T = 0.5 k_e \xi_T/(1 + \xi_T)$; ξ_T is the confining coefficient, and $\xi_T = A_s \bar{f}_{y,\bar{T}_s}/A_c \bar{f}_{ck,\bar{T}_c}$; $k_e = (1-\phi)(\mu_5 - \psi)\backslash(\mu_5 + 50)$; $\bar{\lambda}_{T}^{1^s}$, $\bar{\lambda}_{T}^{C_1 C}$ are equivalent strength of steel and concrete of Eqs.(16) and (18), respectively. Under a fire condition, the inherent weakness of steel under elevated temperature will gradually diminish the confinement effect with time until it can be neglected. In order to assess the confinement effect of CFST columns in fire, the finite element model of CFST under fire from Refs. [25], [26], [27] and [28] was used to calculate the bearing capacities of a solid CFST column (diameter is 500 mm, tube thickness is 5 mm, hollow ratio is 0.0, f_{ck} = 30 MPa, and f_y = 350 MPa) and a hollow CFST column (diameter is 500 mm, tube thickness is 5 mm, hollow ratio is 0.45, f_{ck} = 30 MPa, and f_y = 350 MPa) under fire with and without considering the confinement effect. The obtained numerical results are compared with predictions from Eq.(26) (with confinement effect) and from Eq. (22) (without confinement effect) in Fig. 6. The comparisons show that Eq. (26) exhibits comparable confinement effect at early stage of the fire, and a continuous reduction of the confinement effect to virtually zero. Therefore, Eq. (26) can be used for CFST columns at both room and elevated temperatures.

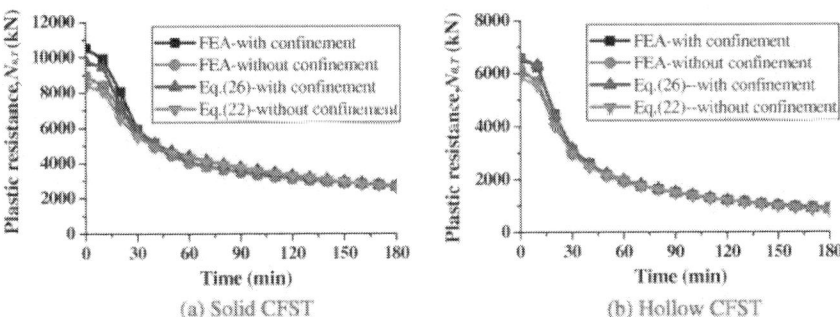

Figure 6: Comparison of the confinement effect.

Similarly, the stability factor of a CFST column at room temperature in Yu et al. [5] can also be extended to include fire conditions as:

$$\varphi_{sc,T} = \frac{1}{2\bar{\lambda}_{sc,T}^2}\left[\bar{\lambda}_{sc,T}^2 + K\bar{\lambda}_{sc,T} + 1 - \sqrt{(\bar{\lambda}_{sc,T}^2 + K\bar{\lambda}_{sc,T} + 1)^2 - 4\bar{\lambda}_{sc,T}^2}\right] \quad (27)$$

where $\bar{\lambda}_{sc,T}$ is the normalized slenderness ratio under fire, $\bar{\lambda}_{sc,T} = L_0/\pi \sqrt{N_{0,T}/(EI)_{sc,T}}$; K denotes initial imperfection coefficient, and $K = 0.25 - 0.09k_e$; $(EI)_{sc,T}$ is the effective flexural stiffness under fire, $(EI)_{sc,T} = \bar{E}_{c,\bar{T}_c} I_c + E_{s,\bar{T}_s} I_s$.

Thus, the stability load bearing capacity of a CFST column under fire conditions is:

$$N_{u,T} = \varphi_{sc,T} N_{0,T} \quad (28)$$

where $\varphi_{sc,T}$ is the stability factor from Eq. (27); $N_{0,T}$ is the plastic resistance from Eq. (26).

Validation of the Fire Resistance Formulas for CFST Columns Subjected to Axial Compression

In order to validate the fire resistance formulas, Eqs. (25) and (28), extensive comparisons with existing fire experimental data were made, including 69 circular CFST columns [6], [7], [8], [9], [19], [20] and [38] and 61 square CFST columns [6], [7], [9], [19] and [39], as shown in Table 2 and Table 3. Because ASTM E119 fire curve was used for some of the fire tests for both carbonate and silicate aggregates, the equivalent exposure time under ISO 834 fire was estimated by [33] and [40]:

$$t_e = \alpha_{agg} \alpha_{ISO} t \quad (29)$$

where t_e is the equivalent exposure time in fire; α_{agg} is an aggregate factor taking 1 for siliceous aggregate and 0.9 for carbonate concrete; α_{ISO} is the fire type factor taking 1 for ISO-834 fire and 0.85 for ASTM-E119 fire.

Table 2: Comparison of formulas and experimental results for circular CFSTs under fire

Ref.	No.	Explanation of numbering	Geometric parameters				B.C.		Material parameters				Test			Based on Eurocode 4 (Method 1)		Unified Formula (Method 2)	
		Number	External diameter D (mm)	Steel thickness t (mm)	Radius of hollow r_o (mm)	Length L_0 (mm)	α_{IS0}	α_{BC}	f_y (MPa)	α_{ag}	(N/P_a)	f_{ck} (MPa)	Applied load N_{test} (kN)	Fire time t (min)	$N_{u,T}$ (kN)	$N_{u,T}/N_{tes t}$	$N_{u,T}$ (kN)	$N_{u,T}/N_{test}$
[6]	1	C-02	141.3	6.55	0	3810	0.85	0.5	350	1	33.1	27.5	110	55	144	1.31	185	1.68
	2	C-04	141.3	6.55	0	3810	0.85	0.5	350	1	31	25.9	131	57	131	1	168	1.28
	3	C-05	168.3	4.78	0	3810	0.85	0.5	350	1	32.7	27.3	150	76	125	0.83	146	0.98
	4	C-06	168.3	4.78	0	3810	1	0.5	350	1	32.7	27.3	150	60	82	0.55	98	0.65
	5	C-08	168.3	4.78	0	3810	0.85	0.5	355	1	35.5	29.3	218	56	227	1.04	264	1.21
	6	C-09	168.3	6.35	0	3810	0.85	0.5	350	1	35.4	29.2	150	81	122	0.82	144	0.96
	7	C-	21	4.	0	3	0	0	3	1	3	2	4	8	3	0.64	335	0.6

Design of Reinforced Concrete

	11	9.1	78		810	0.85	.50	5		1	5.9	92	0	15			8
8	C-13	219.1	4.78	0	3810	0.85	0.50	1	32.3	3	27	384	102	219	0.57	236	0.62
9	C-15	219.1	8.18	0	3810	0.85	1	35	31.9	3	26.7	525	73	207	0.39	252	0.48
10	C-17	219.1	8.18	0	3810	0.85	0.50	1	31.7	3	26.5	525	82	328	0.63	356	0.68
11	C-20	273.1	5.56	0	3810	0.85	0.50	1	28.6	3	23.9	574	112	446	0.78	443	0.77
12	C-21	273.1	5.56	0	3810	0.85	0.50	1	29	3	24.2	525	133	351	0.67	352	0.67
13	C-22	273.1	5.56	0	3810	0.85	0.50	1	27.2	3	22.7	1000	70	727	0.73	707	0.71
14	C-23	273.1	12.7	0	3810	0.85	0.50	1	27.4	3	22.9	525	143	281	0.53	288	0.55
15	C-25	323.9	6.35	0	3810	0.85	0.50	1	27.6	3	23.1	699	145	621	0.89	589	0.84
16	C-26	323.9	6.35	0	3810	0.85	0.50	1	24.3	3	20.3	1050	93	884	0.84	817	0.78
17	C-28	355.6	6.35	0	3810	0.85	0.50	1	23.8	3	19.9	1050	111	1012	0.96	915	0.87
18	C-29	355.6	12.7	0	3810	0.85	0.50	1	25.4	3	21.2	1050	70	652	0.62	613	0.58
19	C-30	406.4	12.7	0	3810	0.85	0.50	1	27.6	3	23.1	1900	71	2400	1.26	2196	1.16

UNIFIED FORMULATION OF FIRE RESISTANCE FOR CFST COLUMNS

20	C-31	141.3	6.55	0	3810	0.85	0.50	0.09	30.2	25.2	80	82	69	0.86	87	1.08
21	C-32	141.3	6.55	0	3810	0.85	0.50	0.09	34.8	28.8	143	64	130	0.91	161	1.13
22	C-34	219.1	4.78	0	3810	0.85	0.50	0.09	35.4	29.2	500	111	239	0.48	256	0.51
23	C-35	219.1	4.78	0	3810	0.85	0.50	0.09	42.7	34.4	560	108	282	0.5	303	0.54
24	C-37	219.1	8.18	0	3810	0.85	0.50	0.09	28.7	24	560	102	255	0.46	277	0.49
25	C-40	273.1	6.35	0	3810	0.85	0.50	0.09	46.5	37.4	1050	106	791	0.75	782	0.74
26	C-41	273.1	6.35	0	3810	0.85	0.50	0.09	50.7	40.5	1050	76	1218	1.16	1173	1.12
27	C-42	273.1	6.35	0	3810	0.85	0.50	0.09	55.4	44	1050	90	1089	1.04	1068	1.02
28	C-44	273.1	6.35	0	3810	0.85	0.50	0.09	38.7	31.5	715	178	310	0.43	313	0.44
29	C-45	273.1	6.35	0	3810	0.85	0.50	0.09	38.2	31.1	712	144	447	0.63	450	0.63
30	C-46	273.1	6.35	0	3810	0.85	0.50	0.09	82.2	64.5	1050	48	2609	2.48	2500	2.38
31	C-50	323.9	6.35	0	3810	0.85	0.50	0.09	42.4	34.2	820	234	502	0.61	481	0.59
32	C-51	323.9	6.35	0	3810	0.85	0.50	0.09	47.5	38	1118	114	1441	1.2	1303	1.1

Ref	No	ID	D (mm)	t (mm)	—	L (mm)	—	—	—	—	f_c'	—	—	—	—	Ratio	P_{calc}	Ratio
	33	C-53	355.6	6.35	0	3810	0.085	0.50	30	0.09	42.4	34.2	133.5	149	1400	1.05	1271	0.95
	34	C-55	355.6	12.7	0	3810	0.085	0.50	30	0.09	40.7	32.9	96.5	274	523	0.54	489	0.51
	35	C-57	406.4	6.35	0	3810	0.085	0.50	30	0.09	44	35.5	140.0	294	1202	0.86	1085	0.78
	36	C-59	406.4	12.7	0	3810	0.085	0.50	30	0.09	37.4	30.6	190.0	125	2234	1.18	1958	1.03
	37	C-60	406.4	12.7	0	3810	0.085	0.50	30	0.09	45.1	36.3	190.0	152	2242	1.18	1970	1.04
[9]	38	77.125 24A	168.3	3.6	0	3600	1	0.5	323	1	43.9	35.4	30.0	56	187	0.62	212	0.71
	39	77.125 24B	219.1	3.6	0	3600	1	0.5	395	1	43.7	35.2	60.0	45	676	1.13	687	1.15
	40	77.125 24D	219.1	3.6	0	3600	1	0.5	385	1	43.9	35.4	60.0	43	710	1.18	720	1.2
	41	77.125 24E	219.1	3.6	0	3600	1	0.5	400	1	43.7	35.2	30.0	102	200	0.67	215	0.72
	42	77.125 24F	219.1	3.6	0	3600	1	0.5	400	1	43.7	35.2	90.0	35	870	0.97	885	0.98
[17]	43	CA	317.7	7	0	35	0.	1	30	1	27.3	94	80	78		0.83	852	0.91

UNIFIED FORMULATION OF FIRE RESISTANCE FOR CFST COLUMNS

]		L1-1	5	.		40	85		4		5		1		4			
	44	CAL1-2	317.5	7	0	3540	0.85	1	304	1	27.5	23	941	80	784	0.83	852	0.91
	45	CAL2	317.5	7	0	3540	0.85	1	304	1	27.5	23	774	150	349	0.45	402	0.52
	46	CAH1	317.5	7	0	3540	0.85	1	304	1	37.8	30.9	1548	28	2462	1.59	2753	1.78
	47	CBL1-1	406.4	9	0	3540	0.85	1	311	1	27.5	23	1676	80	1837	1.1	1793	1.07
	48	CBL1-2	406.4	9	0	3540	0.85	1	311	1	27.5	23	1676	59	2261	1.35	2214	1.32
	49	CBL2	406.4	9	0	3540	0.85	1	311	1	27.5	23	1254	120	1332	1.06	1340	1.07
	50	CBH1	406.4	9	0	3540	0.85	1	311	1	37.8	30.9	2509	47	3359	1.34	3285	1.31
	51	CBH2-1	406.4	9	0	3540	0.85	1	311	1	37.8	30.9	1676	88	2200	1.31	2163	1.29
	52	CBH2-2	406.4	9	0	3540	0.85	1	311	1	37.8	30.9	1676	108	1861	1.11	1871	1.12
[8]	53	C3-1	150	4.6	0	3810	1	1	259	1	55	43.8	920	20	277	0.3	336	0.37
	54	C4-1	219	4.6	0	3810	1	1	381	1	55	43.8	1800	21	966	0.54	1202	0.67
	55	C4-	219	4.6	0	38	1	1	38	1	55	43	18	20	10	0.56	1268	0.7

								10			1			.8	00		15			
[38]	56	C159-6-30-0-40-P-P	159	6	73.5	3810	1		1	337.8	0.9	30.1	25.2	338	18	477	1.41	627	1.86
	57	C159-6-30-0-20	159	6	73.5	3810	1	0.7		337.8	0.9	35.8	29.4	198	42	223	1.12	283	1.43
	58	C159-6-30-0-40	159	6	73.5	3810	1	0.7		337.8	0.9	28.6	23.9	398	25	467	1.17	643	1.61
	59	C159-6-30-0-60	159	6	73.5	3810	1	0.7		337.8	0.9	34.1	28.2	594	14	892	1.5	1262	2.12
	60	C159-6-80-0-20	159	6	73.5	3810	1	0.7		341.4	0.9	71.1	55.8	335	38	327	0.98	399	1.19
	61	C159-6-80	159	6	73.5	3810	1	0.7		341.4	0.9	69	54.1	670	11	1183	1.77	1571	2.34

		-0-40																
[19]	62	C0-F	219	3.8	0	24440	1	1	291.5	1	42.6	34.4	741	29	846	1.14	916	1.24
	63	C1-F	219	3.8	57.8	24440	1	1	291.5	1	42.6	34.4	529	37	470	0.89	489	0.92
	64	C2-F	219	3.8	66.3	24440	1	1	291.5	1	42.6	34.4	443	41	357	0.81	368	0.83
	65	C3-F	219	3.8	78	24440	1	1	291.5	1	42.6	34.4	356	28	440	1.24	455	1.28
[20]	66	0I1	219	1.6	80.75	19999	1	1	318.3	1	57.3	45.4	250	59	175	0.7	173	0.69
	67	0I2	219	1.6	87.225	20001	1	1	318.3	1	57.3	45.4	180	30	316	1.75	298	1.66
	68	2I1	219	1.6	87	20003	1	1	318.3	1	57.3	45.4	250	37	248	0.99	235	0.94
	69	4I2	219	1.63	74.15	20000	1	1	318.3	1	57.3	45.4	250	51	276	1.1	273	1.09

Note: no. 1–62 are solid CFST columns, no. 63–69 are hollow CFST columns.

Table 3: Comparison of formulas and experimental results for square CFSTs under fire

Ref.	No.	Numbering	Length of side B (mm)	Steel thickness t (mm)	Radius of hollow rco (mm)	Length L (mm)	Fire curve	B.C.	f_y (MPa)	α_a/α_g	f_c (MPa)	f_{ck} (MPa)	Applied load $N_{t,est}$ (kN)	Fire Time t (min)	$N_{u,T}$ (kN)	$N_{u,T}/N_{te st}$	Cal Ratio	Unified Formula (Method 2) Ratio $N_{u,T}/N_{test}$
[16]	1	SQ-01	152.4	6.35	0	3810	0.85	0.55	3	1	58.3	46.2	376	66	266	0.71		0.74
	2	SQ-02	152.4	6.35	0	3810	0.85	0.55	3	0.9	46.5	37.4	286	86	174	0.61		0.64
	3	SQ-07	177.8	6.35	0	3810	0.85	0.55	3	1	57	45.2	549	80	338	0.62		0.63
	4	SQ-17	254	6.35	0	3810	0.85	0.55	3	1	5.8	46	1096	62	1748	1.59		1.4

UNIFIED FORMULATION OF FIRE RESISTANCE FOR CFST COLUMNS

					0					3	2						2
	5	SQ-20	254	6.35	0	3810	0.85	0.5	350	0.9	46.5	37.4	931	97	1053	1.13	1.03
	6	SQ-24	304.8	6.35	0	3810	0.85	0.5	350	1	58.8	46.5	1130	131	1632	1.44	1.29
[7]	7	SAL1-1	300	9	0	3540	0.85	1	363	1	27.5	23	843	80	1025	1.22	1.2
	8	SAL1-2	300	9	0	3540	0.85	1	363	1	27.5	23	843	80	1025	1.22	1.2
	9	SAL2	300	9	0	3540	0.85	1	363	1	27.5	23	745	130	575	0.77	0.8
	10	SAH1	300	9	0	3540	0.85	1	363	1	37.8	30.9	1401	44	2275	1.62	1.59
	11	SBL1-1	350	9	0	3540	0.85	1	363	1	27.5	23	1294	80	1702	1.32	1.23
	12	SBL1-2	350	9	0	3540	0.85	1	363	1	27.5	23	1294	80	1702	1.32	1.23
	13	SBL2	350	9	0	3540	0.85	1	363	1	27.5	23	1039	160	882	0.85	0.85
	14	SBH1	350	9	0	3540	0.85	1	363	1	37.8	30.9	1941	108	1670	0.86	0.83
	15	SBH2-1	350	9	0	3540	0.85	1	363	1	37.8	30.9	1558	140	1277	0.82	0.82
	16	SBH2-2	350	9	0	3540	0.85	1	363	1	37.8	30.9	1558	140	1277	0.82	0.82

Ref	No	Specimen	c1	c2	c3	c4	c5	c6	c7	c8	c9	c10	c11	c12	c13	c14	c15
[9]	17	78.U47/T46	150	5	0	3600	1	0.5	389	1	49	39.3	250	81	102	0.41	0.42
	18	78.U48/T47	150	5	0	3600	1	0.5	351	1	48	38.5	250	81	100	0.4	0.41
	19	76.U56/T55	200	5	0	3600	1	0.5	351	1	34.5	28.5	950	36	839	0.88	0.84
	20	78.U46/T45	200	5	0	3600	1	0.5	351	1	49	39.3	740	80	394	0.53	0.52
	21	75.U26/T24	225	3.6	0	3600	1	0.5	368	1	49	39.3	1085	56	969	0.89	0.82
	22	75.U28/T26	225	3.6	0	3600	1	0.5	368	1	49	39.3	1520	42	1258	0.83	0.74
	23	75.U29/T27	225	3.6	0	3600	1	0.5	368	1	49	39.3	430	165	147	0.34	0.31
	24	75.U52/T50	225	8	0	3600	1	0.5	400	1	44.5	35.8	1970	29	1998	1.01	0.97
	25	75.U53/T51	225	8	0	3600	1	0.5	400	1	44.5	35.8	1405	37	1531	1.09	1.02
	26	75.U54/T52	225	8	0	3600	1	0.5	400	1	44.5	35.8	560	144	190	0.34	0.32
	27	77.U50/T50	260	8	0	3600	1	0.5	390	1	41.5	33.5	1500	49	1613	1.08	0.95
	28	77.U51/T51	260	4	0	3600	1	0.5	550	1	41.5	33.5	1500	45	1657	1.1	0.97
	29	77.U5	260	8.8	0	36	1	0.	32	1	34	28	800	10	624	0.78	0.

UNIFIED FORMULATION OF FIRE RESISTANCE FOR CFST COLUMNS

	6/T56				0 0		5	6			.2	2				71
30	77.U60/T60	260	6.3	0	3600	1	0.5	37 0	1	41.5	33.5	800	86	916	1.15	1.03
31	78.U15/T15	260	8	0	3600	1	0.5	39.5	1	41.8	33.8	800	114	634	0.79	0.73
32	74.8202	140	3.6	0	3600	1	0.5	42.5	1	47	37.7	685	24	575	0.84	0.89
33	74.8199	140	3.6	0	3600	1	0.5	42.4	1	51.5	41.1	410	42	273	0.67	0.71
34	74.8203	140	3.6	0	3600	1	0.5	38.4	1	40.1	32.4	190	66	103	0.54	0.57
35	75.9995.S	140	3.6	0	3600	1	0.5	42.9	1	51.6	41.2	530	28	495	0.93	0.98
36	75.9995.T	140	3.6	0	3600	1	0.5	42.9	1	51.6	41.2	530	24	601	1.13	1.2
37	75.9995.W	200	5	0	3600	1	0.5	36.0	1	49.5	39.6	1660	18	2015	1.21	1.18
38	75.9995.X	200	5	0	3600	1	0.5	36.0	1	55.8	44.3	1240	39	1080	0.87	0.81
39	75.9995.Y	200	5	0	3600	1	0.5	36.0	1	55.8	44.3	740	88	362	0.49	0.49
40	75.9995.2	225	3.6	0	3600	1	0.5	31.0	1	40.5	32.7	1000	36	1210	1.21	1.07
41	77.12413	260	6.3	0	3600	1	0.5	37.0	1	41.5	33.5	800	98	784	0.98	0.89
42	75.	26	4	0	3	1	0	3	1	3	2	91	6	94	1.0	0

	2	9995.D	5			600		.5	60		0.2	5.2	0	8	8	4	.93
	43	Fl.34 07	250	10	0	3600	1	0.5	243	1	57.9	45.9	1950	68	1314	0.67	0.06
	44	Fl.34 08	250	10	0	3600	1	0.5	243	1	47.7	38.3	1740	25	2867	1.65	1.52
	45	Fl.31 08	260	6.3	0	3600	1	0.5	370	1	41.5	33.5	800	133	504	0.63	0.58
	46	Fl.34 02	350	10	0	3600	1	0.5	326	1	47.7	38.3	2250	85	2817	1.25	1.04
	47	Fl.34 04	350	10	0	3600	1	0.5	326	1	47.7	38.3	3150	39	4623	1.47	1.24
	48	Fl.34 05	350	10	0	3600	1	0.5	313	1	47.7	38.3	4390	30	5484	1.25	1.08
	49	Fl.34 06	350	10	0	3600	1	0.5	313	1	48.8	39.1	3950	55	3803	0.96	0.8
[39]	50	BAM.77.1	260	6.3	0	3600	1	0.5	370	1	41.5	33.5	800	134	498	0.62	0.57
	51	BR.77.1	260	6.3	0	3600	1	0.7	370	1	41.5	33.5	800	81	810	1.01	0.99
	52	BR.78.1	160	6.3	0	5800	1	0.5	305	1	41.3	33.4	100	68	116	1.16	1.29
	53	BR.78.2	180	6.3	0	5800	1	0.5	273	1	41.3	33.4	200	42	431	2.16	2.31
	54	BR.78.3	200	3.6	0	5800	1	0.5	335	1	41.	33.	300	52	411	1.37	1.4

UNIFIED FORMULATION OF FIRE RESISTANCE FOR CFST COLUMNS

					0			3	4				7
55	BR.78.4	220	6.3	0	5800	1	1	244	1	41.3 33.4	490	16	1178	2.4	2.76
56	BR.78.5	220	6.3	0	5800	1	0.7	244	1	41.3 33.4	800	15	1763	2.2	2.41
57	BR.78.6	220	6.3	0	5800	1	0.5	244	1	41.3 33.4	800	34	1052	1.32	1.3
58	BR.78.7	260	7.1	0	5800	1	0.5	317	1	41.3 33.4	1000	51	1226	1.23	1.18
[19] 59	S1-F	200	3.9	53.3	2440	1	1	322.8	1	42.6 34.4	454	66	278	0.61	0.62
60	S2-F	200	3.9	64.5	2440	1	1	322.8	1	42.6 34.4	365	48	358	0.98	0.96
61	S3-F	200	3.9	79.3	2440	1	1	322.8	1	42.6 34.4	278	38	352	1.26	1.22

Note: no. 1–58 are solid CFST columns, no. 59–61 are hollow CFST columns.

Available fire tests on circular and square CFST columns were compared with the predictions from Eq. (25)(Method 1) in Fig. 7. It is found from the figure that the average ratio of the predictions and test values is 0.941 and 1.027, with variances of 0.143 and 0.179 (the second central moment about the mean) respectively, for circular and square CFST columns. The test results were also compared with the predictions of Eq. (28) (Method 2) in Fig. 8. The average ratio of the predictions and the test values is 1.003 and 0.995, with a variance of 0.186 and 0.203 for circular and square CFST columns, respectively. It is important to notice that Eq. (25) uses the cylinder strength of concrete (f_c), while Eq. (28) uses the standard compressive strength concrete (f_{ck}). However, both can be easily converted from one to the other by [41] and [42]

$$f_{ck} = 0.88\alpha_1\alpha_2 f_{cu}, \quad f'_c = 0.8 f_{cu} \tag{30a}$$

where f_{cu} is the cube strength of concrete; α_1, and α_2 are f_{cu} dependent constants that are calculated as below [41]

$$\begin{aligned} \alpha_1 &= 0.76 \quad \text{when} \quad f_{cu} \leqslant 50 \text{ MPa} \\ \alpha_1 &= 0.82 \quad \text{when} \quad f_{cu} = 80 \text{ MPa}, \\ \alpha_2 &= 1.00 \quad \text{when} \quad f_{cu} \leqslant 40 \text{ MPa} \\ \alpha_2 &= 0.87 \quad \text{when} \quad f_{cu} = 80 \text{ MPa} \end{aligned} \tag{30b}$$

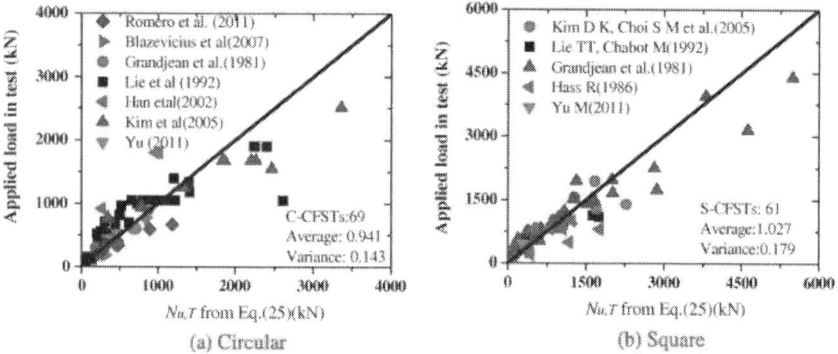

Figure 7: Comparison between the formula based on Eurocode 4 and tests of CFST under fire.

CALCULATION PROCEDURE AND DISCUSSION

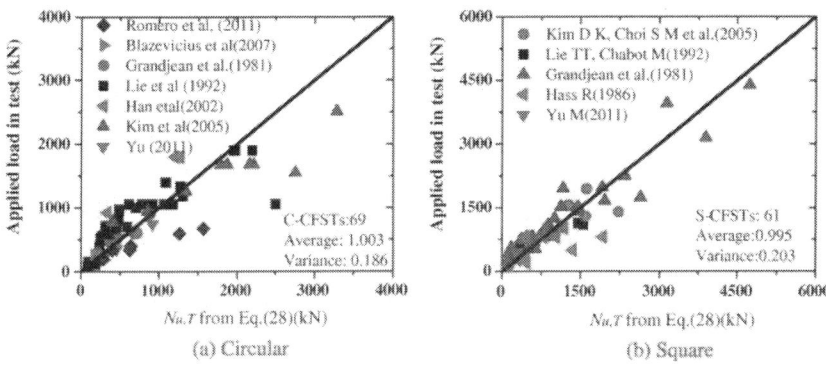

Figure 8: Comparison between the unified formula and tests of CFST under fire.

Linear interpolation technique can be used to approximate the values of α_1, and α_2 are outside the above range of f_{cu}.

The comparisons shown in Table 2 and Table 3 are, respectively, for circular and square CFST columns. Comparing with the work reported by Wang and Kodur [30] and Tan and Tang [33], the average ratio of the predictions from our newly proposed formulas and the test values is generally closer to 1 and the variance is smaller. However, it should be concluded that both methods provide comparable accuracy of prediction.

CALCULATION PROCEDURE AND DISCUSSION

Step-by-Step Calculation Procedure
In order to help engineers to use the proposed formulas in a practical design, a step-by-step procedure is summarized below. The procedure is for calculating fire resistance of solid and hollow CFST columns with circular and polygonal sections.

1. Calculate cross-sectional areas of the steel tube, the concrete core and the hollow part, respectively, if the section of a column is non-circular, otherwise start from step 2 directly.
2. Calculate equivalent thicknesses of the steel tube, the concrete, and the hollow ratio, based on the equivalent area method (see the formulas in the paragraphs below Eqs. (20) and (21)).
3. Calculate average temperatures of the steel and the concrete using Eqs. (20) and (21), respectively.

4. Calculate equivalent strength and elastic modulus of the steel tube and the concrete core at the average temperatures from step 3 by using Eqs. (16), (17), (18) and (19).
5. Calculate strength of the CFST column under fire using Eq. (22) for method 1, and/or Eq. (26) for method 2.
6. Calculate stability factor of the CFST column under fire using Eq. (24) for method 1, and/or Eq. (27) for method 2.
7. Calculate fire resistance by multiplying the strength and the stability factor using Eqs. (25) and (28) for methods 1 and 2, respectively.

Restrictions on the Application of the Formulas

It is evident that for solid and hollow CFST columns with various sectional profiles, method 2 provides a unified formulation to cover a wide range of temperature starting from the room temperature. The restrictions on the application of the unified fire resistance formulas depend on the applicability of using average temperatures of the steel tube and the concrete in the calculation. The proposed calculation method using average temperature can be applied (but may not be limited) to the analysis of concentrically loaded CFST columns satisfying the following conditions:

a. Diameter or equivalent diameter of cross-section: $120 \text{ mm} \leqslant \tilde{D}$ and $\tilde{D} \leqslant 2000 \text{ mm}$.
b. Fire resistance time: $t \leqslant 4 \text{ h}$.
c. Hollow ratio: $0 \leqslant \psi \leqslant 0.75$.
d. Normal weight concrete and structural steel.

CONCLUDING REMARKS

A new fire resistance calculation method using average temperature has been proposed in this paper. It has been shown that the equivalent strength and stiffness can be calculated accurately, on the basis of the average temperature, irrespective of cross-section shapes and variations of temperature gradient in the concrete under fire.

Combined with Eurocode 4 and the unified formula at room temperature from the authors' previous work, a unified formulation for calculating fire resistances of both solid and hollow CFST columns with various section profiles was proposed and validated through comparisons with existing experimental results. A step-by-step procedure was provided and the applicability of the methods was discussed.

Further work is needed to extend the current method to include columns subjected to bending and combined loadings, and also CFST columns with fire protections. The development of the formula for average temperature to include lightweight concrete and some other special type of concrete or steel is also possible.

ACKNOWLEDGEMENT

The authors are grateful for the financial support from the Fundamental Research Funds for the Central Universities (Grant No. 2042014kf0010) and the China Postdoctoral Science Foundation (Grant No.2013M531047).

REFERENCES

1. EN 1994-1-2:2005. Design of composite steel and concrete structures Part 1.2: general rules, structural fire design. Brussels: British Standards Institution; 2005.
2. DBJ13-51-2003. Technical specification for concrete-filled steel tubular structures. Fuzhou: The Construction Department of Fujian Province; 2003 [in Chinese].
3. ASCE/SFPE 29-99. Standard calculation method for structural fire protection. Reston, USA: American Society of Civil Engineers; 1999.
4. ACI 216.1M-07. Standard method for determining fire resistance of concrete and masonry construction assemblies. Detroit, USA: American Concrete Institute; 2007.
5. Yu M, Zha XX, Ye JQ, et al. A unified formulation for circle and polygon concrete-filled steel tube columns under axial compression. Eng Struct 2013;49:1–10.
6. Lie TT, Chabot M. Experimental studies on the fire resistance of hollow steel columns filled with plain concrete. In: NRC-CNRC internal report no. 611, Construction National Research Council Canada; 1992.
7. Kim DK, Choi SM, Kim JH, et al. Experimental study on fire resistance of concrete-filled steel tube column under constant axial loads. Int J Steel Struct 2005;5(4):305–13.

8. Han LH, Xu L, Feng JB, et al. Fire resistance and fire protective cover of concrete-filled steel tubular column. China Civ Eng J 2002;35(6):6–13 [in Chinese].
9. Grandjean G, Grimault JP, Petit L. Determination De La Duree Au Feu Des Profils Creux Remplis De Beton. In: Rapport Final. Commission des Communautes Europeennes, Recherche Technique acier, Luxembourg; 1981.
10. Moliner V, Espinos A, Romero ML, et al. Fire behavior of eccentrically loaded slender high strength concrete-filled tubular columns. J Constr Steel Res 2013;83:137–46.
11. Han LH, Yang YF, Xu L. An experimental study and calculation on the fire resistance of concrete-filled Shs and Rhs columns. J Constr Steel Res 2003;59(4):427–52.
12. Lu H, Zhao XL, Han LH. Fire behaviour of high strength self-consolidating concrete filled steel tubular stub columns. J Constr Steel Res 2009;65(10–11):1995–2010.
13. Espinos A, Romero ML, Hospitaler A. Fire design method for bar-reinforced circular and elliptical concrete filled tubular columns. Eng Struct 2013;56:384–95.
14. Lie TT, Irwin RJ. Fire resistance of rectangular steel columns filled with barreinforced concrete. J Struct Eng 1995;121(5):797–805.
15. Lu H, Zhao XL, Han LH. Testing of self-consolidating concrete-filled double skin tubular stub columns exposed to fire. J Constr Steel Res 2010;66(8–9):1069–80.
16. Lu H, Han LH, Zhao XL. Fire performance of self-consolidating concrete filled double skin steel tubular columns: experiments. Fire Safety J 2010;45(2): 106–15.
17. Yang H, Liu F, Gardner L. Performance of concrete-filled Rhs columns exposed to fire on 3 sides. Eng Struct 2013;56:1986–2004.
18. Yang H, Liu F, Zhang S, et al. Experimental investigation of concrete-filled square hollow section columns subjected to non-uniform exposure [J]. Eng Struct 2013;48:292–312.
19. Yu M. Research on the consecutive theory of concrete filled steel members under normal to high temperature and impact load. Harbin: Harbin institute of technology. Ph.D. dissertation; 2011 [in Chinese].
20. Blaževičius Z, Kvedaras AK. Experimental investigation into fire resistance of H-Cfst columns under axial compression. J Civ Eng Manage 2007;8(1):1–10.
21. Zha XX. FE analysis of fire resistance of concrete filled CHS columns. J Constr Steel Res 2003;54(3):365–86.
22. Yin J, Zha XX, Li LY. Fire resistance of axially loaded concrete filled steel tube columns. J Constr Steel Res 2006;65(12):701–6.

23. Chung KS, Park SH, Choi SM. Fire resistance of concrete filled square steel tube columns subjected to eccentric axial load. Steel Struct 2009;9:69–76.
24. Espinos A, Gardner L, Romero ML, et al. Fire behaviour of concrete filled elliptical steel columns. Thin-Wall Struct 2011;49(2):239–55.
25. Yu M, Zha XX, Ye JQ, et al. Fire responses and resistance of concrete-filled steel tubular frame structures. Int J Struct Stabil Dynam 2010;10(2):253–71.
26. Espinos A, Romero ML, Hospitaler A. Advanced model for predicting the fire response of concrete filled tubular columns. J Constr Steel Res 2010;66(8–9):1030–46.
27. Han LH, Chen F, Liao FY, et al. Fire performance of concrete filled stainless steel tubular columns. Eng Struct 2013;56:165–81.
28. Lu H, Zhao XL, Han LH. Fe modelling and fire resistance design of concrete filled double skin tubular columns. J Constr Steel Res 2011;67(11):1733–48.
29. Kodur VKR. Performance of high strength concrete-filled steel columns exposed to fire. Can J Civ Eng 1998;25(6):975–81.
30. Wang YC, Kodur VKR. An approach for calculating the failure loads of unprotected concrete filled steel columns exposed to fire. Struct Eng Mech 1999;7(2):127–45.
31. Wang YC. A simple method for calculating the fire resistance of concrete-filled CHS columns. J Constr Steel Res 2000;44(3):203–23.
32. Li GQ, He JL, Han LH. Load bearing capacity of fire resistance of concrete filled steel tubular columns. Build Struct 2001;31(01):60–62, 59 [in Chinese].
33. Tan KH, Tang CY. Interaction model for unprotected concrete filled steel columns under standard fire conditions. J Struct Eng 2004;130(9):1405–13.
34. Espinos A, Romero ML, Hospitaler A. Simple calculation model for evaluating the fire resistance of unreinforced concrete filled tubular columns. Eng Struct 2012;42:231–44.
35. Zhao XL, Han LH, Lu H. Concrete-filled tubular members and connections. Spon Press; 2010.
36. Rush D, Bisby L, Melandinos A, et al. Fire resistance design of unprotected concrete filled steel hollow sections: meta-analysis of available furnace test data. Fire Safety Sci 2011;10(1):1549–62.
37. Wang ZH, Tan KH. Green's function solution for transient heat conduction in concrete-filled CHS subjected to fire. Eng Struct 2006;63(7):997–1007.
38. Romero ML, Moliner V, Espinos A, et al. Fire behavior of axially loaded slender high strength concrete-filled tubular columns. J Constr Steel Res 2011;67(12):1953–65.
39. Hass R. Practical rules for the design of reinforced concrete and composite columns submitted to fire. In: Technical rep. no. 69, Institute fur Baustoffe, Massivbau und Brandschutz der Technischen Univ. Braunschweig, Braunschweig, Germany; 1986.

40. Tang CY. An interactive formula for fire resistance of columns. Nanyang Technological University, School of Civil and Environmental Engineering, Master of Engineering; 2002.
41. GB50010-2010. Code for design of concrete structures. Beijing, China: China Building Industry Press; 2011 [in Chinese].
42. Kong FK, Evans RH. Reinforced and prestressed concrete. 3rd ed. London, UK: Chapman and Hall; 1987.

CITATION

Min Yu, Xiaoxiong Zha, Jianqiao Ye ,Baolin Wang, A unified method for calculating fire resistance of solid and hollow concrete-filled steel tube columns based on average temperature, http://dx.doi.org/10.1016/j.engstruct.2014.03.038.

CHAPTER 4

Development of UHPC Mixtures from an Ecological Point of View

N. Randl, T. Steiner S. Ofner, E. Baumgartner, T. Mészöly

ABSTRACT

Reinforced Concrete (RC) is the predominant and most frequently used building material with a worldwide annual material flow of approximately 20–25 billion tons. Consequently, cement as the most used inorganic binding material is responsible for more than 5% of the total anthropogenic CO_2 emissions. Ultra High Performance Concrete (UHPC) is an emerging high-tech building material that – in comparison to normal strength concrete (NSC) – allows for more slenderness and increased durability when designing RC-structures. The ecological impact of UHPC is affected by the high cement content with more than double the amount needed in comparison to normal strength concrete. Substitution of cement in the mixture by less-energy-intensive hydraulic concrete additives is investigated regarding its influence on the concrete properties and its environmental impact parameters calculated for the different UHPC mixtures.

RESEARCH SIGNIFICANCE: SUSTAINABILITY IN CONCRETE CONSTRUCTION

In the European Union about 40% of total energy consumption is attributed to the building and construction sector. In central European countries about 70% of the total material flow is caused by the building industry[1] and [2]. These Figure illustrate the importance of sustainability in the building sector. Therefore, besides the efforts to improve construction materials, the issue of sustainability has gained more and more attention in recent years and has become a primary focus in the construction materials industry.

The ecological targets include the minimizing of the exploitation of non-renewable resources, thereby ensuring the regeneration of renewable resources and the reduction of building waste and residues. Furthermore, the efficient use of raw materials for the production of building materials and concepts for the reuse and the recycling of building waste are necessary to keep up with future demand as laid out in the Brundtland Report of 1987, where the term "sustainability" was first defined [3].

Reinforced Concrete (RC) is well known as the most important construction material worldwide. Recent success in the formation of superplasticizers has given way to the development of the new concrete family of Ultra High Performance Concrete (UHPC), which is reaching a level in compressive strength that was earlier only possible with steel. Several guidelines dealing with the material properties and design concepts for UHPC have meanwhile been elaborated [4], [5] and [6].

The world's annual overall material flow for concrete is estimated to be approximately 20–25 Gt [7] and [8]. This amount of concrete would correspond to a cube with a side length of more than 2 km filled with concrete. Cement is the most used inorganic binding material. According to the literature its worldwide production in 2012 amounted to about 3.6 Gt [9], which has a significant ecological impact due to its production technology. The current rate of growth in cement production is about 3–5% per year. The cement industry is responsible for 5–8% of the total anthropogenic CO_2 emissions [10]. This high Figure comes predominantly from the de-acidification of limestone, the main raw material in cement production and in addition from the energy compounds necessary to reach the calcination temperature of 1450 °C. Therefore a considerable potential reduction of the environmental impact of concrete lies in the partial substitution of cement by less-energy-intensive hydraulic concrete additives. This has an even greater significance in concrete materials like UHPC with a high cement content.

In the first part of the present study, UHPC mixtures with steel fibers using different supplementary cementitious materials (SCMs) are investigated in comparison with a reference UHPC mixture. The goal is to reach similar properties of fresh and hardened concrete with a lower impact on the environment. To quantify this effect, in a second step the primary energy input (PEI) and the following environmental impact indicators were considered in a quasi-life-cycle assessment (LCA) approach for UHPC:

- Global warming potential (GWP).
- Acidification potential (AP).
- Eutrophication potential (EP).

The influence of ozone in the stratosphere (ODP) and the photochemical creation process (POCP) is not taken into account. Data reflecting the energy and environmental impact indicators were taken from the literature [11], [12] and [13].

Substitution of cement in UHPC mixtures by SCM
A main focus of this research was to develop new mixtures for UHPC with the substitution of high-energy-intensive cement by locally available supplementary cementitious materials like granulated blast furnace slag (GBS) or fly ash (FA). Due to the high cement content of about 800 kg/m^3 in its mixture proportions, UHPC has a critical impact on the environment if compared with NSC. By substituting the cement content with SCMs, attention was directed to the workability of fresh concrete and the mechanical properties of hardened concrete. To visualize the effect the properties were studied in comparison with a reference mixture using only cement as a binder. Since the highest achievable compressive strength was not within the focus of this research, no heat treatment was applied to the UHPC specimens.

Degree of Substitution
The substitution of cement of >30% by weight with quartz filler material was investigated at the Royal Institute of Technology, Stockholm [14] for different types of high strength concrete. The mixtures with reduced cement content had similar workability and compressive strength. The increase in packing density by the ultra-fine filler material and the large

content of unreacted cement due to the low water-binder ratio was discussed as being responsible for this behavior.

Results of another study with a similar focus were presented by Heinz [15], substituting Portland cement by using GBS at a different percentage by volume. The effect on workability and mechanical properties of the UHPC mixtures is discussed. For non-heat-treated mixtures, the best results were obtained at a substitution range between 35% and 55% by volume.

The degree of substitution of Portland cement by SCMs (fly ash, granulated blast furnace slag) in UHPC mixtures was also studied based on the concept of the particle packing density by Puntke [16]. An optimum substitution rate for GBS and FA in this respect was obtained at 31% by weight [17].

In the present study, Portland cement was substituted by GBS in fine and extra fine quality, as well as by FA. The results, gained on the basis of a substitution rate in the UHPC mix design of 45% by weight, are discussed in Section 3.

Mixture Proportions

The reference mixture is a fine grain mixture, UM-5 with a maximum grain size of 0.5 mm. As binder material a CEM I 42.5 R, SR 0 (free of C_3A) was used. The range of the grain sizes was 0.1–0.5 mm for quartz sand, below 40 μm for quartz powder and for the finest grain, microsilica (97% SiO_2), 0.1–0.3 μm. The steel fibers had a length of 15 mm and a diameter of 0.20 mm. As superplasticizer a special formulation provided by SIKA-Austria was applied. The mix design of all mixtures (reference mixture, mixtures with SCMs) is presented in Table 1. The mixture proportion of the reference mix UM-5 was strongly based on the maximization of the packing density of the fine grain, thereby reducing the required amount of water. The methodology used was the set-up developed by Puntke [16], identifying the voids in a powder-filled small container by slowly adding water until the level of the powder surface drops and thus indicates the point of water saturation. The maximum packing density corresponds to the minimum required amount of water.

Table 1: Constituents of the different UHPC mixtures

Components	UM-5 (kg/m³)	UM-5-FA	UM-5-GBSf	UM-5-GBSef
Cement CEM I 42.5 R	729	401	401	401
Microsilica ($k = 1.0$)	124	124	124	124
FA ($k = 0.4$)	–	328	–	–
GBSf ($k = 0.8$)	–	–	328	–
GBSef ($k = 0.8$)	–	–	–	328
Quartz powder	397	397	397	397
Quartz sand	833	833	833	833
Total water (incl. SP)	200	200	200	200
Superplasticizer (SP)	30	30	30	30
Fibers (Stratec 0.2/15)	155	155	155	155
w/c_{eq}	0.234	0.305	0.254	0.254
w/f	0.47	0.44	0.45	0.45

The w/c_{eq} value in Table 1 is the equivalent water to binder ratio and has been derived on the basis of the k-value concept according to EN 206-1 [18]. Thereby the hydraulic activity of SCMs is taken into account via the k-factor ($k = 0.4$ for FA and $k = 0.8$ for GBS). In addition the volume based water/fines ratio, w/f is defined as an indirect measure for the packing density. With respect to this decisive role of the fines (particles <125 µm) [6] and [19], the w/f ratio was kept nearly constant in the mixture proportions (see Table 1).

Characterization of Supplementary Cementitious Materials Used

The material characterization of the SCMs was performed using specific surface analysis (Blaine value, cm²/g), material density and grain size distribution by laser granulometry. The material properties for the SCMs used in the UHPC mixtures are shown in Table 2.

Table 2: Material properties of cement and SCMs

	CEM I	FA	GBSf	GBSef
Density (g/cm^3)		3.24	2.51	2.74	2.9
Blaine value (cm^2/g)		4387	4410	4790	5620
D_{50}:MMD (mass-median-diameter) (μm)		11.05	14.29	14.71	8.47

Cement: CEM I 42.5 R, SR 0.
FA: fly ash.

GBSf: granulated blast furnace slag fine.

GBSef: granulated blast furnace slag extra fine.

The grain size distribution of the SCMs and the cement is shown in Figure. 1. Due to their latent hydraulic properties, GBS and FA provide favorable properties for the substitution of cement. Both are locally available in Austria as by-products of the blast furnace process of steel or from caloric power stations. Therefore the environmental impact of these SCMs is accounted for in the industry where they first appear and is not taken into account for the environmental impact balance of concrete (this approach being in line with the recommendations in [20]).

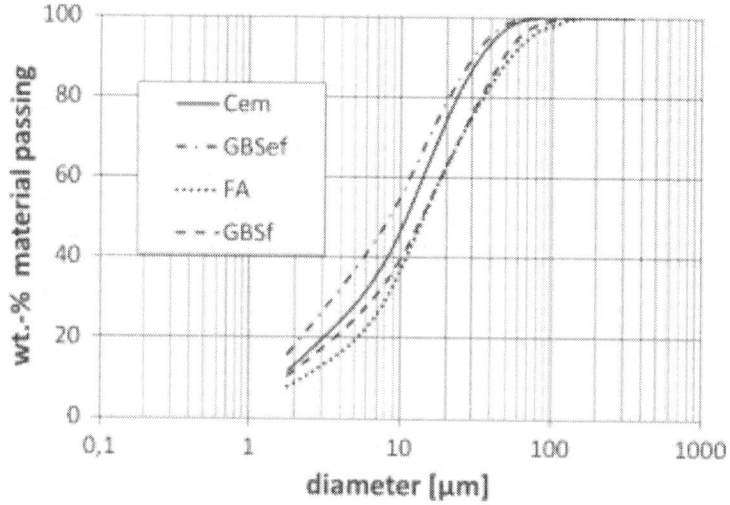

Figure 1: Grain size distribution by laser granulometry.

RESEARCH SIGNIFICANCE: SUSTAINABILITY IN CONCRETE CONSTRUCTION

Alternative approaches for the allocation of the environmental impact generated by the industrial processes to main products and by-products or waste differ between primary and secondary process, the latter one representing the required specific treatment of waste or by-products for further use [21] and [22]. Different allocation methods, e.g. based on the mass ratio between product and by-product or related to the currently added economic value, can lead to different and sometimes even higher environmental burdens of the by-product than the replaced material; however, none of the procedures are incontestable [21]. Moreover other advantages like resource savings should then be taken into account in the total balance.

Material Properties of UHPC with Supplementary Cementitious Materials

Fresh Concrete Properties of UHPC Mix Design with Reduced Cement Content

Taking into consideration the manufacturing technique, sufficient time should be allowed before the UHPC stiffening process starts. For the mixtures under investigation it was found that the workability was appropriate approximately 20 min from the addition of water, thus enabling the casting process from placing the concrete until release of entrapped air within this time slot. To provide a basis for judging the workability and identifying the optimum viscosity of the UHPC mix, the slump-flow test for mortars was performed on the basis of the European Guidelines for Self-Compacting Concrete [23]. However, with respect to the quick stiffening process, the slump flow test was modified in terms of measuring the spread of the fresh concrete already after 2 min (see results in Figure. 2). Thereby a diameter of 270 mm turned out to be the lower limit of the slump flow to enable proper handling of the UHPC mix. The temperature of the mixture plays an important role and should not exceed 30 °C during the mixing process.

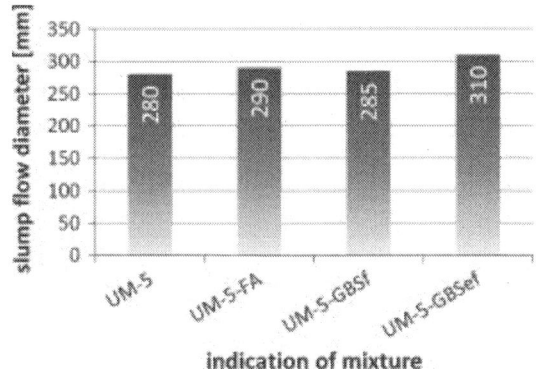

Figure 2: Results of slump flow test after 2 min.

Hardened Concrete Properties

Curing and storing conditions of specimens were in accordance with the Austrian standard ONR 23303 [24](remove from mold after 24 h, then up to the 7th day storage under water in curing tank, afterwards further curing in air under laboratory conditions up to the 28th day). The compression tests were performed on 100 mm cubes made of fiber reinforced UHPC on the 28th day after preparation. As shown in Figure. 3, the compressive strength of the reference mixture UM-5 was 166.1 MPa. A similar result with only 2.6 MPa below was obtained for the mixture UM-5-GBSef, fiber reinforced UHPC with the substitution of 45% by weight of the cement by extra fine GBSef. The other two substitution mixtures reached values of 139.4 MPa (UM-5-GBSf) and 124.7 MPa (UM-5-FA) respectively, which is 83% and 75% of the compressive strength of the reference mixture.

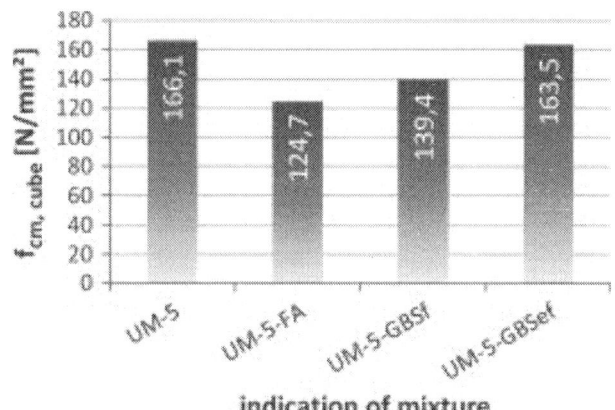

Figure 3: Compressive strength.

RESEARCH SIGNIFICANCE: SUSTAINABILITY IN CONCRETE CONSTRUCTION

The best results in terms of workability (see Figure. 2) as well as compressive strength (see Figure. 3) were obtained from the substitution of cement by GBSef with a Blaine value close to 6000 cm²/g. For the evaluation of the packing density of the different mix proportions Puntke tests [16] were performed. The results of these tests (representing average values of 3 tests each) are listed in Table 3. The packing density of the fine grain (n_f) corresponds to the amount of water (n_w) required to fill the voids ($n_f = 1 - n_w$). The packing densities of the mixtures with SCMs are slightly above the value of the reference mixture, the highest one with 61.1% for GBSef.

Table 3: Puntke test results – packing density

UHPC mix	n_w (%)	n_f (%)
UM-5	39.7	60.3
UM-5-FA	39.2	60.8
UM-5-GBSf	39.1	60.9
UM-5-GBSef	38.9	61.1

Comparison of the Ecological Properties of Different UHPC Mixtures

Based on the promising mechanical properties, the developed UHPC mixtures using SCMs were evaluated in terms of environmental impact indicators. In radar charts, usually used to indicate environmental impact categories of construction materials [25], the results of the influence of the substitution of cement in UHPC and the position of UHPC in relation to the concept of "green concrete" according to [26] are shown.

Comparison of UHPC with NSC

The main topic of this section is the comparison between the relevant UHPC mixtures and NSC on the basis of their ecological properties. These were calculated from the primary energy input parameter and environmental impact indicators for the constituents of the different mixtures. The respective data have been derived from sources [13] and

[27]. The procedure applied is a simplified LCA approach according to EN ISO 14040 [28], focusing on the materials required for 1 m³ compacted concrete. For the sake of better comparability to NSC, for the UHPC mixtures the influence of potential steel fibers was not considered. The environmental impact parameters taken into account are listed in Table 4, including the scaling factors to be applied when interpreting the graphs in Figure 4 and Figure 5.

Table 4: Energy and environmental impact indicators

Environmental impact indicators	Unit	Scaling factor
Primary energy input – renewable, PEI_{re}	(MJ/m³)	102
Primary energy input – non-renewable, PEI_{non-re}	(MJ/m³)	104
Global warming potential, GWP	(kgCO$_2$-eq/m³)	103
Acidification potential, AP	(kgSO$_2$-eq/m³)	1
Eutrophication potential, EP	(kgPO$_4$-eq/m³)	1

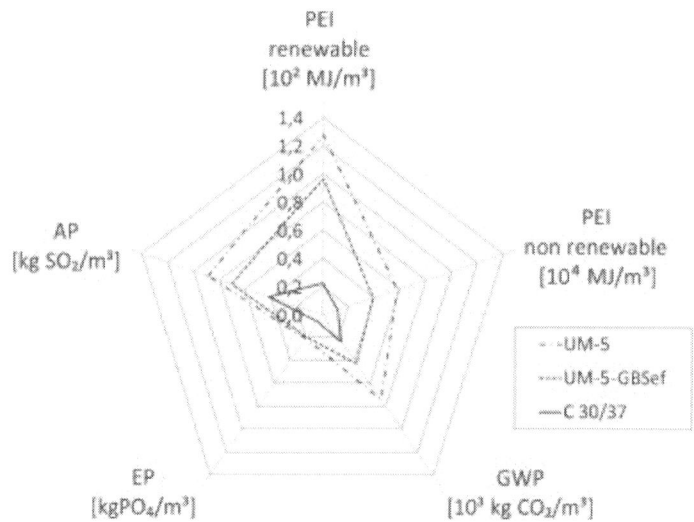

Figure 4: Omparison of ecological indicators in UHPC mix design.

RESEARCH SIGNIFICANCE: SUSTAINABILITY IN CONCRETE CONSTRUCTION

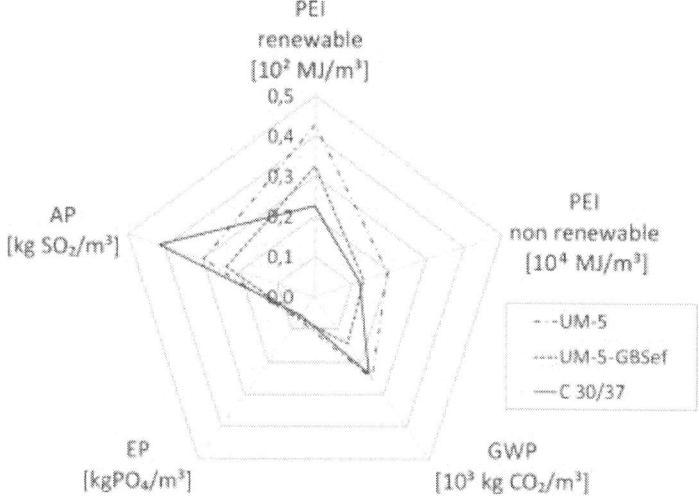

Figure 5: Comparison of environmental impact parameters between 1 m³ of C30/37, UHPC reference mixture UM-5 and UHPC with GBS extra fine (considering a reduction of the cross-section and increased durability of UHPC).

Figure. 4 shows the effect of the environmental impact indicators in the mix design of 1 m³ compacted UHPC. The ecological data of the individual ingredients were assessed and weighted according to their percentage in each mixture. The results were generated for the three mixtures discussed, using the scaling factors listed in Table 4 for illustration reasons (see Figure. 4 and Figure. 5).

In comparison to normal strength concrete C30/37, the data show a substantial increase for UHPC in all parameters. Comparing the two UHPC mixtures UM-5 and UM-5-GBSef, a significant reduction in the parameters thanks to the substitution of cement can be seen as the result: in detail a reduction of about 32% of PEI non-renewable, 24% of PEI renewable, 42% of GWP and 20% of AP is achieved. The results in Figure. 4 thus demonstrate clearly the effect in the UHPC mix design towards mixtures of less ecological impact when substituting cement with SCMs. In addition, in order to provide a realistic evaluation and make use of the full ecological potential of UHPC, the possible reduction in the amount of material used to reach the same load bearing capacity and the increase of the durability has to be taken into account.

Comparison of Building Members Made of UHPC with NSC

Due to its extraordinary compressive strength and the increased tensile strength (approximately 3 times higher than for NSC) UHPC allows for a reduction of the cross section compared to standard RC members, see e.g. the study presented in [1]. The reduction potential depends on the kind and the geometry of a building member, the relevant load scenarios and the decisive failure modes. While compression members allow for significantly increased slenderness when using UHPC, the reduction is rather limited when considering members subject mainly to bending. In the latter case the amount and the properties of the reinforcing steel and the inner lever arm, to some extent influenced by the compressive strength of the concrete, are decisive for the achievable slenderness. By adequately reducing the width of web sections and increasing the inner lever arm according to the shifting of the center of the compression zone, in the case of flexural members the cross sectional reduction potential may range from less than 10% to about 20%.

On the other hand, building columns are slender compression members where buckling is the predominant failure mode and cast-in reinforcement bars overtake usually substantial parts of the compression force. In this case, when assuming standard reinforcement degrees between 2% and 4%, reductions of the cross section by 30–50% can be achieved. Concerning rather compact members under compression without risk of buckling failure, the possible material savings are even larger and nearly proportional to the enhancement of the concrete strength.

In order to take the optimization of the cross section into account, in the present study a reduction of one third, i.e. 33% was considered as representative. In that context, it should be borne in mind that also the requirements on fire resistance could lead to a higher reduction. For the comparison with NSC, a reference concrete C30/37 is chosen.

Another important aspect is the increased durability and lifetime of UHPC members. Regarding experimental investigations on durability parameters like chloride ion penetration, carbonation, abrasion and freeze–thaw resistance, a substantial increase of the durability can be deduced. Based on experimental investigations at Kassel University [29], compared to standard NSC, the carbonation process under outdoor conditions is 3–6 times slower in UHPC. Several other studies report similar beneficial durability properties of UHPC [30],[31] and [32]. In general a very low level of migration of chloride ions into the UHPC can be observed. According to [33] the chloride diffusion is retarded (based on rapid chloride migration tests) with a time factor of larger than 4 compared to ordinary concrete.

In order to consider the increased lifetime of UHPC compared to NSC structures, in the present study a factor of 2 is applied (Figure. 5). The chosen ratio corresponds to [30] where the authors expect, based on a variety of performed durability tests, that UHPC outperforms NSC by at least twice as much in service life. While the durability tests reported in the above mentioned studies [29], [30], [31] and [32] would justify even higher durability factors (at least a ratio of 3–4 can be argued), current codes on the other hand do not require a corresponding extension of the design life of buildings and structures so that it would be difficult to argue the actual benefit when applying such high factors.

Taking into account both cross-sectional reduction and enlarged lifetime in the mentioned way, the generated radar chart in Figure. 5 shows that the ecological impact is significantly reduced and thus UHPC building members may finally cause less environmental burden than NSC. Additional subsidiary factors like reduced cross sections of foundations or savings in floor space due to the use of, e.g., slender columns [1] are thereby not taken into account.

In addition the consideration of reinforcing steel and/or steel fibers is another important aspect when evaluating the ecological impact of building members. RC-structures usually contain at least a minimum amount of steel reinforcement bars while UHPC due to its brittleness is preferably equipped with a certain amount of steel fibers. Based on tensile tests with Ultra High Performance Fiber Reinforced Concrete (UHPFRC), a steel fiber amount of at least 2% by volume may lead to a strain-hardening tensile behavior of the UHPFRC rather than strain-softening [34]. However, in many cases for structural applications a fiber amount of 0.5–1% by volume may already be sufficient to avoid brittle failure. In addition UHPFRC members will usually also contain a reduced amount of steel reinforcement bars. The incorporation of both fibers and steel rebars will increase the environmental impact factors substantially due to the energy-consuming production process and may thus become one of the most dominant factors when considering all UHPFRC ingredients [35]. However, considering the environmental impact of the steel ingredients makes only sense with reference to real building members with a given reinforcement layout and is therefore not taken into account in the present study.

CONCLUSIONS

The present study investigates the substitution of cement in UHPC by less energy-intensive latent hydraulic concrete additives, focusing on its effect on the mechanical properties and the environmental impact categories. The production-related CO_2 emissions of such alternative additives are not considered in this context, as they are by-products of industrial processes, in which their environmental impact is accounted for. The outcome of the investigations can be summarized as follows:

1. The substitution of cement by appropriate less energy intensive cementitious materials is possible up to about 45% by weight without significant degradation of mechanical properties and workability parameters.
2. The results indicate that achieving an adequate packing density when using ultra-fine materials like extra-fine granulated blast furnace slag (GBSef) is even more decisive for the UHPC properties than the hydraulic reactivity of such materials.
3. Comparing the environmental impact categories of UHPC with that of NSC, the substitution of cement by SCMs is only a first step towards improving the sustainability of UHPC from the ecological point of view. However, when considering building members and also taking into account the reduction of material consumption and the increased durability and lifetime, the overall picture improves substantially.
4. Further optimization of the partial substitution of the cement and the use of alternative fiber materials are required to increase the acceptance and competitiveness of UHPFRC from the environmental point of view.

ACKNOWLEDGEMENTS

This study is part of the research project "HiPerComp – High Performance Composite Structures" and the authors like to express their sincere thanks to the Austrian Research Foundation (FFG) for the funding, furthermore to the Austrian material supplier w&p Zement GmbH for providing the supplementary cementitious materials.

REFERENCES

1. Aaleti S, Petersen B, Sritharan S. Design guide for precast UHPC waffle deck panel system, including connections. Tech report no. FHWA-HIF-13-032. Washington, DC: Federal Highway Administration. U.S. Department of Transportation; 2013.
2. AFGC. Recommendation: Ultra high performance fibre-reinforced concretes, revised ed. Association Française de Génie Civil; Service d'études techniques des routes et autoroutes; 2013.
3. Ahlborn TM, Peuse EJ, Misson DL. Ultra-high-performance-concrete for Michigan bridges material performance – Phase I. Research report. RC-1525. Tech rep. Lansing, MI: Michigan Department of Transportation Construction and Technology Division; 2008.
4. Aßbrock O, Becke A, Bernhofen T, Hauer B, Kaczmarek T, Lotz U, et al. Nachhaltiges Bauen mit Beton; Ein Fachbeitrag für Architekten, Planer und Bauherren. Beton Marketing Deutschland GmbH; 2011.
5. Bundesministerium für Verkehr. Bau und Stadtentwicklung. Ökologisches Baustoffinformationssystem WECOBIS. Germany; 2013. <http://www.wecobis.de/portale/nachhaltiges-bauen.html>.
6. Bundesverband der Deutschen Ziegelindustrie e.V. Green Building Challenge Handbuch. <http://www.ziegel.at/gbc-ziegelhandbuch>.
A. Chen, G. Habert, Y. Bouzidib, A. Jullien, A. Ventura, LCA allocation procedure used as an incitative method for waste recycling: an application to mineral additions in concrete, Resour Conserv Recycl, 54 (2010), pp. 1231–1240
7. Cembureau. Activity report 2012. Tech rep. European Cement Association; 2013.
8. DAfStb. Sachstandsbericht Ultrahochfester Beton; 2008.
9. Ecoinvent – swiss centre for life cycle inventories; 2013. <http://www.ecoinvent.ch/>.

10. EFNARC. The European guidelines for self-compacting concrete – specification, production and use. experts for specialised construction and concrete systems; 2005.
11. Fehling E, Schmidt M, Teichmann T, Bunje K, Bornemann R, Middendorf B. Entwicklung, Dauerhaftigkeit und Berechnung ultrahochfester Betone (UHPC). DFG-Forschungsbericht, Schriftenreihe Baustoffe und Massivbau, Structural Materials and Engineering Series; 2005 p. 1.
12. Fib (fédération internationale du béton/International Federation for Structural Concrete). fib bulletin 67: Guidelines for green concrete structures; 2012.
13. Graybeal BA. Material property characterization of ultra-high performance concrete. Tech report no. FHWA-HRT-06-103. McLean, VA: Research, Development, and Technology Turner-Fairbank Highway Research Center; Federal Highway Administration, U.S. Department of Transportation; 2006.
14. Haist M, Müller HS. Betontechnologie im Spannungsfeld zwischen Ökobilanz und Leistungsfähigkeit. In: Nachhaltiger Beton. Symposium: Nachhaltiger Beton – Werkstoff, Konstruktion und Nutzung. Karlsruhe: KIT Scientific Publishing; 2012.
15. Heinz D. UHPC mit alternativen Bindemitteln. 9. Münchner Baustoffseminar der Technische Universität München; 2011.
16. JSCE. Recommendations for design and construction of ultra-high strength fiber reinforced concrete structures (draft); JSCE guidelines for concrete no. 9 ed. Japan Society of Civil Engineers; 2006.
17. K. Sakai, Sustainability in fib model code 2010 and its future perspective, Struct Concr, 14 (4) (2013), pp. 301–308 http://dx.doi.org/10.1002/suco.201300012
18. K. Sakai, T. Noguchi, The sustainable use of concrete, CRC Press, Taylor & Francis Group (2013)
19. M. Schmidt, E. Fehling, T. Teichmann, K. Bunje, R. Bornemann Ultra-high performance concrete: perspective for the precast concrete industry – discussion of the composition and the properties of UHPC, Concr Plant + Precast Technol, 69 (3) (2003), pp. 16–29
20. Müller HS, Vogel M, Haist M. Service life design – a tool for sustainable application of concrete. In: First international conference on concrete sustainability (ICCS13). ICCS13, Japan; 2013, p. 56–69 [S-2-5-1].
21. Müller HS, Wiens U. Beton. In: Bergmeister K, Fingerloos F, Wörner JD, editors. Beton-Kalender 2014. Berlin: Ernst & Sohn; 2013.
22. ÖNORM EN 206-1. Concrete – Part 1: specification performance, production and conformity; 2005.
23. ÖNORM EN ISO 14040:2006. Environmental management – life-cycle assessment – principles and framework; 2006.

24. ONR 23303. Test methods for concrete – national application of testing standards for concrete and its source materials. Austrian Standards Institute; 2010.
25. Puntke W. Wasseranspruch von feinen Kornhaufwerken. Beton Schriftenreihe 5; 2002.
26. Racky P. Wirtschaftlichkeit und Nachhaltigkeit von UHPC. Schriftenreihe Baustoffe und Massivbau, Heft 2. Ultra-Hochfester Beton, Planung und Bau der ersten Brücke mit UHPC in Europa. Tagungsbeiträge zu den 3. Kasseler Baustoff- und Massivbautagen. Kassel: Kassel University Press GmbH; 2003.
27. Randl N, Däuber F. Material properties of fiber reinforced UHPC. In: 8th RILEM international symposium on fibre reinforced concrete: challenges and opportunities. PRO 88; Guimarães, Portugal: RILEM; 2012.
28. Rafiee Computer modeling and investigation on the steel corrosion in cracked ultra high performance concrete, No. 21 in Structural materials and engineering seriesKassel University Press (2012) ISBN print: 978-3-86219-388-2.
29. Schmölzer C. Kriterien der Mischungsentwicklung von UHPC aus regional verfügbaren Ausgangsstoffen. Master's thesis. Spittal/Drau, Austria: Carinthia University of Applied Sciences; 2011.
30. T. Proske, S. Hainer, M. Rezvani, C.A. Graubner, Eco-friendly concretes with reduced water and cement contents – mix design principles and laboratory tests, Cement Concr Res, 51 (2013), pp. 38–46
31. T. Stengel, P. Schießl Life cycle assessment of UHPC bridge constructions, Archit Civil Eng Environ, 1 (2009), pp. 109–118
32. Vogt C. Ultrafine particles in concrete. Ph.D. thesis. Stockholm, Sweden: School of Architecture and the Built Environment, Royal Institute of Technology; 2010.
33. Von Weizsäcker EU, Hargroves K, Smith M. Faktor fünf – Die Formel für nachhaltiges Wachstum. München: Droemer Verlag; 2010.
34. WCED (World Commission on Environment and Development). Our common future (Brundtland Report). Oxford University Press; 1987.

CITATION

N. Randl , T. Steiner S. Ofner , E. Baumgartner, T. Mészöly, Development of UHPC mixtures from an ecological point of view, dx.doi.org/10.1016/j.conbuildmat.2013.12.102

CHAPTER 5

Numerical Study of FRP Reinforced Concrete Slabs at Elevated Temperature

Masoud Adelzadeh, Hamzeh Hajiloo, and Mark F. Green

Civil Engineering Department, Queen's University, Kingston, K7L 3N6, Canada

ABSTRACT

One-way glass fibre reinforced polymer (GFRP) reinforced concrete slabs at elevated temperatures are investigated through numerical modeling. Serviceability and strength requirements of ACI-440.1R are considered for the design of the slabs. Diagrams to determine fire endurance of slabs by employing "strength domain" failure criterion are presented. Comparisons between the existing "temperature domain" method with the more representative "strength domain" method show that the "temperature domain" method is conservative. Additionally, a method to increase the fire endurance of slabs by placing FRP reinforcement in two layers is investigated numerically. The amount of fire endurance gained by placing FRP in two layers increases as the thickness of slab increases.

INTRODUCTION

During the past decade, the civil engineering community has been very interested in applications of fibre reinforced polymer (FRP) reinforcement in concrete structures as an alternative to steel reinforcement [1].

Outstanding characteristics of FRP materials such as high strength-to-weight ratios and resistance to corrosion make FRPs suitable for structures subjected to harsh environments. Progress in FRP manufacturing technology has reduced the material cost and increased the confidence in applications of FRP for civil engineering applications.

Among the many areas of application of FRP, strengthening of existing buildings has gained the most attention, given advantages for fast construction. Nevertheless, applications of FRPs as internal reinforcement in new construction of concrete structural members are growing rapidly. FRP reinforcing bars are now available in different forms for both flexural and shear reinforcement. Demand for FRP internal reinforcement in highly-corrosive environments such as bridges, barrier walls, parking lots, buildings in coastal areas and industrial structures has increased [2]. Apart from notable advantages of FRPs, application of FRP in structures has a few drawbacks. One of the potential disadvantages of FRP materials is their performance in fire. Degradation of strength and stiffness of FRPs induced by high temperatures could cause substantial loss in load carrying capacity of concrete structures, specifically where they are the only form of reinforcement. On the contrary, conventional concrete structural members with internal steel reinforcement generally exhibit good performance in fire [3].

Kodur and Bisby [4] studied the behaviour of FRP reinforced concrete slabs in fire through numerical modelling. They discovered significant differences in the behaviour of FRP reinforced compared to steel reinforced concrete slabs. They have investigated the effects of the type of FRP bars, the overall thickness of the slab, and the cover of the slab on performance of slabs. Bisby and Kodur [5] developed a simple numerical method to investigate the differences of FRP reinforced slabs with steel reinforced ones. Nigro et al. [6] conducted fire tests on 6 FRP reinforced slabs. They recommended to use continuous FRP bars from end to end of the concrete member, and to protect a portion of member at the ends from the effects of fire to avoid bond failure. In North America, design guidance for FRP reinforcement in building is given in ACI 440.1R [7] and CAN/CSA S806-02 [8]. Many other countries have similar documents. Although CSA S806 presents recommendations for estimating fire endurance of FRP reinforced concrete elements, the approach is based only on the temperature of the reinforcement and does not directly consider the strength of the FRP reinforced slab in fire. In view of this fact, this study focuses on numerical modelling of the structural behaviour of FRP reinforced slabs in fire. Specific attention has been given to details of reinforcement placement in order to increase the fire endurance.

MATERIAL BEHAVIOUR AT HIGH TEMPERATURES

The behaviour of concrete at elevated temperatures is well understood [9,10,11]. On the other hand, FRP behaviour at high temperature is more problematic. Firstly, the properties of commercially available materials can vary widely. Additionally, the time-dependent visco-elastic behaviour of matrix or adhesive makes experimental characterization difficult, and few tests have been conducted. In general, the mechanical properties of FRP degrade at high temperatures depending mainly on the properties of the matrix. Blontrock *et al.* [12] suggested the tensile strength of carbon fiber reinforced plastic (CFRP) and aramid fiber reinforced plastic (AFRP) remains unaffected up to 100 °C but that of GFRP bars decreased consistently with the increase of the temperature. In this paper, two models proposed by Saafi [13] and Bisby [14] for degradation of GFRP bars at elevated temperatures are considered. Figure 1 compares the strength and elastic modulus degradation for GFRP composites in Saafi and Bisby's models. Saafi's model produces conservative results compared to Bisby's model, particularly for strength retention. Both models suffer from the dearth of experimental data on FRP bars that are currently employed for reinforced concrete. Nevertheless, Bisby's model is more consistent with recent tests on commercially available FRP bars for reinforcing concrete [15].

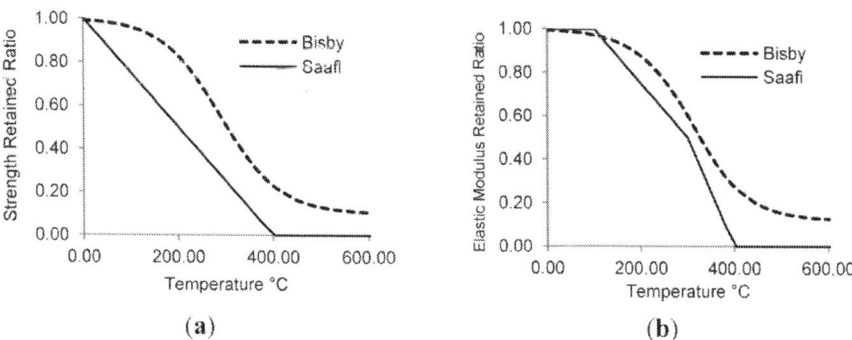

Figure 1: Comparing Bisby's and Saafi's models; (**a**) Strength reduction; (**b**) Elastic modulus degradation for GFRP at elevated temperatures.

HEAT CONDUCTION SIMULATION IN REINFORCED CONCRETE MEMBERS

In this study, the ASTM E119 [16] time temperature curve has been used to simulate the temperature rise due to compartment fire in heat transfer model. The heat transfer model formerly developed [17] is a finite-volume code, which is capable of predicting temperature in an insulated concrete section. The partial differential equation of heat conduction can be expressed as:

$$\rho c \frac{\partial T}{\partial t} = \nabla \cdot (k \nabla T) = \frac{\partial}{\partial x}\left(k \frac{\partial T}{\partial x}\right) + \frac{\partial}{\partial y}\left(k \frac{\partial T}{\partial y}\right) \quad (1)$$

where T is temperature, t is time, k is thermal conductivity, ρ is density and c is heat capacity. These field variables are functions of temperature and spatial variables. For the predictions in this paper, the thermal properties of the concrete are taken from the recommendations of Lie [9] for concrete with carbonate aggregates. The thermal properties of the GFRP are assumed to be the same as that of concrete because of the relatively small amount of GFRP compared to the volume of concrete. The temperature-time curve of ASTM E119 temperature predictions at different concrete depths are presented [18] in Figure 2.

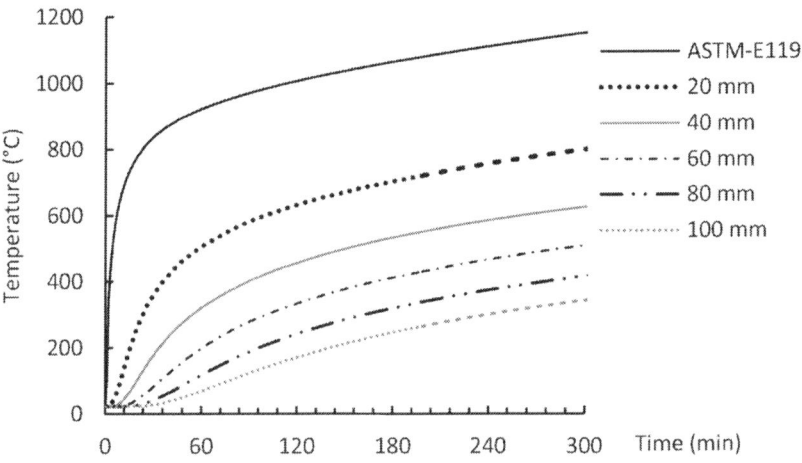

Figure 2: Temperature predictions at different concrete depths *vs.* exposure time.

LOAD CAPACITY MODEL

Several studies carried out to determine temperature profile in cross section of concrete members [19]. Once the distribution of temperatures throughout the slab is known at each time step during fire exposure, the flexural capacity of slab can be calculated using Euler-Bernoulli beam theory. The following assumptions are made in the model:

1) Slabs are exposed to fire from the bottom of the slab only
2) Slabs carry loads in bending in one direction only (one-way slab)
3) Slabs are simply-supported and no axial restraint or axial forces are present
4) The bond of FRP bars to concrete is unaffected by heat, and
5) Plane sections remain plane throughout the analysis.

The fourth assumption regarding the bond of FRP is particularly important because recent tests have shown that bond often governs the failure of FRP reinforced slabs [2,20]. However, this work by Nigro *et al.* [2] also demonstrated that such bond failures can be prevented by special anchoring details such as bending the bars at the ends of the slabs. Thus, the information in this paper is applicable only to designs that have such suitable anchoring details.

In order to calculate the flexural capacity of the section during fire, concrete and FRP characteristics have been adjusted at each step to account for the loss of strength due to fire exposure. However the loss of strength in concrete has been neglected during fire exposure. The reason for this simplification is that the temperature of the compressive portions of the slab is fairly low even after 6 h of standard fire exposure, Figure 3. Therefore, the compressive strength of concrete is not affected by fire. The degradation of modulus of elasticity of concrete would not affect the flexural capacity of the section although it affects the deflection of the member. Therefore serviceability design criteria used for the initial design of the slabs will not be predicted during the fire.

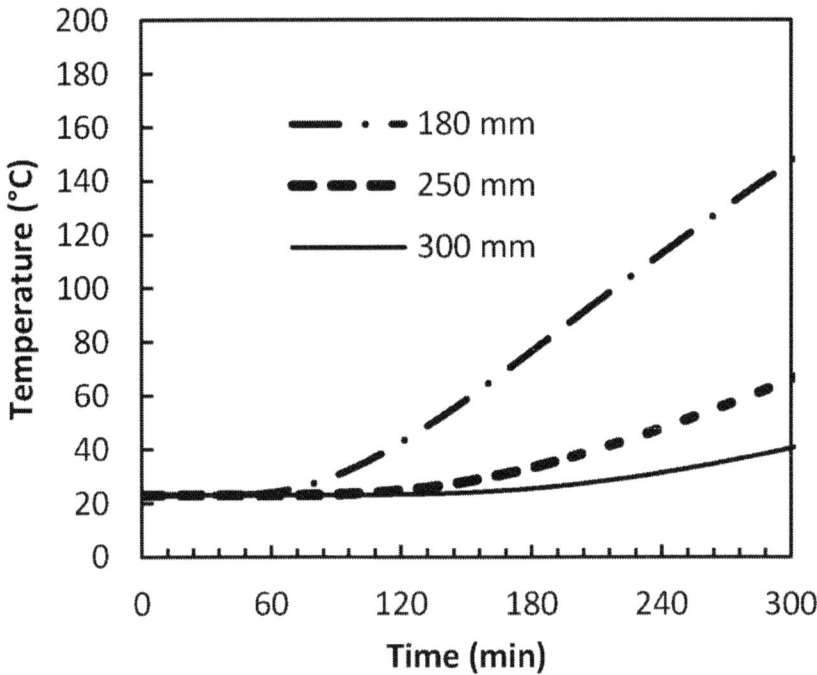

Figure 3: Temperatures at unexposed surface of slabs.

The procedure for determining the ultimate capacity of FRP reinforced slab in fire is more complicated compared to the steel reinforced slab. In the case of steel reinforced slab, the nonlinear behaviour of steel after its yield point makes the calculation simpler. However, for FRP reinforced concrete slabs, the strain compatibility diagram changes continuously as the FRP loses strength at high temperature and the balanced FRP reinforcement ratio ($\rho_{FRP,b}$) increases continuously. Thus, at each time step, $\rho_{FRP,b}$ is recalculated using temperature adjusted characteristics of FRP.

Properties of FRP bars at elevated temperature are obtained based on strength and stiffness degradation models given in Figure 1. Once $\rho_{FRP,b}$ is calculated, the existing FRP ratio (ρ_{FRP}) is compared to $\rho_{FRP,b}$ to identify whether failure will be by crushing of concrete or by FRP rupture. In the case of slabs, with two or more layers of FRP the possibility of rupture in each FRP layer is considered at every time step, especially when the ultimate strain of FRP decreases as the temperature increases. For example in a slab with two layers of reinforcement, in some occasions the section reaches a higher flexural capacity because the bottom layer ruptures before the concrete crushes. The ultimate strain of concrete is set to be constant and equal to 0.0035. However the mechanical rupture strain of the FRP bars varies throughout the fire exposure as shown in Figure 4. The curves in this figure were obtained by a combination of the strength and modulus curves presented in Figure 1.

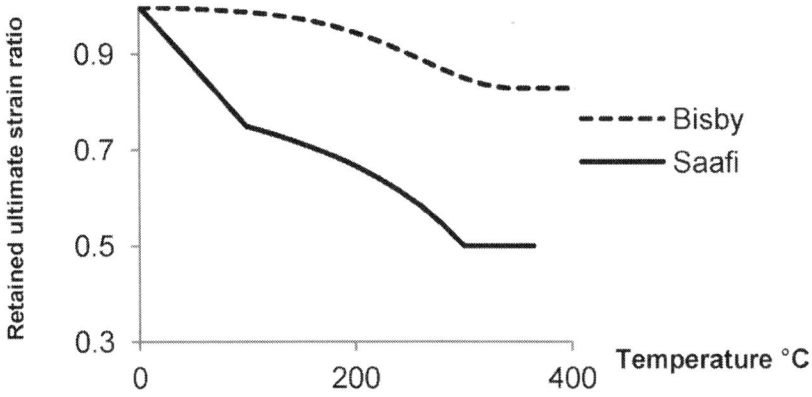

Figure 4: Comparison of the models proposed by Bisby and Saafi for ultimate strain of GFRP bars at elevated temperatures.

When ρ_{FRP} is greater than $\rho_{FRP,b}$, the governing failure mechanism is concrete crushing and the strain at top compression layer is 0.0035. The strain in FRP bars is determined using strain compatibility. As the strength of the FRP decreases due to temperature effects, the neutral axis depth decreases resulting in a smaller concrete stress block.

When ρ_{FRP} is less than $\rho_{FRP,b}$, the FRP will rupture before the concrete crushes. The strain in FRP bars is set to their ultimate strain at that temperature. Hence, the stress in the bar is known and the resultant tensile force can be calculated. A subroutine checks many different neutral axis depths to satisfy equilibrium of tensile and compressive forces in the slab cross-section. Once the equilibrium criterion is fulfilled and the neutral axis is determined, strains and stresses are calculated for any point in the section. The resultant forces and moment resistance are calculated once all the stresses are determined and this gives the moment resistance of concrete slab at any instant of fire exposure. The known parameters of the analysis are FRP reinforcement ratio, which is calculated beforehand by conforming to ACI 440.1R [7] serviceability limitations, as well as dimensions of the slabs, and material properties of reinforced concrete at room temperature.

The reliability of the above mentioned analyses is highly dependent upon the model being used to predict the thermal and mechanical properties of the constituent materials. Although adequate tests and research on concrete and steel are available, a lack of information for FRP bars calls for more research in this area. Figure 5 illustrates the significance of FRP material behaviour model in overall response of the slab. The solid line is the calculated moment capacity curve using the model proposed by Saafi's and the dashed line is moment capacity calculated using Bisby's model. The slab is 180 mm thick and the concrete cover is 50 mm over GFRP reinforcement. For this case, the Saafi model estimates the fire resistance as only 85 min compared to 140 min for Bisby's model. Bisby's model is employed in the remaining calculations in this paper because it appears more accurate based on recent test data. Nevertheless, the FRP material property models still need more tests and thus the results in this paper are intended to demonstrate the trends that changing different slab configurations will have on fire endurance rather than as accurate predictions of fire endurance.

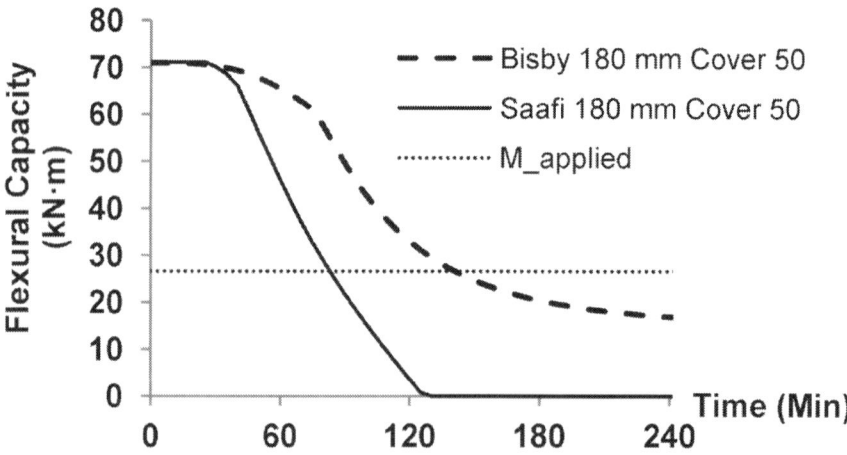

Figure 5: Moment curves using different FRP degradation model in high temperature.

To verify the results of the model, it was used to estimate the fire endurance of GFRP reinforced concrete beams tested by Abbasi*et al.* [21]. They tested two 4400 mm long beams with a cross section of 400 × 350 mm. The average cube compressive strength of concrete was reported as 42 MPa. Each beam was reinforced by 7 of 12.7 mm in diameter GFRP rebars in tension. For Beam "b1" the GFRP rebars' ultimate strength and elastic modulus were 690 MPa and 40.8 GPa respectively, and for beam "b2" 1000 MPa and 41 GPa. During the fire test, the beams experienced spalling in the corners of the cross section and extensive cracking and failed under the sustained load after 124 min in the case of beam "b1" and 94 min in the case of beam "b2". The fire endurance estimates from the model developed here (see Figure 6) are 90 min for beam "b1" and 110 min for beam "b2". Thus the model is conservative for beam "b1". Considering that the model does not account for spalling, the fire endurance results from the model give a reasonable estimate of the fire endurance.

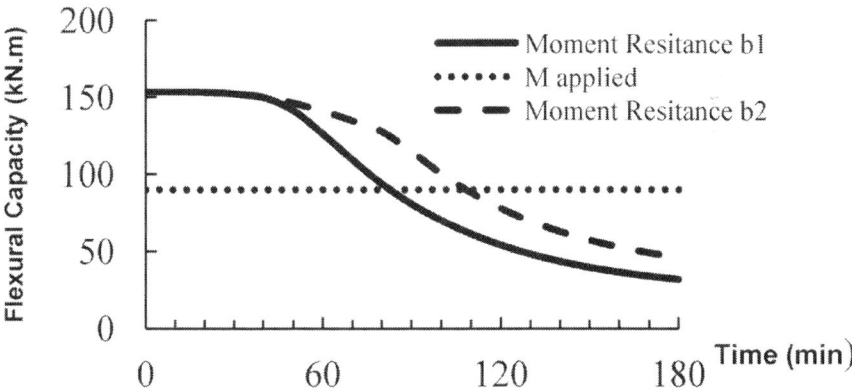

Figure 6: Model prediction for tests of the beams tested by Abbasi et al. [21].

SLABS WITH ONE LAYER OF FRP

To study the effect of fire on slabs, several different slab configurations have been considered as shown in Table 1. Since serviceability limitations are the governing criteria in FRP reinforced slabs, the slabs are designed for crack width and deflection serviceability criteria. The crack width of the slabs has been limited to 0.7 mm for interior exposure as suggested by ACI-440.1R [7]. The permissible deflection is $L/360$. For the two above defined limits, the required FRP reinforcement ratio is found for the all slabs. Slab thicknesses are 180, 250, and 300 mm and concrete cover ranges from 30 to 70 mm.

Table 1: Characteristics of FRP reinforced slabs investigated in this study

Slab number	Thickness (mm)	Rebar type	f'_c (MPa)	cover (mm)	L (mm)	Spacing (mm)	$A_{f,req}$	M_u^{**} (kN·m)	M_{cr}	Deflection (mm) $M_a/M_{cr} = 1.5$
1	180	GFRP	30	30	3600	150	1006	77.6	17.8	4.1
2	180	GFRP	30	40	3600	150	1243	74.1	17.8	4.1
3	180	GFRP	30	50	3600	150	1576	70.2	17.8	4.1
4	180	GFRP	30	60	3600	150	2051	65.6	17.8	4.0
5	180	GFRP	30	70	3600	150	2754	59.9	17.8	4.0
6	180	GFRP	30	80	3600	150	3845	53.1	17.8	3.9
7	250	GFRP	30	30	5000	150	1235	156.4	34.2	5.7
8	250	GFRP	30	40	5000	150	1446	153.5	34.2	5.7
9	250	GFRP	30	50	5000	150	1719	149.7	34.2	5.6
10	250	GFRP	30	60	5000	150	2072	145.4	34.2	5.6
11	250	GFRP	30	70	5000	150	2530	140.2	34.2	5.5
12	250	GFRP	30	80	5000	150	3136	134.0	34.2	5.4
13	300	GFRP	30	30	6000	150	1409	222.6	49.3	6.9
14	300	GFRP	30	40	6000	150	1617	227.2	49.3	6.8
15	300	GFRP	30	50	6000	150	1879	223.9	49.3	6.7
16	300	GFRP	30	60	6000	150	2206	220.2	49.3	6.7
17	300	GFRP	30	70	6000	150	2614	215.7	49.3	6.6
18	300	GFRP	30	80	6000	150	3122	210.0	49.3	6.5

For the purposes of illustration, consider a 250 mm thick concrete slab with a 28 day concrete compressive strength of 30 MPa with carbonate aggregate. If the concrete cover to the centre of the FRP bars is 50 mm, the required reinforcement area assuming $M_a/M_{cr} = 1.50$ is 1719 mm². Placing the required amount of reinforcement to satisfy serviceability criteria gives a nominal moment resistance 150 kN·m. The cracking moment (M_{cr}) of the slab is 34 kN·m. Exposed to fire from below, the slab loses its moment capacity as a consequence of thermal degradation of the mechanical properties of the FRP. The initial flexural capacity of the slab drops to the applied moment (M_a = 51 kN·m) at 140 min. It should be mentioned that the resistance model given by the model does not include member reduction factors as recommended by ACI 216 [22]. The maximum likely crack width is calculated using the following equation:

$$w_{cr} = 2 \frac{f_f}{E_f} \beta k_b \sqrt{d_c^2 + (\frac{s}{2})^2} \tag{2}$$

in which w = maximum crack width; f_f = reinforcement stress; β = ratio of distance between neutral axis and tension face to distance between neutral axis and centroid of reinforcement; d_c = thickness of cover from tension face to center of closest bar; and s = bar spacing. Since crack width is a function of stress in the FRP bars, the design is affected by the service load level as expressed by the M_a/M_{cr} ratio. Three common ratios of 1, 1.25, and 1.5 are selected for the M_a/M_{cr} ratio.

As expected, by increasing the concrete cover, the initial moment capacity decreases. Because the defining design criteria relate to serviceability rather than strength, all slabs fulfill the strength requirements at room temperature. Fire performance of slabs considerably increases by increasing their concrete cover as shown in Figure 7, but this performance improvement comes at the expense of higher reinforcement ratio due to crack width limitations. For example, a 180 mm slab with 30 mm of cover has approximately 1 h of fire endurance while the slab with 60 mm of concrete cover has in excess of 4 h of fire endurance. However, the required reinforcement ratio is 2 times higher in the slab with 60 mm of concrete cover to meet serviceability design criteria.

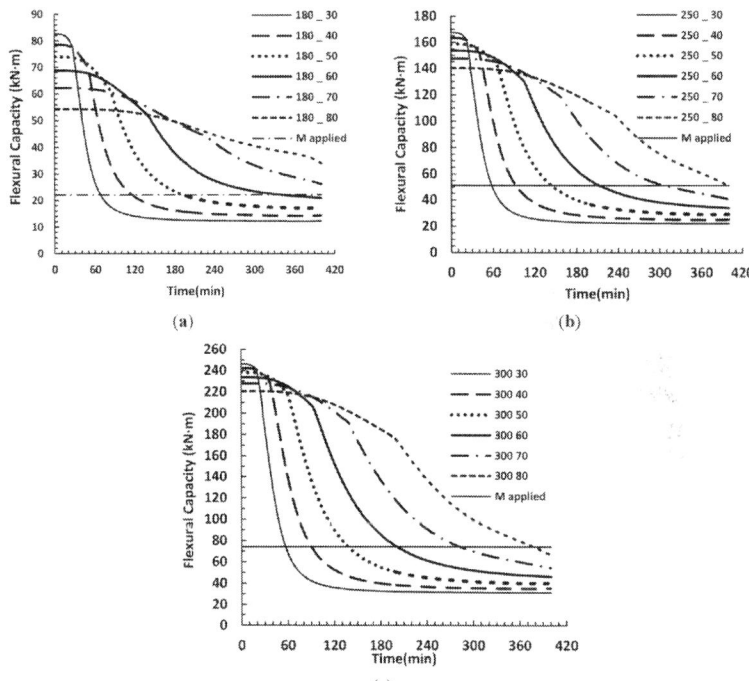

Figure 7: Moment capacities of 180 mm thick slabs with various cover depths in fire, $M_a/M_{cr} = 1.5$, (a) 180 mm thick slab; (b) 250 mm; (c) 300 mm.

An interesting observation is that the fire endurance of slabs is independent of their level of applied service load. In other words, changing the M_a/M_{cr} ratio during the design process does not significantly affect the failure time of the slab within the range of M_a/M_{cr} between 1.0 and 1.5. This effect is illustrated in Figure 8 where moment resistance (M_r) curves are normalized *versus* applied load or service load (M_a). While slabs with different M_a/M_{cr} ratios behave differently in the beginning, they approach each other when the moment capacity reaches the service load level. For example, a slab with an initial $M_a/M_{cr} = 1$, has approximately the same fire endurance as a slab with $M_a/M_{cr} = 1.5$. Obviously in a slab with $M_a/M_{cr} = 1.5$ the amount of reinforcement is higher due to crack width requirements but this extra reinforcement does not increase the fire endurance.

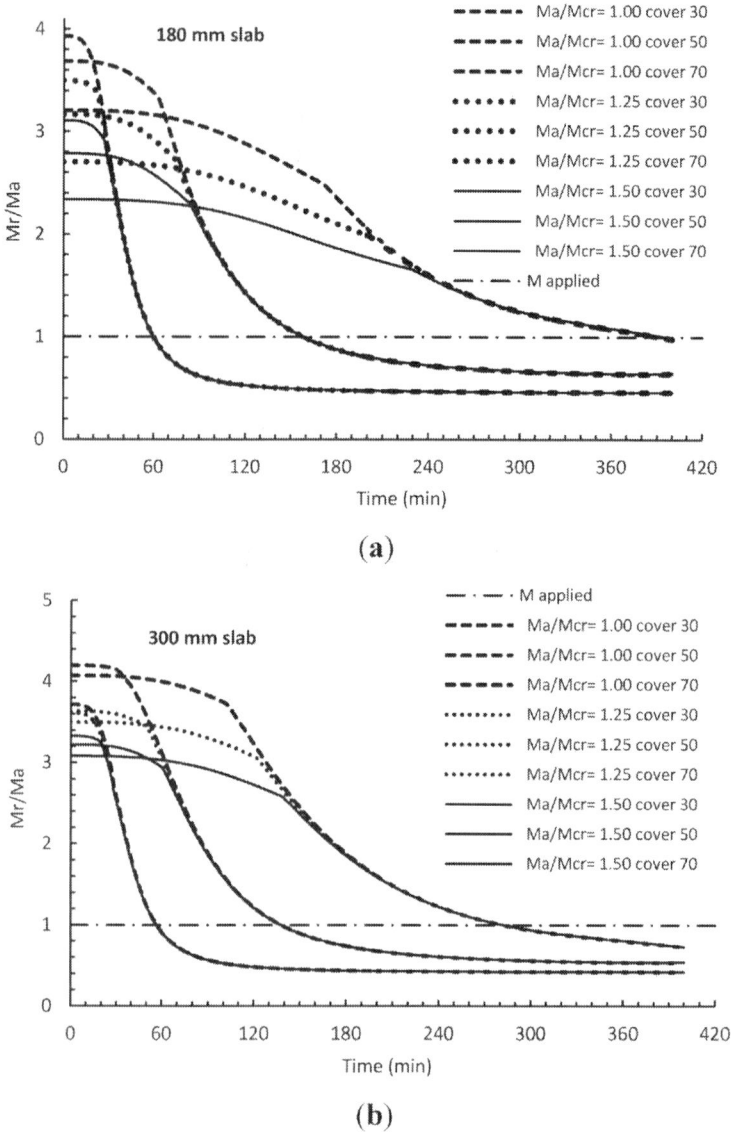

Figure 8: Ratios of M_r/M_a in fire for various concrete covers for: (**a**) 180 mm and (**b**) 300 mm.

STRENGTH-DOMAIN AND TEMPERATURE-DOMAIN FAILURE

Another approach in determining fire endurance is temperature-domain approach. In the temperature-domain approach, the fire endurance of the slabs is specified based on the critical temperature of the reinforcement defined as the temperature at which the bar loses 50% of the its tensile strength at room temperature. For steel reinforcement, the critical temperature is 593 °C [23]. Wang and Kodur [24] have reported 325 °C as the critical temperature for GFRP bars. Robert and Benmokrane reported 46% of tensile strength loss at 300 °C [15]. The fire endurance prediction of the strength model described in previous sections is presented in Figure 9 in comparison with results of temperature-domain method of CSA-S806. In the strength-domain method, the slab fails once its nominal flexural capacity drops below the applied moment (M_a). Figure 9 illustrates a considerable conservative prediction of temperature domain model. For instance, for a 180 mm slab, the fire endurance prediction of temperature domain failure for cover of 30 mm is 41% of the strength-domain failure. This is even lower for a slab with higher concrete cover. For the same slab with cover of 60 mm, the ratio is 24%. Bisby's model for GFRP behaviour at fire is used here to predict the fire performance of the slabs. The results of strength-domain result would be closer to temperature-domain results if a more conservative FRP material model (e.g., Saafi's model) is used in the calculation. For example, for a 180 mm thick slab with 50 mm of cover, Figure 5 shows that Saafi's model would estimate the fire endurance as 85 min compared to 75 min for the temperature-domain approach as shown in Figure 9.

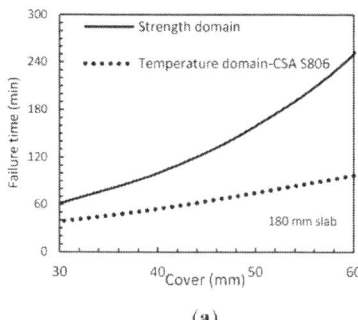
(a)

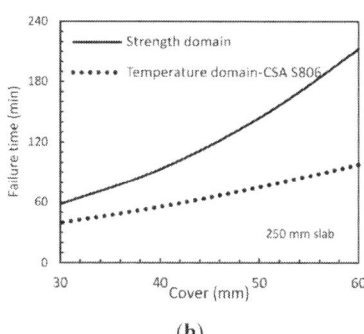
(b)

Figure 9: Strength-domain fire rate *versus* temperature-domain of CSA-S806 for (**a**) slab 180; and (**b**) 250 mm.

Given that the slabs in this study have been designed conforming to ACI-440-1R serviceability requirements, the nominal flexural capacities of the slabs are greater than that needed to resist the applied loads. This fact highlights the difference between FRP and steel reinforced slabs. The design of most steel reinforced slabs is governed by strength considerations. Since FRP reinforced slabs have reserve flexural strength because of serviceability considerations, a temperature-domain approach will always underestimate their fire endurance. This conclusion is also consistent with observations of Bisby and Kodur [5].

SLABS WITH TWO LAYERS OF FRP

Simulation results for slabs with one layer of FRP show that to fulfill the requirements of serviceability in FRP reinforced slabs, the amount of FRP is considerably larger than the amount needed considering the strength requirements. Since the serviceability requirements during a fire event do not need to be fulfilled (especially crack width criteria), the FRP could be employed more effectively to increase the fire endurance by placing the FRP reinforcement in two or more layers. During fire, a slab with two layers of FRP will perform better because the inner layer has more protective cover. To further investigate the effectiveness of this approach (*i.e.*, placing FRP in two layers), simulations are performed on slabs with two FRP layers and their behaviour is compared to that of slabs with one layer of FRP with the same amount of FRP reinforcement. The two types of slabs were designed to meet the same criteria at room temperature. Slab thicknesses are 180, 250 and 300 mm. Half of the reinforcement is placed in one layer and the remaining half in the other layer. The covers chosen for the bottom FRP layers are 30, 40, 50, and 60 mm and the distances between FRP layers are 30, 40, and 50 mm. The cross-sectional area of FRP reinforcement is determined according to ACI-440.1R serviceability criteria, similar to the design procedure used for slabs with one layer of FRP. The relations for calculating crack width and deflection are modified for slabs with two layers of FRP. For strength design purposes, the FRP in each layer is considered separately rather than as a single bundled FRP layer. The crack width limit is set to be equal to 0.7 mm for both slabs. M_a/M_{cr} in all simulations is equal to 1.5 since fire endurance was found to be independent of the service load level. Sample moment capacity curves during fire are shown in Figure 10. As expected, a slab with two layers of GFRP reinforcement outperforms a slab with same amount of reinforcement placed in one layer in terms of fire endurance. For example, a slab with GFRP placed at two layers with covers of 30 and 60 mm achieves approximately 2 h of fire endurance while the same slab with one layer of GFRP has a fire endurance of only 100 min. While the initial strength of the slab is higher for the one-layer slab, the decline in strength is faster during the fire exposure. Thus, the slab with two layers of reinforcement has more gradual fire degradation than the slab reinforced with one layer.

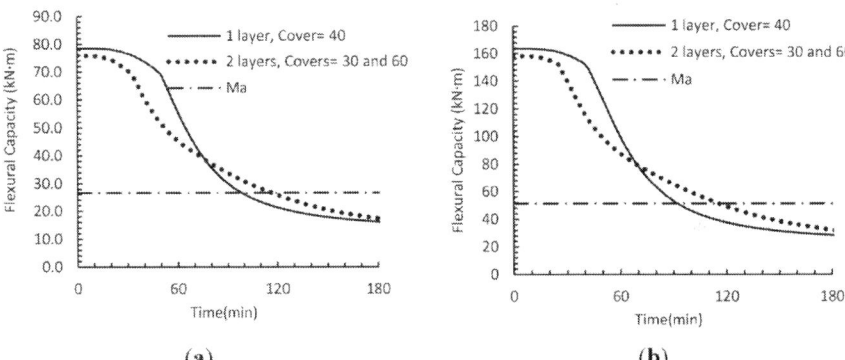

Figure 10: Prediction of the flexural capacity in fire of a slab with two layers; (a) Slab thickness of 180 mm; (b) Slab thickness of 250 mm.

The obtained increase in fire endurance by placing reinforcements in two layers varies by slab thickness. For example, in a slab with 180 mm thickness, the fire endurance gain was approximately 15 min on average, which is not significant considering the amount of effort needed for placing FRP bars in two layers. On the other hand, for a slab with 250 mm thickness, the average gain is 35 min and for a 300 mm slab it is 45 min. Based on these observations, placing FRP in two layers is more effective in terms of fire endurance for thicker slabs.

Figure 11 shows fire endurance results for slabs with two layers of reinforcement and corresponding results for slabs with one layer of reinforcement. The slab thicknesses are 180, 250, and 300 mm and the results are plotted against the reinforcement ratio. Therefore two points on a vertical line have the same amount of GFRP reinforcement and their vertical separation is the amount of increased fire endurance in min. There is a strong linear relation between reinforcement ratio and fire endurance in one layer slabs. The same linear dependency is generally present for two layer slabs. The fluctuations in two layer data are because of a sudden change in the distance between two layers. For example the distance between two layers in the first three points from left in Figure 11 is 30 mm and for the next three points it is 40 mm. Figure 12 shows flexural behavior of slabs in fire. With the same amount of reinforcement in one layer slabs and two layers reinforced slabs, initial strength of one layer reinforced slabs are slightly higher. However, they have gained higher fire endurance rate.

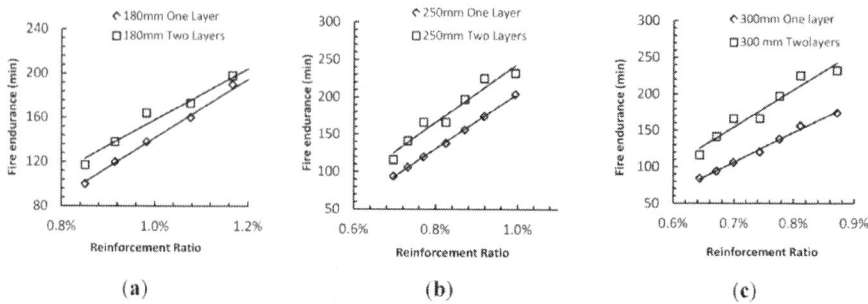

Figure 11: Fire endurance of one-layer compared to two-layer FRP reinforced concrete slab (a) 180 mm, (b) 250 mm, and (c) 300 mm.

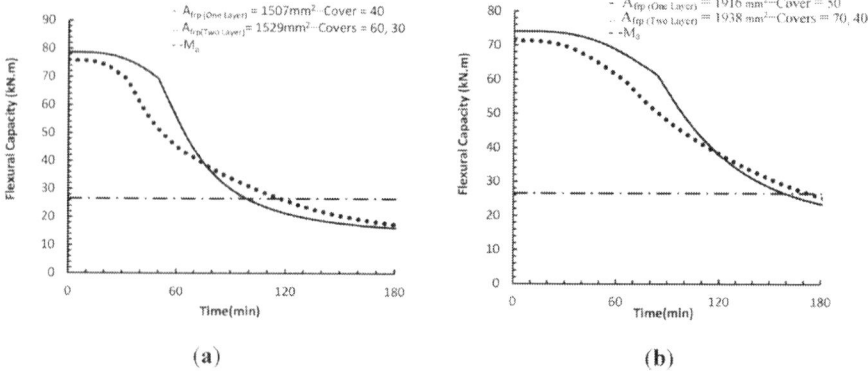

Figure 12: Comparison of flexural capacities of one-layer slab (180 mm) with two-layer with the same area of reinforcement for three cover thicknesses.

CONCLUSIONS

This paper investigated the fire endurance of FRP reinforced concrete slabs using numerical modelling. The model only considers the strength of the FRP and does not model degradation of the bond of FRP bars to concrete. Thus, the results are only applicable to designs where the bars are anchors to prevent bond failure in a fire. The results and recommendations of this study are summarized as follows:

- Concrete cover thickness drastically influences the fire endurance of the slabs.
- The validity of the temperature-domain method for fire endurance (*i.e.*, specifying a critical temperature for the reinforcement to represent failure) is investigated against the strength-domain method. Since the temperature-domain method is developed from the behaviour of steel reinforced concrete members in fire, it is not entirely applicable to FRP reinforced concrete members and represents a lower bound of the expected fire endurance.
- The results for two-layer FRP reinforced slabs show that by changing the distribution of the required reinforcement from one layer to two layers (using the same amount of FRP reinforcement), fire endurance of the slabs increases. The increase is more notable for thicker slabs.
- The results are greatly influenced by the implemented thermal strength degradation of FRP materials. Comprehensive tensile tests at various temperatures are needed to improve the FRP material degradation models especially with different FRP materials available in the market.
- The model should be extended to simulate the effects of temperature on the bond of FRP to concrete.

ACKNOWLEDGEMENTS

The authors would like to thank the Natural Sciences and Engineering Research Council of Canada (NSERC), Queen's University, and the Ministry of Transportation of Ontario (MTO) for support of this paper.

REFERENCES

1. Rafi, M.; Nadjai, A.; Ali, F. Fire resistance of carbon FRP reinforced-concrete beams. *Mag. Concr. Res.* 2007, *59*, 245–255.
2. Nigro, E.; Bilotta, A.; Cefarelli, G.; Manfredi, G.; Cosenza, E. Performance under fire situations of concrete members reinforced with FRP rods: Bond models and design nomograms. *J. Compos. Constr.* 2011, *16*, 395–406.

3. Kodur, V.K.R.; Baingo, D. *Fire Resistance of FRP Reinforced Concrete Slabs*; Institute for Research in Construction: Ottawa, Canada, 1998.
4. Kodur, V.; Bisby, L.A. Evaluation of fire endurance of concrete slabs reinforced with fiber-reinforced polymer bars. *J. Struct. Eng.* 2005, *131*, 34–43.
5. Bisby, L.; Kodur, V. Evaluating the fire endurance of concrete slabs reinforced with FRP bars: Considerations for a holistic approach. *Compos. Part B Eng.* 2007, *38*, 547–558.
6. Nigro, E.; Cefarelli, G.; Bilotta, A.; Manfredi, G.; Cosenza, E. Fire resistance of concrete slabs reinforced with FRP bars. Part I: Experimental investigations on the mechanical behavior. *Compos. Part B Eng.* 2011, *42*, 1739–1750.
7. American Concrete Institute. *Guide for the Design and Construction of Concrete Reinforced with FRP Bars*; ACI-440.1R-06. American Concrete Institute: Detroit, MI, USA, 2006.
8. Canadian Standards Association. *Design and Construction of Building Components with Fiber-Reinforced Polymers*; CSA-S806. Canadian Standards Association: Mississauga, Ontario, Canada, 2002.
9. Lie, T.T. *Structural Fire Protection*; Manuals and Reports on Engineering Practice No. 78. American Society of Civil Engineers: New York, NY, USA, 1992.
10. Buchanan, A. *Structural Design for Fire Safety*; John Wiley&Sons: New York, NY, USA, 2001.
11. Bazant, Z.P.; Kaplan, M.F.; Haslach, H. Concrete at high temperatures: Material properties and mathematical models. *Appl. Mech. Rev.* 1997, *50*, B75–B75.
12. Blontrock, H.; Taerwe, L.; Matthys, S. Properties of fiber reinforced plastics at elevated temperatures with regard to fire resistance of reinforced concrete members. *ACI Spec. Publ.* 1999, *1999*, 188.
13. Saafi, M. Effect of fire on FRP reinforced concrete members. *Compos. Struct.* 2002, *58*, 11–20.
14. Bisby, L. Fire behaviour of fibre-reinforced polymer (FRP) reinforced or confined concrete. Ph.D Thesis, Department of Civil Engineering, Queen's University, Kingston, ON, Canada, 2003.
15. Robert, M.; Benmokrane, B. Behavior of GFRP reinforcing bars subjected to extreme temperatures. *J. Compos. Constr.* 2009, *14*, 353–360.
16. American Society for Testing and Materials. *Standard Methods of Fire Test of Building Construction and Materials*; ASTM E119–07. American Society for Testing and Materials: West Conshohocken, PA, USA, 2007.

17. Adelzadeh, M.; Green, M.F.; Bénichou, N. Behaviour of fibre reinforced polymer-strengthened T-beams and slabs in fire. *Proc. ICE Struct. Build.* 2012, *165*, 361–371.
18. Adelzadeh, M.; Hajiloo, H.; Green, M.F. FRP Reinforced Concrete Slabs in Fire: A Parametric Analysis. In Proceedings of Second Conference on Smart Monitoring, Assessment, and Rehabilitation of Civil Structures, Istanbul, Turkey, 9 September 2013.
19. Kodur, V.; Yu, B.; Dwaikat, M. A simplified approach for predicting temperature in reinforced concrete members exposed to standard fire. *Fire Saf. J.* 2013, *56*, 39–51.
20. Weber, A. Fire-Resistance Tests on Composite Rebars. In Proceedings of CICE2008, Zurich, Switzerland, 22 July 2008.
21. Abbasi, A.; Hogg, P.J. Fire testing of concrete beams with fibre reinforced plastic rebar. *Compos. Part A Appl. Sci. Manuf.* 2006, *37*, 1142–1150.
22. American Concrete Institute. *Guide for Determining the Fire Endurance of Concrete Elements*; ACI-216. American Concrete Institute: Detroit, MI, USA, 1994.
23. Lie, T. Calculation of the fire resistance of composite concrete floor and roof slabs. *Fire Technol.* 1978, *14*, 28–45.
24. Wang, Y.; Kodur, V. Variation of strength and stiffness of fibre reinforced polymer reinforcing bars with temperature. *Cem. Concr. Compos.* 2005, *27*, 864–874.

CITATION

Masoud Adelzadeh, Hamzeh Hajiloo, and Mark F. Green, Numerical Study of FRP Reinforced Concrete Slabs at Elevated Temperature, doi:10.3390/polym6020408.

CHAPTER 6

Achievements of Truss Models for Reinforced Concrete Structures

Panagis G. Papadopoulos, Hariton Xenidis, Panos Lazaridis, Andreas Diamantopoulos, Periklis Lambrou, Yannis Arethas

Department of Civil Engineering, Aristotle University of Thessaloniki, Thessaloniki, Greece

KEYWORDS

Reinforced Concrete Structure; Truss Model; Constitutive Law; Material and Geometric Nonlinearities; Concrete Cracking; Reinforcement Yield; Concrete Ultimate Compressive Strength; Plastic Hinge; RC Column Confinement; Buckling of Inner Concrete Struts; Global Instability

ABSTRACT

Achievements are presented for truss models of RC structures developed in previous years: 1) Two constitutive models, biaxial and triaxial, are based on regular trusses, with bars obeying nonlinear uniaxial σ-ε laws of material under simulation; both models have been compared with test results and show a dependence of Poisson ratio on curvature of σ-ε law; 2) A truss finite element has been used in the nonlinear static and dynamic analysis of plane RC frames; it has been compared with test results and describes, in a simple way, the formation of plastic hinges; 3) Thanks to the very simple geometry of a truss, the equilibrium equations can be easily written and the stiffness matrix can be easily updated, both with respect to the deformed truss, within each step of a static incremental loading or within each time step of a dynamic analysis, so that to take into account geometric nonlinearities. So the confinement of a RC column is interpreted as a structural stability effect of concrete. And a significant role of the transverse reinforcement is revealed, that of preventing, by its close spacing and sufficient amount, the buckling of inner longitudinal concrete struts, which would lead to a global instability of the RC column; 4) The proposed truss model is statically indeterminate, so it exhibits some features, which are not met by the "strut-and-tie" model.

INTRODUCTION

In 1967, in a pioneering work [1], D. Ngo and A. C. Scordelis presented a detailed finite element model for a RC beam, in which separate finite elements are used for concrete and steel reinforcement. The material nonlinearities of the reinforcement can be easily described by the nonlinear uniaxial σ-ε law of a bar element. However, it is difficult to represent the nonlinear biaxial or triaxial stress-strain behavior of concrete or any other material. The relevant problems are discussed in two state-of-theart reports on nonlinear finite element analysis of RC structures, one by P. G. Bergan and I. Holand in 1979 [2] and another in a special publication of ASCE in 1982 [3], written by specialists on this field, under the co-ordination of A. C. Scordelis. Also, the difficulties appearing in the application of finite elements to nonlinear problems have been discussed in the series of three Conferences F. E. No. Mech. (Finite Elements in Nonlinear Mechanics), organized by J. H. Argyris in the Institute of Statics and Dynamics, University of Stuttgart, Germany in the years 1978, 1981, 1984 [4].

In order to describe the nonlinear biaxial or triaxial stress-strain behavior of a structure by the Finite Element Method, constitutive models for the structural materials have to be developed in order to be embodied in the individual finite elements. Efforts to develop such constitutive models have been made by many researchers, e.g. plasticity models by W. F. Chen [5] and Z. Mroz [6], the plastic-fracturing model of Z. P. Bazant [7], as well as the more practical contributions by D. Darwin [8] and K. I. Willam [9], for nonlinear, biaxial and triaxial, respectively, stress-strain behavior of concrete.

In 1977 [10], N. J. Burt and J. W. Dougill presented a random network constitutive model, in order to describe the nonlinear biaxial stress-strain law of a material, and noticed that equivalent results can be obtained by use of simple regular networks. By applying this idea, P. G. Papadopoulos developed in 1984 and 1986 [11,12] a biaxial and a triaxial network constitutive model, based on a regular plane octagon and a regular space rhombic dodecahedron, respectively, in which sides and diagonals are bars

INTRODUCTION

obeying the nonlinear uniaxial σ-ε laws of the material under simulation. Results from the above network constitutive models have been found in satisfactory approximation with corresponding published test results [13-15].

Trusses have been used not only in constitutive models, but also in finite elements of structures. In 1978 [16], E. Absi, in his "theorie des equivalences" stated that simple truss finite elements give equivalent results with the usual more complicated continuum finite elements. This idea was extended to problems with material nonlinearities and to the nonlinear static and dynamic analysis of plane RC frames by P. G. Papadopoulos [17,18]. A simple truss finite element was proposed, based on a plane rectangle in which all sides and diagonals are bars obeying nonlinear uniaxial σ-ε laws of concrete or steel. So, the nonlinear biaxial stress-strain behavior of the element is, in a simple way, described, thus the embodying of a constitutive law in the individual finite elements is no more needed. Results from nonlinear static analysis for cyclic loading, as well as nonlinear seismic dynamic analysis of simple plane RC frames, by the proposed truss RC element, were compared with corresponding published test results and found in a satisfactory approximation with them [19,20]. As the bars of the proposed finite element include the main material nonlinearities of concrete and steel, that is concrete tensile cracking and ultimate compressive strength, as well as tensile yield of reinforcement, the proposed truss model can, in a simple way, describes the formation of plastic hinges in a RC frame.

Afterwards, some other versions of E. Absi ideas for truss finite elements were developed for plane structures, under various names but all similar to each other, e.g. "truss analogy" in 1997 [21] for steel structures, "lattice model" in 1997 [22] and "lumped stress model" in 2002 [23], the latter two for RC structures.

In 1987 [24] J. Schlaich invented the so called "strutand-tie" model, which is a statically determinate truss model, consisting of concrete and steel bars. These bars include the main material nonlinearities of a RC structure. So, the "strut-and-tie" model can effectively describe the main stress-strain states of a RC structure, that is bending, shear and even torsion in 3D, thus it has been proved as a very useful practical tool in analysis and design of RC structures.

The "strut-and-tie" model has been further developed by other researchers, as by T. T. Hsu in 1993 [25], by F. J. Vecchio and M. P. Collins in 1993

[26], as well as by ASCE-ACI Committee on shear and torsion in 1998 [27].

The proposed here truss model is a statically indeterminate structure, so it exhibits some features that are not met by the statically determinate "strut-and-tie" model:

1) It can describe lateral expansion (Poisson ratio) effect.
2) It takes into account geometric nonlinearities, by writing the equilibrium equations and updating the stiffness matrix, both with respect to the deformed truss, within each step of a static incremental loading or within each time step of a dynamic analysis. This is easily achieved thanks to the very simple geometry of a truss.

By this proposed truss model which includes geometric nonlinearities, the confinement of a RC column is interpreted as a structural stability effect of concrete [28-30].

And, beyond the already known roles of the transverse reinforcement [31-33] (that is, shear transfer, reduction of concrete spalling, preventing of buckling of longitudinal reinforcement, increase of compressive stiffness, strength and ductility of the confined concrete core), another significant role of the transverse reinforcement is revealed by the proposed truss model with structural instability, that of retarding and even preventing, by its close spacing and sufficient amount, the buckling of inner longitudinal concrete struts, which would lead to a global instability of the RC column.

Results from the application of this proposed model with structural instability on RC column confinement have been found in a satisfactory approximation to Codes requirements [34-36], regarding the spacing and amount of transverse reinforcement, which, in turn, are based on test results, too.

In the following, some of the achievements of the above proposed truss models for nonlinear analysis of structures, mainly RC structures, proposed in previous years, will be described in more detail.

TRUSS CONSTITUTIVE MODELS

A biaxial and a triaxial constitutive model for the nonlinear stress-strain law of a material have been developed [11,12], based on a regular plane octagon and a regular space rhombic dodecahedron, respectively, in which sides and diagonals are bars obeying the nonlinear uniaxial σ-ε law of the material under simulation. So, in a simple way, by the nonlinear uniaxial σ-ε laws of the bars, the nonlinear biaxial or triaxial stress-strain behavior of the whole truss is described. Results from the above truss constitutive models, for various loading histories, have been found in satisfactory approximation to corresponding published test results [13-15]. Both above truss constitutive models show a dependence of the Poisson ratio value ν on the curvature κ of the nonlinear uniaxial σ-ε law of the material under consideration, as shown in Figure 1.

TRUSS FINITE ELEMENT FOR PLANE RC FRAME

A truss finite element is proposed for beams of a plane RC frame, based on a rectangle, in which all sides and diagonals are bars, obeying nonlinear uniaxial σ-ε laws of concrete or steel, as shown in Figure 2. The σ-ε law of concrete bars includes tensile cracking, ultimate compressive strength, as well as loading-unloading rules after compressive yield. Whereas, the σ-ε law of a steel bar includes ultimate tensile and compressive strengths, as well as loading-unloading rules after tensile or compressive yielding.

DETERMINATION OF BAR SECTIONS

In the above proposed truss finite element for beams of plane RC frames, the cross-section areas of steel bars are reasonably and easily determined as sums of sections of the corresponding steel reinforcing bars. Whereas, in order to determine the cross-sections areas A_1, A_2, A_3 of the concrete bars of the truss element as shown in Figure 3(b), we have to compare it to the corresponding continuum concrete beam element of Figure 3(a), as regards three representative stress-strain states in the linear elastic region. And we chose, as such characteristic states, the pure bending, the confined axial deformation as well as the confined transverse deformation.

For the pure bending shown in **Figure** 3(c), the curvature angle of the beam element is $\Delta\varphi = Ml/EJ$ where $J = wh^3/12$, whereas for the truss element $\Delta\varphi = 2\Delta l/h$

where $\Delta l = \dfrac{M}{h} l/A_1$. By combining the above equations we obtain $A_1 = wh/6$.

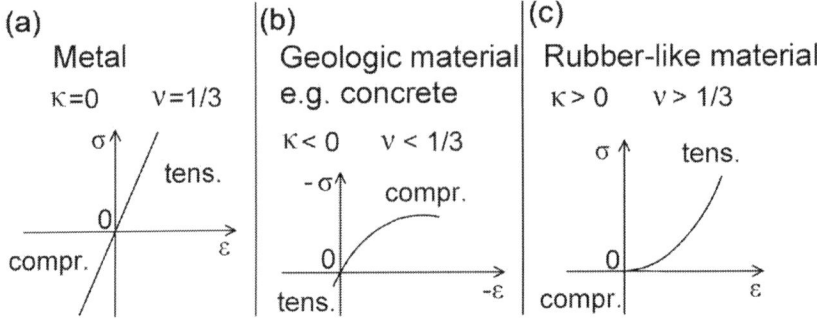

Figure 1: Dependence of Poisson ratio ν on the curvature κ of nonlinear uniaxial σ-ε law of the material. (a) Metal κ = 0 g ν = 1/3; (b) Geologic material e.g. concrete κ < 0 g ν < 1/3; (c) Rubber-like material κ > 0 g ν > 1/3.

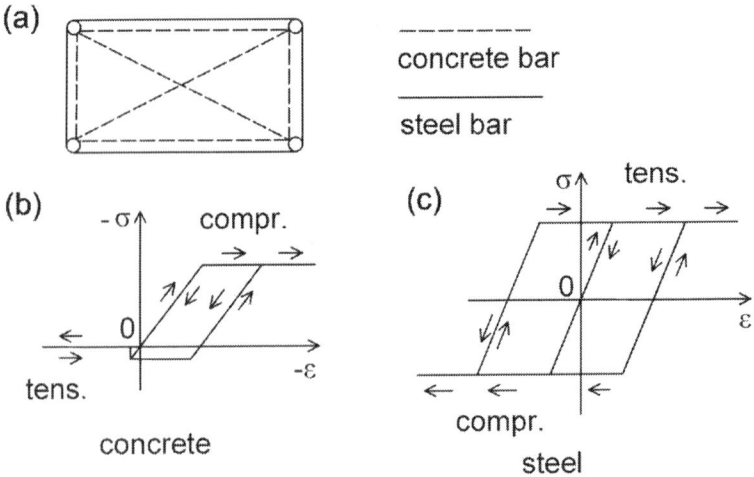

Figure 2: (a) Truss finite element for beam of a plane RC frame, with concrete and steel bars; (b) Nonlinear uniaxial σ-ε law of concrete bars; (c) Nonlinear uniaxial σ-ε law of steel bars.

DETERMINATION OF BAR SECTIONS

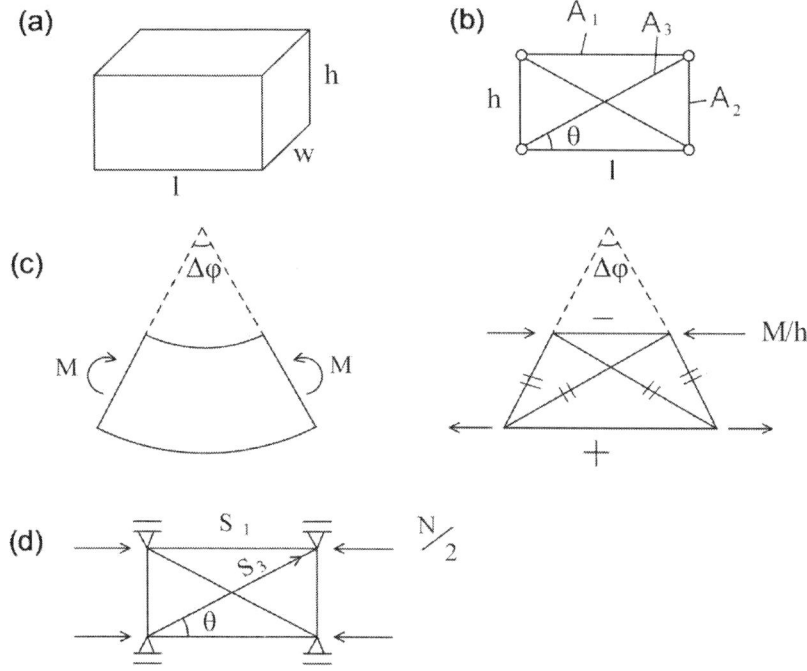

Figure 3: Comparison between characteristic stress-strain states of the concrete beam element and the corresponding truss element in order to determine the concrete bar sections. (a) Concrete beam element; (b) Corresponding truss element; (c) Pure bending; (d) Confined axial compression.

For the confined axial deformation, the elasticity theory gives $\sigma_x = \frac{E}{1-v^2}\varepsilon_x$.

For $v \approx 0.2$, $1-v^2 \approx 1$, thus $\sigma_x \approx E\varepsilon_x$ where $\sigma_x = N/wh$ and $\varepsilon_x = \Delta l/l$. In the corresponding state of the truss element shown in **Figure 3**(d), we have $S_1 + S_3 \cos\theta = N/2$

where $S_1 = \frac{EA_1}{l}\Delta l$ and $S_3 = \frac{EA_3}{l/\cos\theta}\Delta l\cos\theta$.

From combination of above equations, we obtain

$$A_3 = wh/3\cos^3\theta$$

From similar considerations for confined transverse deformation, we obtain $A_2 = wl/2 - (wh/3)3\mathrm{tg}^3\theta$.

Obviously, when the angle θ tends to zero, $\theta \to 0$, the sections tend to $A_3 \to wh/3$ and $A_2 \to wl/2$.

NONLINEAR STATIC ANALYSIS

The incremental loading of the structure is preferably performed by strain control, which is a more stable procedure than stress control. The material nonlinearities are taken into account by the variations of the elasticity moduli E of the bars during the loading. Whereas, in order to take into account geometric nonlinearities, the equilibrium equations are written and the global stiffness matrix updated, both with respect to the deformed truss, within each step of incremental loading. The local stiffness matrix of a bar in 2D, consisting of elastic and geometric part, is:

$$\mathbf{k} = \mathbf{k}_E + \mathbf{k}_G = \frac{EA}{l_o} \begin{pmatrix} c_x^2 & c_x c_y \\ c_x c_y & c_y^2 \end{pmatrix} + \frac{N}{l} \begin{pmatrix} c_y^2 & -c_x c_y \\ -c_x c_y & c_x^2 \end{pmatrix}$$

where A section, l_o undeformed length, l present length, N axial force and c_x, c_y direction cosines of the bar.

Whereas, the global stiffness matrix of the truss is:

$$\mathbf{K} = \mathbf{B}\operatorname{diag}(\mathbf{k}_i)\mathbf{B}^t \quad i = 1 \cdots n_b$$

where B Boolean linkage matrix and n_b number of bars of the truss.

Based on the proposed algorithm, a very short computer program, with only about 200 FORTRAN instructions, has been developed, for the nonlinear static analysis of a truss model of a plane RC frame.

NONLINEAR DYNAMIC ANALYSIS

A lumped mass is assigned to every free node of the truss. Zero damping and zero initial velocities are assumed. The resulting initial value problem: $\dot{\mathbf{y}} = \mathbf{q}(t, \mathbf{y})$, $\mathbf{y}(0) = \mathbf{y}_0$ where the state vector is $\mathbf{y} = \{\mathbf{r}, \mathbf{v}, \mathbf{c}\}$ with r, v positions and velocities of nodes and c constitutive variables of the bars, is solved by the step-by-step algorithm of trapezoidal rule, which coincides with the Newmark's algorithm of constant average acceleration:

$$y_{n+1} = y_n + \frac{1}{2}\left[q(t_n, y_n) + q(t_{n+1}, y_{n+1})\right]\Delta t$$
combined with a predictor-corrector technique with two corrections per step, PE(CE)² [37]. So, there is no need to solve an algebraic system within each step of the algorithm.

The stability criterion of the algorithm is $\omega_{max}\Delta t < 2.0$ rad and the accuracy criterion is $\omega_{max}\Delta t < 0.5$ rad, that is $\Delta t < T_{min}/4\pi$, which dictates the choosing of the time step-length Δt of the algorithm.

An upper bound for the normal frequencies can be found from the norm of the matrix M⁻¹K, where M mass matrix and K stiffness matrix of the structure:

$$\omega_{max} < \|M^{-1}K\|$$

Based on the proposed algorithm, a very short computer program has been developed, with only about 150 FORTRAN instructions, for the nonlinear dynamic analysis of a truss model of a RC frame.

APPLICATIONS TO ANALYSIS OF SIMPLE PLANE RC FRAMES

The above proposed truss finite element for plane RC frames, as well as the proposed algorithms for nonlinear static and dynamic analysis, have been applied to the nonlinear static analysis of a simple plane RC frame for cyclic loading [17], as well as to the nonlinear dynamic seismic analysis of a simple plane RC frame [18]. The results of these analyses have been found in satisfactory approximation with corresponding published test results [19,20].

As the nonlinear uniaxial σ-ε laws, of the bars of the proposed truss model, include all the main material nonlinearities of a RC structure, that is tensile cracking and ultimate compressive strength of concrete, as well as tensile yielding of steel reinforcement, the formation of plastic hinges in a RC frame is, in a simple way, described, as shown in Figure 4.

APPLICATION TO CONFINEMENT OF A RC COLUMN

In order to take into account geometric nonlinearities, by the proposed truss model, the equilibrium equations are written and the stiffness matrix is updated, both with respect to the deformed truss, within each step of a static incremental loading or within each time step of a dynamic analysis. This is easily achieved thanks to the very simple geometry of a truss.

As the proposed truss model includes geometric nonlinearities, it interprets the confinement of a RC column as a structural stability effect of concrete [28-30].

In Figure 5(a), the compressive axial σ-ε diagram of a confined RC column is shown. An early small drop of the stress $-\Delta\sigma$ is observed, which is due to a local instability because of spalling (buckling) of outer concrete. As the loading further increases, preferably by strain control, for a significant value of the compressive axial deformation, the stress σ suddenly drops to zero, which is an obvious mark of global structural instability, observed in experiments and verified by the proposed truss model, too.

In Figure 5(b), a part of a confined RC column, between two successive sets of transverse reinforcement, is shown. The longitudinal reinforcement is omitted, for simplicity. As the compressive axial loading N gradually increases, a lateral expansion of concrete occurs. For a significant compressive axial deformation, because of the large lateral expansion of concrete, a tensile yielding of the transverse reinforcement occurs, which implies a

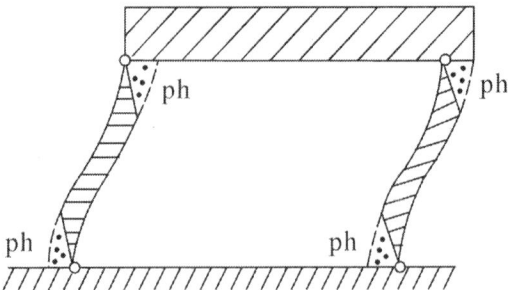

Figure 4. Description of formation of plastic hinges, in a RC frame, by the proposed truss model. "......" cracked concrete. "------" reinforcement in tensile yielding. "//////" rigid parts. "ph" plastic hinges.

APPLICATION TO CONFINEMENT OF A RC COLUMN

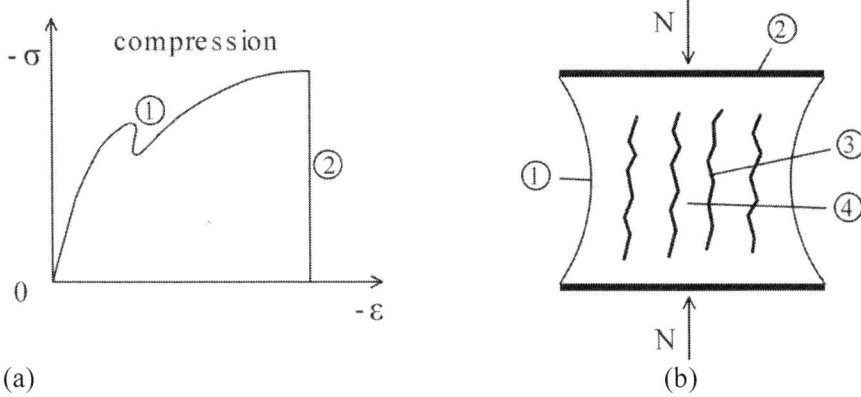

(a)   (b)

Figure 5. (a) Compressive axial σ-ε diagram of a confined RC column. 1. Early small drop of stress due to spalling of outer concrete. 2. For a significant value of compressive axial deformation, the stress σ suddenly drops to zero, which is a mark of global structural instability; (b) A part of a confined RC column between two successive sets of transverse reinforcement. 1. Spalling of outer concrete. 2. Transverse reinforcement in tensile yielding. 3. Longitudinal concrete cracks. 4. Longitudinal concrete struts.

further lateral expansion of concrete. So, wide longitudinal-vertical concrete cracks are formed, and between such successive concrete cracks, inner longitudinal-vertical concrete struts are formed, which tend to buckle, leading to a global instability of the RC column.

Beyond the already known roles of transverse reinforcement in a RC column [31-33] (that is shear transfer, reduction of concrete spalling, preventing of buckling of longitudinal reinforcement, increase of compressive axial stiffness, strength and ductility of the confined concrete core), another significant role of the transverse reinforcement is revealed by the proposed truss model with structural instability, that of retarding and even preventing, by its close spacing s and sufficient amount ρ (mechanical ratio), the buckling of the above inner longitudinal-vertical concrete struts, which would lead to a global instability of the RC column.

Results, from application of the proposed truss model with structural instability to the confinement of RC columns, have been found in satisfactory approximation with corresponding requirements of codes [34-36], regarding the spacing s and the mechanical ratio ρ of transverse reinforcement; these requirements are, in turn, based on test results, too.

CONCLUSIONS

Some achievements have been presented for truss models of structures, mainly RC structures, which have been developed in previous years and found in satisfactory approximation with test results and Codes requirements:

1) N. J. Burt and J. W. Dougill developed in 1977 [10] random network constitutive models and stated that equivalent results can be, in a simple way, obtained by regular networks. This idea was realized in 1984 [11] and 1986 [12] by two network constitutive models, a biaxial and triaxial one, based on a regular plane octagon and on a regular space rhombic dodecahedron, respectively, in which sides and diagonals are bars obeying the nonlinear uniaxial σ-ε law of the material under simulation. Both models show a dependence of Poisson ratio on the curvature of the nonlinear uniaxial σ-ε law of the material.

2) E. Absi in 1978 [16], in his "theorie des equivalences", stated that simple truss finite elements give equivalent results with the usual more complicated continuum finite elements. This idea was extended in 1988 [17,18] to structures with material nonlinearities, and applied particularly to the nonlinear static and dynamic analysis of plane RC frames. As the individual bars of the proposed truss finite element include, in their uniaxial σ-ε laws, the main material nonlinearities of a RC structure, that is the concrete tensile cracking, the reinforcement tensile yield, as well as the ultimate compressive strength of concrete, the formation of plastic hinges in a RC frame can be, in a simple way, described.

3) Compared to the "strut-and-tie" model for RC structures, invented by J.Schlaich in 1987 [24] and further developed by other researchers, which proved as a very effective tool in the analysis of RC structures, the proposed here truss model exhibits the difference that it is a statically indeterminate, whereas the "strut-and-tie" model is statically determinate. So, the proposed truss model has some features that are not met by the "strut-and-tie" model: a) It can describe lateral expansion (Poisson ratio) effect. b) It takes into account geometric nonlinearities, by writing equilibrium equations and by updating stiffness matrix, both with respect to the deformed truss, within each step of a nonlinear static or dynamic analysis. So, it interpreted in 1999 [31] the confinement of a RC column as a structural stability effect of concrete. And revealed a significant role of transverse reinforcement, that of retarding and even preventing, by its close spacing and sufficient amount, the buckling of inner longitudinal

concrete struts, which would lead to a global instability of the RC column.

REFERENCES

1. D. Ngo and A. C. Scordelis, "Finite Element Analysis of Reinforced Concrete Beams," ACI Journal, Vol. 64, 1967, pp. 152-163.
2. P. G. Bergan and I. Holand, "Nonlinear Finite Element Analysis of Concrete Structures," Computer Methods in Applied Mechanics and Engineering, Vol. 17-18, 1979, pp. 443-467. doi:10.1016/0045-7825(79)90027-6
3. A. C. Scordelis, Editor, ASCE Task Committee on Concrete and Masonry Structures, "State-of-the-Art Report on Finite Element Analysis of Reinforced Concrete," ASCE Special Publication, 1982.
4. J. H. Argyris, Organizer, International Conferences F.E.No.Mech. (Finite Elements in Nonlinear Mechanics). Institute for Statics and Dynamics, University of Stutgart, Germany, I.30 August-1 September 1978, II. 25-28 August 1981. III. 10-13 September 1984.
5. W. F. Chen and E. C. Ting. "Constitutive Models for Concrete Structures," Journal of Engineering Mechanics Division ASCE, Vol. 106, No. 1, 1980, pp. 1-19.
6. Z. Mroz, V. A. Norris and O. C. Zienkiewicz, "Application of an Anisotropic Hardening Model in the Analysis of Elastic-Plastic Deformation of Soils," Geotechnique, Vol. 29, 1979, pp. 1-34. doi:10.1680/geot.1979.29.1.1
7. Z. P. Bazant and S. S. Kim, "Plastic-Fracturing Theory for Concrete," Journal of Engineering Mechanics Division ASCE, Vol. 105, No. 3, 1979, pp. 407-428.
8. D. Darwin and D. A. Pecknold, "Analysis of Cyclic Loading of RC Structures," Computers and Structures, Vol. 7, No. 1, 1977, pp. 137-147. doi:10.1016/0045-7949(77)90068-2
9. K. J. Willam and E. P. Warnke, "Constitutive Model for the Triaxial Behavior of Concrete," Proceedings of IABSE, Structural Engineering Report 19, Section III, 1975, pp. 1-30.
10. N. J. Burt and J. W. Dougill, "Progressive Failure in a Model Heterogeneous Medium," Journal of Engineering Mechanics Division ASCE, Vol. 103, 1977, pp. 365-376.
11. P. G. Papadopoulos, "Biaxial Network Constitutive Model," Journal of Engineering Mechanics ASCE, Vol. 110, No. 3, 1984, pp. 449-464. doi:10.1061/(ASCE)0733-9399(1984)110:3(449)

12. P. G. Papadopoulos, "A Triaxial Network Constitutive Model," Computers and Structures, Vol. 23, 1986, pp. 497-501. doi:10.1016/0045-7949(86)90093-3
13. H. B. Kupfer, H. D. Hilsdorf and H. Rusch, "Behavior of Concrete under Biaxial Stresses," ACI Journal, Vol. 66, No. 8, 1969, pp. 656-666.
14. R. Palaniswamy and S. P. Shah, "Fracture and StressStrain Relationships of Concrete under Triaxial Compression," Journal of Structural Division ASCE, Vol. 100, 1974, pp. 901-916.
15. R. Scavuzzo, T. Stankowski, K. Gerstle and H.-Y. Ko, "Stress-Strain Curves for Concrete under Multiaxial Load Histories," University of Colorado, Boulder, 1983.
16. E. Absi, "Méthodes des Calcus Numerique en Elasticité," Eyrolles, Paris, 1978.
17. P. G. Papadopoulos, "Nonlinear Static Analysis of Reinforced Concrete Frames by Network Models," Advances in Engineering Software, Vol. 110, No. 3, 1988, pp. 114- 122.doi:10.1016/0141-1195(88)90010-1
18. P. G. Papadopoulos and C. G. Karayannis, "Seismic Analysis of R/C Frames by Network Models," Computers and Structures, Vol. 28, No. 4, 1988, pp. 481-494.doi:10.1016/0045-7949(88)90022-3
19. K. Stylianidis and G. Penelis, "Experimental Study of, bare and Infilled by Wall, One Story Frames under Cyclic shear Loading," 7th Greek Conference on Concrete, Vol. 2, Patra, 1985, pp. 47-55.
20. P. Hidalgo and R. W. Clough, "Earthquake Simulator Study of a Reinforced Concrete Frame," EERC Report 74-13, University of California, Berkeley, 1974.
21. S. C. Goel, B. Stojadinovicz and K. H. Lee, "Truss Analogy for Steel Moment Connections," Engineering Journal, Second Quarter 1997, pp. 43-53.
22. E. Schlangen and E. J. Garboczi, "Fracture Simulations of Concrete Using Lattice Models: Computational Aspects," Engineering Fracture Mechanics, Vol. 57, No. 2-3, 1997, pp. 319-332. doi:10.1016/S0013-7944(97)00010-6
23. F. Fraternali, M. Angelilo and A. Fortunato, "A Lumped Stress Method for Plane Elastic Problems and the Discrete Continuum Approximation," International Journal of Solids and Structures, Vol. 39, 2002, pp. 6211-6240. doi:10.1016/S0020-7683(02)00472-9
24. J. Schlaich, K. Schäfer and M. Jennewein, "Towards a Consistent Design of Structural Concrete," PCI Journal Special Report, Vol. 32, No. 3, 1987, pp. 75-150.
25. T. T. C. Hsu, "Unified Theory of Reinforced Concrete," CRC Press, 1993.
26. F. J. Vecchio and M. P. Collins, "Compression Response of Cracked Reinforced Concrete," Journal of Structural Engineering ASCE, Vol. 113, 1993, pp. 3590-3610.doi:10.1061/(ASCE)0733-9445(1993)119:12(3590)

REFERENCES

27. ASCE-ACI Committee 445 on Shear and Torsion, "Recent Approaches to Shear Design of Structural Concrete. State-of-the-Art Report," Journal of Structural Engineering ASCE, Vol. 119, No. 12, 1998, pp. 1375-1417.
28. P. G. Papadopoulos and H. C. Xenidis, "A Truss Model with Structural Instability for the Confinement of Concrete Columns," Journal of EEE (European Earthquake Engineering), Part 2, 1999, pp. 57-79.
29. P. G. Papadopoulos, H. Xenidis, C. Karayannis, A. Diamantopoulos and P. Lambrou, "Confinement of Concrete Column Interpreted as a Structural Stability Effect," 6th GRACM (Greek Association of Computational Mechanics) Conference, Thessaloniki, 19-21 June 2008.
30. P. G. Papadopoulos, H. Xenidis, D. Plasatis, P. Kiousis and C. Karayannis, "Concrete Stability Achieved by Confinement in a RC Column," 12th International Conference on Civil, Structural and Environmental Engineering Computing, Coordinator B.H.V. Topping, Madeira, Portugal, 1-4 September 2009.
31. K. Park, M. J. N. Priestley and W. D. Gill, "Ductility of Square Confined Concrete Columns," Journal of Structural Division ASCE, Vol. 108, No. 4, 1982, pp. 929-950.
32. S. Watson, F. A. Zahn and R. Park, "Confining Reinforcement for Concrete Columns," Journal of Structural Engineering ASCE, Vol. 120, No. 6, 1984, pp. 1798-1849.
33. J. B. Mander, M. J. N. Priestley and R. Park, "Theoretical Stress-Strain Model for Confined Concrete," Journal of Structural Engineering ASCE, Vol. 114, No. 8, 1988, pp. 1804-1826.doi:10.1061/(ASCE)0733-9445(1988)114:8(1804)
34. Uniform Building Code 2, "Structural Engineering Design Provisions," Chapter 19. Concrete, 19.2.1. Reinforced Concrete Structures Resisting Forces Induced by Earthquake Motions 19.2.14. Frame Members Subjected to Bending and Axial Load, 1994, pp. 237-239.
35. New Zealand Standards 3101, "Code of Practice for the Design of Concrete Structures," Chapter 17, Members Subjected to Flexure and Axial Loads, Additional Seismic Requirements, 1989.
36. Eurocode 8, "Earthquake Resistant Design of Structures," Part 1-3. General Rules and Rules for Buildings. 2, Specific Rules for Concrete Buildings. 2.8. Provisions for Columns, Brussels, 1993, pp. 35-46.
37. P. G. Papadopoulos, "A Simple Algorithm for the Nonlinear Dynamic Analysis of Networks," Computers and Structures, Vol. 18, No. 1, 1984, pp. 1-8. doi:10.1016/0045-7949(84)90074-9

CITATION

P. Papadopoulos, H. Xenidis, P. Lazaridis, A. Diamantopoulos, P. Lambrou and Y. Arethas, "Achievements of Truss Models for Reinforced Concrete Structures," Open Journal of Civil Engineering, Vol. 2 No. 3, 2012, pp. 125-131. doi: 10.4236/ojce.2012.23018.

CHAPTER 7

Exploring Mechanical and Durability Properties of Ultra-High Performance Concrete Incorporating Various Steel Fiber Lengths and Dosages

Safeer Abbas[a], Ahmed M. Soliman[b], Moncef L. Nehdi[b],

[a] Department of Civil Engineering, University of Engineering & Technology, Lahore, Pakistan
[b] Department of Civil and Environmental Engineering, Western University, London, Ontario, Canada

ABSTRACT

Ultra-high performance concrete (UHPC) is a new generation of steel fiber-reinforced concrete with superior mechanical and durability properties. However, limited data is available on the influence of the steel fiber length and dosage on UHPC mechanical and durability performance. Therefore, in this study, a number of UHPC mixtures with varying steel fiber lengths (8 mm (0.31 in), 12 mm (0.47 in) and 16 mm (0.62 in)) and dosages (1%, 3% and 6%) by mixture volume were tested. Mechanical properties of UHPC including compressive, splitting tensile and flexural strengths were assessed. Moreover, its resistance to chloride ions penetration and mechanical degradation under various chloride exposures (i.e. 3.5% and 10%) were evaluated. Results showed an increase in mechanical properties as the fiber dosage increased. UHPC mixtures incorporating short steel fibers exhibited enhanced flexural properties compared to that of mixtures with similar volume of longer steel fibers. At higher fiber dosage, UHPC mixtures exhibited relatively improved durability. Moreover, no degradation in UHPC mechanical properties was observed after exposure to various chloride ions solutions.

INTRODUCTION

Ultra-high performance concrete (UHPC) is a cement-based composite with a compressive strength typically higher than 150 MPa (22 ksi) and tensile strength greater than 7 MPa (1 ksi) [1], [2], [3] and [4]. It has almost zero porosity, leading to superior resistance to the penetration of aggressive and corrosive materials into its hardened matrix [1] and [3].

The key factor in producing UHPC is to improve the micro and macro properties of its mixture ingredients to ensure mechanical homogeneity, maximum density and dense particle packing [5], [6] and [7]. In UHPC, a high proportion of cement is used compared to that of normal strength (NS) and high-performance concrete (HPC) [5]. A very low w/b is used in UHPC mixtures due to which, only part of the total cement hydrates in UHPC and the un-hydrated cement can be replaced with crushed quartz, fly ash or blast furnace slag [3], [8] and [9]. The reduced workability of UHPC due to its very low w/b can be resolved by adding an effective superplasticizer (SP) [10]. Moreover, the addition of silica fume can improve the workability of UHPC and fill voids between coarser particles due to its finer size and spherical shape, thus enhancing the strength properties through pozzolanic reactions [3] and [11]. Using fine aggregate such as quartz powder (instead of coarse aggregates) plays an important role in reducing the maximum paste thickness (MPT), leading to lower porosity in the matrix [12], which is a key factor in the mixture design of UHPC. Due to its very high strength and homogeneity, UHPC becomes very brittle; yet it can be made ductile by adding steel fibers [1] and [13].

Various researchers have studied the material characterization of different commercially available and locally produced UHPC mixtures. For instance, Graybeal and Davis [14] studied the size effect on the compressive strength of UHPC and recommended 70 mm cube specimens where testing machine capacity and specimen grinding is a concern. Furthermore, 40% average increase in compressive strength was observed for 90 °C (194 °F) heat treatment compared to that of untreated control specimens [9],[15] and [16]. Some studies [17] and [18] reported that the UHPC compressive strength was not influenced by the addition of high dosages of steel fibers. Other studies [9], [15], [19] and [20] reported a slight increase in compressive strength due to fiber addition if proper thermal treatment is applied. The casting direction had no significant effect on the compressive strength of fiber-reinforced UHPC; yet, it significantly affected the flexural capacity [21], [22], [23] and [24]. Beam specimens incorporating fibers showed multiple cracks and exhibited steadier drop in load carrying capacity rather than a sudden drop in load after formation of the first crack [25]. The failure was characterized by a single vertical macro-crack with multiple micro-cracks for UHPC incorporating steel fibers [26]. Various researchers e.g. [27] and [28] conducted durability testing of UHPC and concluded that the chloride ions penetration was highly dependent on the exposure solution and duration, w/b and curing regimes.

UHPC has been successfully employed in various construction projects around the world [29], [30],[31] and [32]. However, applications of UHPC in the North America construction industry are lagging due to the lack of adequate design provisions, qualified contractors and experienced pre-casters, relatively higher initial cost and the requirement of special types of high energy mixers. Prior research showed that the addition of steel fibers in UHPC mixtures enhanced its mechanical properties significantly. However, only scant data is available on the effect of the steel fiber length and dosage on the overall mechanical and durability performance of UHPC. Therefore, this research program was planned to cover this gap in the existing knowledge.

RESEARCH SIGNIFICANCE

Achieving greater awareness of the mechanical and durability properties of ultra-high performance concrete (UHPC) amongst stakeholders in the construction industry is paramount for its wider implementation. Various researchers e.g. [1] and [3] studied the mechanical and durability properties of UHPC. However, very limited data is available in the open literature on the effect of the steel fiber length and dosage on such properties, especially for UHPC cured under normal precast plant conditions. Moreover, the durability properties of UHPC were investigated in previous studies [1] and [33] using a 3% chloride ions penetration test. Thus, in this study, the effect of fiber properties (length and dosage) on UHPC durability properties was investigated under highly corrosive environments (i.e. 10% chloride ions solution). The knowledge acquired on structural and durability properties of various UHPC mixtures incorporating different steel fiber lengths and dosages and cured under regular regimes used in precast plants should assist in producing full-scale precast UHPC structural elements (e.g. tunnel lining segments) with superior mechanical, durability and sustainability features.

EXPERIMENTAL PROGRAM

Materials Composition and Proportions

In this study, various UHPC mixtures incorporating 8 mm (0.31 in), 12 mm (0.47 in) and 16 mm (0.62 in) steel fibers at dosage of 1%, 3% and 6% by mixture volume were investigated for their mechanical and durability properties. The used fibers were copper coated steel having a constant diameter (0.2 mm (0.0078 in)) and tensile strength greater than 2850 MPa (413.35 ksi). The tested UHPC mixtures consist of portland cement, silica fume, quartz sand, quartz powder and a polycarboxylate based super-plasticizer in addition to steel fibers. The chemical and physical properties of these materials are summarized in Table 1. A typical UHPC mixture composition is shown in Table 2. The tested UHPC mixture (Table 2) was selected after various trial mixes and considering recommendations of previous study [7]. Moreover, the steel fiber dosage in various UHPC mixtures was adjusted by altering the quartz sand equivalent volume in order to make the total UHPC mixture volume consistent, in agreement with previous studies [7] and [34].

Table 1: Chemical and physical properties of used materials

Components	Cement	Silica fume	Quartz sand	Quartz powder
SiO_2 (%)	19.6	95.3	>99.50	>99.80
CaO (%)	61.5	0.49	0.01	0.01
Al_2O_3 (%)	4.8	0.17	0.05	0.05
Fe_2O_3 (%)	3.3	0.08	0.03	0.04
MgO (%)	3	0.27	–	–
K_2O (%)	1.2	0.48	–	–
SO_3 (%)	3.5	0.24	–	–
Na_2O (%)	0.1	0.19	–	–
TiO_2 (%)	0.3	0.3	0.03	0.02
Loss on ignition (%)	1.9	1.99	0.11	0.1
Specific surface area (m²/kg)	371	19,530	255	20,000
Specific gravity	3.17	2.12	2.65	2.65
C_3S (%)	55	–	–	–
C_2S (%)	15	–	–	–
C_3A (%)	7	–	–	–
C_4AF (%)	10	–	–	–

Table 2: UHPC mixture proportions

Quantities	Mass/cement mass
Cement	1
Silica fume	0.2
Quartz powder	0.3
Quartz sand	1.2
Superplasticizer	3.50[a]
Steel fibers length	[b]
Steel fibers dosage	[c]
Water	0.23

[a] Polycarboxylate superplasticizer (% by cement mass).
[b] 8 mm (0.31 in), 12 mm (0.47 in) and 16 mm (0.62 in) length.
[c] 1%, 3% and 6% by total volume of mixture.

Mixing of UHPC Constituents

A high shear pan mixer (Fig. 1) with 120 L (32 gal) capacity was used for mixing the UHPC constituents. Quartz sand and silica fume were dry mixed for 3–5 min. Thereafter, cement and quartz powder were added and mixing resumed for another 3 min. Half of the superplasticizer (SP) was mixed with the mixing water and added gradually to the dry mixture and mixing continued for another 3 min. Afterwards, the other half of SP was added over 3 min of mixing. At the end, steel fibers were added and mixing continued until fibers were fully dispersed.

Figure 1: High shear pan mixer for mixing UHPC.

Specimen Preparation and Environmental Conditions

UHPC cylindrical and beam specimens were immediately casted after the end of the mixing process. All specimens were casted in three layers. Each layer was compacted using a vibrating table. Afterwards, specimens were transported to an environmental chamber at 45 °C (113 °F) and relative humidity (RH) > 95%. After 5 h, molds were stripped and specimens were stored in a moist curing room at RH ⩾ 95% and 20 °C (68 °F) for 5 days. Subsequently, all specimens were stored at ambient laboratory conditions (i.e. 20 ± 2 °C (68 ± 4 °F)). This curing regime was selected to realistically resemble that used in industrial precast plants [35].

Mechanical Testing

The compressive strength for various UHPC mixtures was determined in accordance with ASTM C39/C39M (Standard Test Method for Compressive Strength of Cylindrical Concrete Specimens) [36]. The loading rate was 1 MPa/s (145 psi/s) in agreement with previous work [1].

EXPERIMENTAL PROGRAM

For all tested specimens, the stress and corresponding strain under compression load was recorded. The UHPC stress–strain curve showed almost linear behavior up to the peak compressive stress [1] and [37]. Therefore, the initial slope of the stress–strain curve (up to 85% of maximum stress) was used to evaluate the modulus of elasticity.

The splitting tensile strength was determined according to ASTM C496/C496M (Standard Test Method for Splitting Tensile Strength of Cylindrical Concrete Specimens) [38]. A displacement control load at a rate of 0.025 mm/min (0.001 in/min) was applied on the UHPC specimens in agreement with previous study [34].

The flexural performance of UHPC specimens was conducted according to ASTM C1609/C1609M (Standard Test Method for Flexural Performance of Fiber-Reinforced Concrete (Using Beam with Third-Point Loading)) [39]. A controlled displacement was applied at a rate of 0.05 mm/min (0.002 in/min) in agreement with previous work [25]. The first crack load is the point at which the initial linear elastic portion of the flexural load–deflection curve ends, while the peak load is the value of maximum load on the load–deflection curve. Crack width at various loading stages was measured using a crack width ruler. Table 3 shows the number of UHPC specimens, sizes and coefficient of variance (COV) for the various tests conducted.

Table 3: Number of specimens and coefficient of variance for various tests

Test	Specimen description			Coefficient of variance for various fiber length and dosage (%)									
				—	8 mm (0.31 in)			12 mm (0.47 in)			16 mm (0.62 in)		
	Size (mm)	Age (d)	Number	0%	1%	3%	6%	1%	3%	6%	1%	3%	6%
Workability (flow diameter)	—	—	4	4.21	3.56	2.68	1.92	3.5	2.96	2.12	3.78	3.43	2.28
Compressive strength	C: 75 × 150	7	3	2.11	1.31	1.43	1.11	1.05	1.78	0.89	1.03	0.39	1.97
		14	3	1.64	1.81	1.36	1.21	1.14	1.59	0.68	1.23	0.98	1.65

Test	Specimen	Age (days)	n										
Splitting tensile strength	C: 75 × 150	28	3	1.36	1.09	1.65	1.35	1.28	1.9	1.32	1.45	1.13	2.11
		56	3	1.75	0.79	1.87	1.45	1.43	1.96	1.76	1.14	2.12	1.08
		7	3	1.31	2.13	1.14	0.73	1.43	2.31	1.23	2.23	2.14	1.07
		14	3	1.45	2.03	1.28	0.94	2.32	3.12	1.61	2.06	1.53	2.63
		28	3	2.32	1.84	1.85	1.21	2.53	2.1	1.56	1.76	1.29	2.55
		56	3	2.04	1.21	1.69	1.74	2.12	2.13	2.43	1.78	1.96	2.05
Flexural strength	P: 100 × 100 × 400	28	3	3.12	2.43	2.22	3.42	3.12	2.76	3.72	3.12	2.26	2.65
VPV	C: 75 × 150	28	4	1.87	2.13	1.88	2.65	2.32	1.25	1.97	2.34	2.43	2.67
Sorptivity	C: 100 × 50	28	3	2.31	2.12	1.76	1.23	2.34	2.32	1.68	2.32	2.56	2.67
RCPT	C: 100 × 50	28	5	3.12	2.34	1.65	2.31	1.88	2.54	2.56	2.24	2.58	1.37
		56	5	2.32	1.24	2.21	1.45	1.67	2.34	1.31	2.43	1.23	1.65
Salt ponding	C: 100 × 75	90	3	0.85	2.12	1.12	1.05	2.01	1.08	0.78	1.85	1.25	0.95
		180	3	1.34	1.75	0.78	0.89	1.03	0.56	1.11	1.18	1.31	1.21
Salt immersion	C: 75 × 150	30	3	1.81	1.65	1.55	2.21	2.13	1.57	2.34	2.54	1.75	1.81
		90	3	2.12	2.45	1.87	1.94	2.05	1.68	2.52	2.25	1.94	2.17

	150	3	2.53	2.13	1.43	1.64	2.76	2.78	2.12	3.12	3.27	2.23
	180	3	1.45	2.84	2.36	2.87	3.23	3.12	2.32	3.31	3.54	3.65
Porosity	##	30	3	0.76	1.23	2.13	1.43	1.08	0.87	0.43	1.42	1.64	2.11

d = days; C = cylinder; P = prism;
1 mm = 0.04 in; ## small fragments.

Durability Testing

The volume of permeable voids for the UHPC mixtures incorporating various steel fiber dosages (1%, 3% and 6% by mixture volume) were tested in accordance with ASTM C642 (Standard Test Method for Density, Absorption, and Voids in Hardened Concrete) [40]. The volume of permeable voids can be evaluated using Eq. (1):

$$p = \left(\frac{g_2 - g_1}{g_2}\right) \times 100 \tag{1}$$

where p is the permeable pore space (%), g_1 and g_2 are the dry bulk and apparent densities, respectively.

The sorptivity test was performed according to the ASTM C1585 (Standard Test Method for Measurement of Rate of Absorption of Water by Hydraulic Cement Concretes) [41]. The initial weight of the specimen was first measured and the specimen was then immersed in 3–5 mm (0.1–0.2 in) deep water. The specimen was removed and weighed frequently according to the time duration defined in ASTM C1585 [41]. The sorptivity coefficient (S) was evaluated as the slope of the best fit line between the water absorption (I) and the square root of time (t), as given by Eq. (2).

$$S = \frac{I}{\sqrt{t}} \tag{2}$$

The rapid chloride ion penetrability test was conducted on UHPC specimens in accordance with ASTM C1202 (Standard Test Method for Electrical Indication of Concrete's Ability to Resist Chloride Ion Penetration) [42]. In addition to the 3% sodium chloride (NaCl) solution recommended by ASTM C1202, 3.5% and 10% NaCl solutions were also examined to replicate more severe exposure conditions [43], [44], [45] and [46]. The test continued for 6 h according to ASTM C1202 [42] and the number of coulombs passed were noted every 30 min.

The salt ponding test was conducted on UHPC specimens according to AASHTO T259 (Standard Method of Test for Resistance of Concrete to Chloride ion Penetration) [47]. Chloride solutions with various concentrations (i.e. 3%, 3.5% and 10%) were ponded 12–15 mm (0.5–0.6 in) on the specimen's top surface. In order to accelerate the chloride ions penetration into UHPC specimens, all tested specimens were stored at $40 \pm 1\,°C$ ($104 \pm 2\,°F$) inside a walk-in environmental chamber [48]. The chloride contents after 90 and 180 days of exposure were assessed according to ASTM C114-97 (Standard Test Methods for Chemical Analysis of Hydraulic Cement, Section 19-Chlorides) [49] and FHWA-RD-72-12 (Sampling and Testing of Chloride Ion in Concrete) [50].

The salt immersion test was performed over six months in accordance with ASTM D870 (Standard Practice for Testing Water Resistance of Coatings using Water Immersion) [51] and ASTM G31-72 (Standard Practice for Laboratory Corrosion Testing of Metals) [52]. UHPC cylinder and beam specimens were completely immersed in various chloride solutions (i.e. 3.5% and 10%) in plastic containers. All specimens were stored at $40 \pm 1\,°C$ ($104 \pm 2\,°F$) inside a walk-in environmental chamber. Weekly wetting and drying cycles for UHPC specimens were conducted in agreement with previous studies [53], [54] and [55]. The chloride solutions were changed at least every month and whenever it became cloudy. Representative cylinder and beam specimens were taken out from the plastic containers after 1, 3, 5 and 6 months of exposure. All specimens were gently cleaned and dried with paper towel. After one hour of drying, compressive strength, splitting tensile strength and flexural strength tests were performed. In addition, visual inspection including scaling, chipping and crack patterns for all the specimens were conducted.

Micro-structural Analysis

Micro-structural analysis was conducted on thin polished sections and small fragments from selected UHPC specimens using a Hitachi S-4500 scanning electron microscope (SEM). The pore size distribution was determined using a Micromeritics AutoPore IV 9500 Series porosimeter allowing a range of pressures from 0 to 414 MPa (0–60,000 psi).

RESULTS AND DISCUSSION

Fresh UHPC Properties

It was observed that the flowability of UHPC decreased marginally with increased fiber length and dosage (Table 4), as expected. For instance, 13% reduction in UHPC flowability was observed for mixtures incorporating 6% by mixture volume of 16 mm (0.62 in) fibers compared to that of the mixture without fiber addition. The temperature of the freshly mixed UHPC mixtures ranged between 24 and 28 °C (75 and 82 °F), in agreement with previous studies [25] and [34].

Table 4: Flowability, compressive strength and modulus of elasticity results

Mixture	Steel fiber Length (mm)	Dosage (%)	Flow[a] (mm)	Compressive strength				Modulus of elasticity			
				7 d (MPa)	14 d (MPa)	28 d (MPa)	56 d (MPa)	7 d (GPa)	14 d (GPa)	28 d (GPa)	56 d (GPa)
1	–	–	810	131	142	151	160	35.6	39.9	42.6	45.9
2	8	1	800	135	147	156	165	35.2	39.3	42.1	45.3
3	8	3	760	142	154	164	174	35.9	39.6	42.5	45.1
4	8	6	740	148	160	171	181	35.6	39.1	42.8	45.8
5	12	1	790	135	149	158	165	35.8	39.4	42.3	45.5
6	12	3	750	141	155	166	176	35.1	39.8	42.8	46.1
7	12	6	730	147	159	173	184	34.9	39.2	42.3	45.4
8	16	1	780	137	148	159	167	35.6	40.1	43.4	45.7
9	16	3	730	143	152	165	175	35.1	39.5	42.1	46.2
10	16	6	705	149	163	170	185	35.8	39.2	42.9	45.3

[a] Flow diameter; d = days; 1 mm = 0.04 in; 1 MPa = 0.145 ksi; 1 GPa = 145 ksi.

Compressive Strength and Modulus of Elasticity

Fig. 2(a) shows typical 28 days compressive stress–strain curve for UHPC incorporating 8 mm (0.31 in) steel fibers at various dosages (1%, 3% and 6% by mixture volume). The initial slope of stress–strain curves for UHPC mixtures with various steel fiber dosages were comparable. Therefore, the compressive stress-strain curves were horizontally offset (Fig. 2(a)) in order to avoid their overlap, similar to previous study [25]. The mixture without steel fiber showed a brittle failure with an abrupt drop in load carrying capacity. However, mixtures incorporating steel fibers exhibited somewhat more ductile behavior with steadier drop in load carrying capacity (Fig. 2(a)). This was attributed to restricting lateral expansion by steel fibers, leading to higher tolerance for axial deformation [25]. The slope of the descending branch of the compressive stress–strain curves (Fig. 2(a)) decreased with higher dosage of steel fibers.

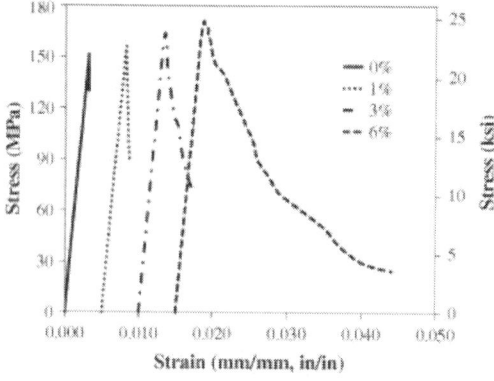

(a) Typical compressive stress-strain curve for UHPC specimens

(b) Cylindrical fiber-reinforced UHPC specimens remained nearly intact after compressive strength testing

Figure 2: Compressive strength behavior of UHPC specimens.

Table 4 shows the compressive strength results for the various UHPC mixtures. The 7 days compressive strength and modulus of elasticity were 131 MPa (19 ksi) and 35.6 GPa (5163 ksi) respectively, for the UHPC mixture without fibers. However, after 28 days, the compressive strength and modulus of elasticity increased to 151 MPa (22 ksi) and 42.6 GPa (6179 ksi), respectively. This progression of UHPC compressive strength and modulus of elasticity continued at later age. For instance, an increase in compressive strength and modulus of elasticity by 6% and 8%, respectively, was observed at 56 days compared to that at 28 days. This was attributed to hydration reaction of pozzolanic cementitious materials at later ages [56].

An increase in compressive strength of 3%, 9% and 13% was observed for mixtures incorporating 1%, 3% and 6% of 8 mm (0.31 in) steel fibers, respectively, compared to that of the mixture without steel fibers. This increase in compressive strength due to steel fiber addition is in agreement with other studies [9],[15] and [57]. The homogenously distributed fibers restrict the internal material deterioration and crack propagation by absorbing the developed stresses at the fiber's tip and consequently enhanced compressive strength [26]. It should be understood that fiber reinforcement is primarily used to enhance the tensile behavior and toughness characteristics of cementitious systems, and not to increase compressive strength. However, since very high compressive strength is a key feature of UHPC, the improved effect of fibers on compressive strength merits further investigation.

No significant effect of the fiber length on compressive strength was observed (Table 4). For instance, increasing the fiber length from 8 mm (0.31 in) to 16 mm (0.62 in) had induced difference less than 2%, regardless of the fiber dosage.

Furthermore, fiber length and dosage had a minimum effect on the modulus of elasticity of UHPC (i.e. <2%). It was observed that the cylinder specimens without fibers were severely damaged through a sudden and explosive behavior, similar to previous work [37]. However, specimens incorporating steel fibers did not show any splitting or breakage into pieces after failure (Fig. 2(b)). This was ascribed to the effect of the fibers, which did not allow the concrete to explode or break into pieces [56].

Splitting Tensile Strength

As expected, the splitting tensile strength increased with the curing time (Table 5). The splitting tensile strength of UHPC was significantly affected by the steel fiber addition (Table 5). For instance, a 48% increase in the 28-days splitting tensile strength was observed for the mixture incorporating 1% of the 16 mm (0.62 in) steel fiber compared to that of the control mixture without fibers. At a higher fiber dosage of 6%, the 28 days splitting tensile strength was approximately 4 times greater than that of the control mixture. Furthermore, it was observed that the fiber length considerably influenced the splitting tensile strength. For example, the mixture incorporating 3% of the 8 mm (0.31 in) steel fiber showed 16% increase in the 28 days splitting tensile strength compared to that of a similar mixture incorporating the same dosage of the 16 mm (0.62) steel fiber. It was observed that cylinder specimens including steel fibers did not split into two parts due to crack bridging action of fibers and the strong bond between steel fibers and the concrete matrix.

Table 5: Splitting tensile strength of UHPC

Mixture	Steel fiber		Tensile strength			
	Length (mm)	Dosage (%)	7 days (MPa)	14 days (MPa)	28 days (MPa)	56 days (MPa)
1	–	–	8.1	8.9	9.4	10.1
2	8	1	16	16.9	17.6	18.3
3		3	20.2	21.1	21.9	22.1
4		6	38.2	39.1	39.8	40.1
5		1	14.6	15.2	15.6	16.2
6	12	3	18.7	19.1	19.7	20.2
7		6	35	35.5	36.2	37.1
8		1	12.8	13.5	13.9	14.5
9	16	3	17.9	18.1	18.9	19.5
10		6	32.7	33.1	33.8	34.6

1 mm = 0.04 in; 1 MPa = 0.145 ksi.

Flexural Strength

The flexural behavior of the UHPC beam specimens incorporating steel fibers and that of the control specimens without fibers were significantly different. For instance, a 37% increase in the peak load was observed for the beam specimens incorporating 1% of the 16 mm (0.62 in) fibers compared to that of the control specimens (Table 6). Furthermore, the beam specimens without fiber addition had a brittle failure and sudden drop in load carrying capacity after reaching the peak load.

Table 6: Flexural properties of UHPC mixtures

Mixture	Steel fiber Length (mm)	Steel fiber Dosage (%)	First crack load (kN)	Cracking deflection (mm)	Peak load (kN)	Peak deflection (mm)	Toughness (kN-mm) Cracking	Toughness (kN-mm) Failure
1	–	–	20.07	0.45	20.45	0.47	9.03	5
2	8	1	28.68	0.81	32.52	1.35	23.23	52
3	8	3	45.87	0.95	54.32	1.65	43.6	108
4	8	6	76.32	1.1	88.24	2.08	83.95	181
5	12	1	26.76	0.68	29.87	1.08	18.2	58
6	12	3	40.43	0.81	49.12	1.4	32.75	120
7	12	6	64.34	0.94	77.54	1.8	60.5	201
8	16	1	24.76	0.61	27.97	0.95	15.1	63
9	16	3	35.38	0.75	43.26	1.26	26.53	130
10	16	6	53.76	0.88	66.13	1.63	47.3	221

1 mm = 0.04 in; 1 kN = 0.224 kip; 1 kN-mm = 0.009 kip-in.

The first crack load is the load at which the initial linear elastic slope of the load-displacement plot ends, while the peak load represents the maximum load of the load-displacement curve. Furthermore, the failure point corresponds to the point just before the complete failure or breakage of the segment into two pieces.

Load-deflection Behavior

The initial linear elastic phase (Portion 0C in Fig. 3(a)) for the UHPC beams incorporating short fibers (8 mm (0.31 in)) and long fibers (16 mm (0.62 in)) was comparable for all UHPC specimens. However, the short 8 mm (0.31 in) fibers better improved the strain hardening behavior (Portion CP in Fig. 3(a)). It is believed that at similar fiber dosage, short fibers better captured the growth of micro-cracks, yielding more stable micro-crack formation [34] and [58].

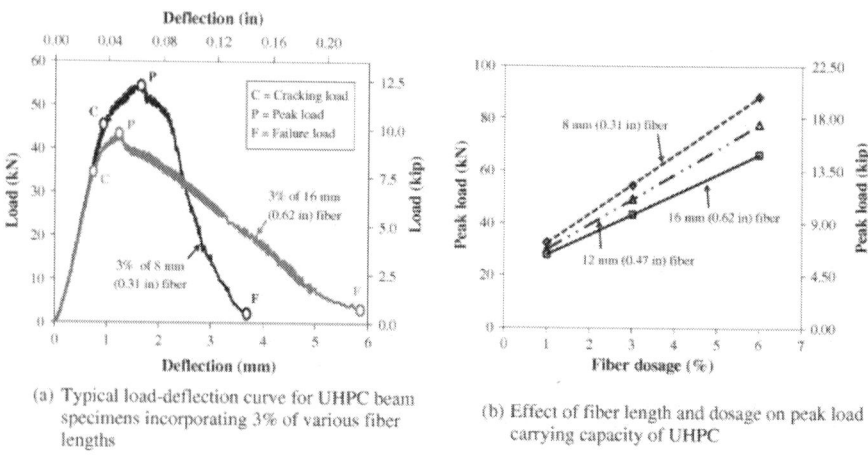

(a) Typical load-deflection curve for UHPC beam specimens incorporating 3% of various fiber lengths

(b) Effect of fiber length and dosage on peak load carrying capacity of UHPC

Figure 3: Flexural testing results of UHPC beam specimens.

UHPC beam specimens incorporating short fibers exhibited larger cracking and ultimate deflections compared to that of beam specimens with long fibers (Fig. 3(a)). For example, an increase in cracking and ultimate deflections by 27% and 31%, respectively was observed for beams incorporating 3% of 8 mm (0.31 in) fibers compared to that of similar specimens with 16 mm (0.62 in) fibers. It seems that the development of multiple micro-cracks delayed the formation of macro-cracks and led to higher first crack and peak load carrying capacities for the beam specimens incorporating short fibers. For instance, beam specimens incorporating 3% of 8 mm (0.31 in) fibers exhibited 30% and 26% higher first crack and peak loads respectively, compared to that of the beam specimens with 3% of 16 mm (0.62 in) fibers.

The difference between the behavior of mixtures incorporating short and long fibers increased at higher fiber dosage. For example, beams incorporating 6% of the 8 mm (0.31 in) fiber exhibited 42% and 34% increase in first crack and peak loads, respectively, compared to that of the similar beams with 16 mm (0.62 in) fibers. Higher cracking and peak load capacities of beams incorporating short fibers was attributed to the fiber pinching force applied at the crack tips, which suppressed the propagation and development of cracks [34], [56] and [58]. Furthermore, at the same dosage, more short fibers are available for crack bridging compared to that of the longer fibers, thus increasing the load carrying capacity [58]. The increased crack load capacity of UHPC beams incorporating short fibers is beneficial against the penetration of aggressive materials responsible for corrosion of steel fibers.

The length of the softening phase for short fibers (Portion PF in Fig. 3(a)) was smaller and exhibited a steeper drop in load carrying capacity since short fibers can pull-out or de-bonded relatively easier from the matrix because of smaller embedment length. Conversely, the beam specimens incorporating longer fibers (16 mm (0.62 in)) exhibited improved strain softening (post peak) behavior (Fig. 3(a)). Long fibers especially at low fiber concentration (i.e. 1% or 3%) have relatively larger inter-fiber spacing; therefore, they became effective only after the development of macro-cracks. Furthermore, the load-deflection curves for beam specimens incorporating the long fibers exhibited steadier drop in load carrying capacity after the peak load compared to steeper load drop in the case of short fibers. This is due to the increased energy required for de-bonding or pull-out of the longer steel fibers.

Moreover, it was observed that the first crack load and peak load increased with higher fiber dosage. For instance, 51% and 64% increase in the first crack and peak loads were observed for beam specimens incorporating 3% of the 12 mm (0.47 in) steel fibers, respectively, compared to that of similar beam specimens with 1% fiber dosage. At higher dosage, fibers are closely spaced, providing more effective and localized control for the growth of micro-cracks into macro-cracks [56], [58], [59] and [60]. This delay of macro-cracks propagation increases the ultimate load carrying capacity.

It was observed that the initial stiffness of the load-deflection curves of UHPC beams was not affected by the fiber dosage (Fig. 3(a)). The deflection at crack and ultimate loads increased with higher fiber dosage. For example, 38% and 67% increase in deflection at crack and ultimate loads, respectively, were observed for beam specimens incorporating 6% of the 12 mm (0.47 in) fibers compared to that of similar specimen with 1% of the 12 mm (0.47 in) fibers (Table 6).

Toughness
The toughness of UHPC beams incorporating various fiber lengths and dosages was calculated by estimating the area under the load-deflection

curve up to the cracking and failure points. It was observed that beam specimens incorporating the long 16 mm (0.62 in) fibers exhibited higher toughness compared to that of identical beams with short fibers. For instance, a 20% increase in toughness (up to failure point) was observed for the beams incorporating 3% of the 16 mm (0.62 in) fibers compared to that of similar beams with the 8 mm (0.31) fibers. This can be ascribed to improved post-peak (strain softening) behavior of the beam specimens incorporating the longer fibers. Moreover, the steel fiber dosage considerably influenced the toughness of UHPC beam specimens. For instance, beam specimens incorporating 3% and 6% of the 16 mm (0.62 in) fibers showed approximately 2.0 and 3.5 times higher toughness (up to failure point) than that of beam specimen with 1%, respectively, regardless of the fiber length (Table 6).

Crack and Failure Pattern

As expected, no cracks were observed during the initial linear elastic loading (Portion 0C in Fig. 3(a)). At the end of the linear elastic phase, cracks were initiated in the form of various hairline micro-cracks, barely visible with the naked eye. New micro-cracks in between the previously developed cracks were observed as the load increased. Most of these cracks continued to propagate in both directions. Closely spaced cracks developed perpendicular to the flexural tensile stresses on the beam specimens (Fig. 4(a)), demonstrating the ability of fiber-reinforced UHPC to redistribute the stresses through several micro-cracks until fiber pull-out or fracture occur [56].

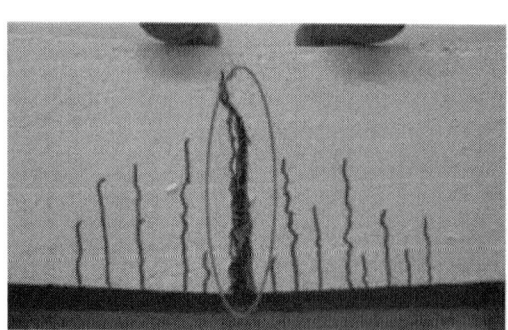
(a) Localization of strain at a single crack in UHPC beam specimen leading to failure

(b) Failure surface of beam specimen

Figure 4: Flexural failure of UHPC beam specimen.

RESULTS AND DISCUSSION

Fiber pull-out from the matrix began as the load carried by individual fibers exceeded the ability of the UHPC matrix to hold the fiber. The pulled-out of fibers increased the load in neighboring fibers across cracks, pulling-out more fibers from the matrix. As the load increased on the beam specimens, stresses continued to increase with more formation of cracks [59] and [60].

At the ultimate load, the highly stressed fibers started pulling-out from the matrix at one particular cross-section, which represented the weakest point (Fig. 4(a)). This was attributed to the localization of maximum strain higher than the strain capacity of concrete matrix [37]. Afterwards, the softening branch began with steadier or steeper drop in the load carrying capacity depending on the fiber length and dosage. The failure of UHPC beam specimens was characterized by a local failure of the fiber–matrix bond with significant increase in crack width (Table 7).

Table 7: Measured crack width during flexural test of UHPC beam specimens

Mixture	Steel fiber		Average crack width at		
	Length (mm)	Dosage (%)	First crack load (mm)	Peak load (mm)	Failure point (mm)
2	8	1	0.22	0.5	14
3		3	0.08	0.25	9
4		6	0.04	0.15	7
5	12	1	0.32	0.6	18
6		3	0.12	0.3	13
7		6	0.06	0.18	10
8	16	1	0.48	0.73	22
9		3	0.2	0.36	16
10		6	0.1	0.22	11

1 mm = 0.04 in.

The crack width after first crack, at peak load and failure load decreased with higher dosage of fibers (Table 7). For instance, the average crack width after first crack, at peak load and failure load decreased by 81%, 71% and 50% for beam specimens incorporating 6% of 8 mm (0.31 in) fibers compared to that of beam specimens with 1% of 8 mm (0.31 in) fibers. Furthermore, a decrease in crack width was observed for beam specimens incorporating short fibers compared to that similar beam specimens made with a similar dosage of the long fibers (Table 7).

The average spacing between cracks decreased for beam specimens incorporating short 8 mm (0.31 in) fibers in comparison with that in beam specimens made with longer 16 mm (0.62 in) fibers as shown in Fig. 5. A similar decreasing trend in average crack spacing was observed at higher fiber dosage compared to that at lower fiber dosage (Fig. 5).

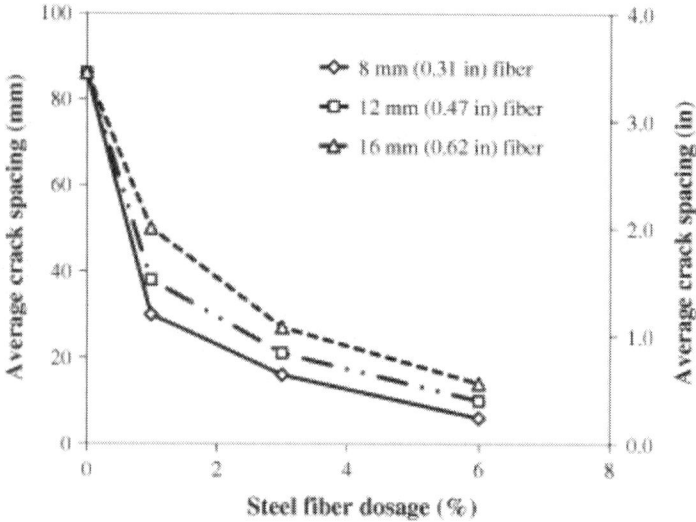

Figure 5: Average crack spacing for UHPC beam specimens incorporating various fiber dosages.

Visual inspection of the failure surfaces of beam specimens revealed that fiber pull-out was dominant compared to fiber fracture (Fig. 4(b)). The fracture surfaces of beam specimens were also analyzed under scanning electron microscope (SEM), indicating very dense micro-structure (Fig. 6(a)) and interface between aggregates and the cementitious matrix (Fig. 6(b and c)). This confirms the ability of fiber reinforced UHPC to transfer stresses through cracks and achieve enhanced toughness [26] and [56].

RESULTS AND DISCUSSION

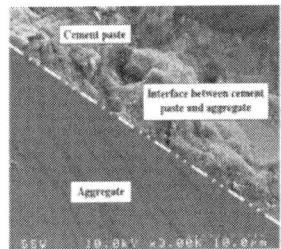
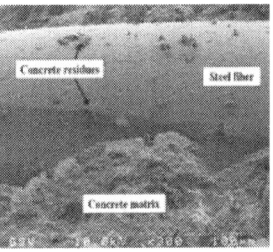

(a) Strong microstructure (b) Interface between aggregate and matrix (c) Fiber-cementitious intimate contact

Figure 6: SEM images of UHPC specimen.

Permeable Voids

The volume of permeable voids (VPV) in UHPC was less than 4% (Table 8), which is lower than that in high-strength concrete (around 14%) [61]. This is attributed to the dense micro-structure of UHPC and the improved interfacial zone between the cementitious matrix and aggregates (Fig. 6). There was no significant influence of steel fiber length on the VPV (Table 8). However, the steel fiber dosage in UHPC mixtures reduces the VPV leading to improved durability properties. For instance, 12% and 36% decrease in VPV was observed for the mixtures incorporating 3% and 6% of steel fibers, respectively compared to that of the control mixture without fibers. The steel fiber addition disturbs the continuity of capillary pores and reduced the VPV [61] and [62]. This was confirmed through mercury intrusion porosimetry (MIP) analysis (Fig. 7). For instance, UHPC mixtures incorporating 3% of the 8 mm (0.31 in) steel fiber had 17% reduction in the total porosity compared to that of the control UHPC mixture without fiber addition. The lower capillary pore volume in fiber-reinforced UHPC (Fig. 7) can reduce the capillary suction and intrusion of aggressive solutions (e.g. chlorides), leading to improved durability properties [63].

Table 8: VPV and sorptivity coefficient results for UHPC mixtures

Mixture	Steel fiber		VPV[a] (%)	Initial sorptivity (kg/m²/h^0.5)	Secondary sorptivity (kg/m²/h^0.5)
	Length (mm)	Dosage (%)			
1	–	–	3.52	0.0631	0.0432
2	8	1	3.28	0.0582	0.0382
3	8	3	3.11	0.0534	0.031
4	8	6	2.25	0.047	0.0271
5	12	1	3.3	0.0589	0.0387
6	12	3	3.1	0.054	0.0314
7	12	6	2.27	0.0477	0.0275
8	16	1	3.31	0.0591	0.039
9	16	3	3.13	0.0544	0.0316
10	16	6	2.29	0.0479	0.0277

[a] VPV = volume of permeable voids.

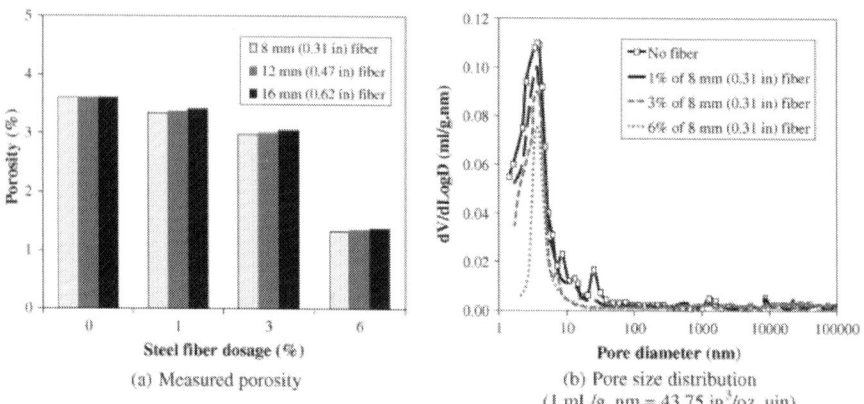

(a) Measured porosity

(b) Pore size distribution (1 mL/g, nm = 43.75 in³/oz, μin)

Figure. 7: Measured porosity for UHPC mixtures incorporating various fiber lengths and dosages.

Sorptivity Voids

Table 8 shows the initial (i.e. measured during the first 6 h of the test) and secondary (i.e. measured after the first 6 h and up to 8 days) sorptivity coefficient results for UHPC specimens. The initial and secondary sorptivity values for the tested UHPC specimens without fibers were 0.0631 and 0.0432 kg/m²/h^{0.5} respectively, which is lower than that of high-performance concrete and normal-strength concrete.

It was observed that the fiber length had no appreciable effect on the sorptivity coefficients. However, the initial and secondary sorptivity coefficients decreased by 7% and 12%, respectively, due to the addition of 1% of the 8 mm (0.31) steel fibers. This difference was more evident at higher dosage of steel fibers. For instance, the initial and secondary sorptivity coefficients decreased by 26% and 37%, respectively for UHPC mixtures incorporating 6% of the 8 mm (0.31 in) steel fibers compared to that of the control mixture without fibers. This indicates that the addition of steel fibers resulted in relatively less connected pores, leading to denser microstructure and consequently enhanced durability properties. This was verified through MIP analysis (explained earlier). Fig. 8(a) shows a typical sorptivity curve for the UHPC mixture incorporating 3% of the 8 mm (0.31 in) steel fibers.

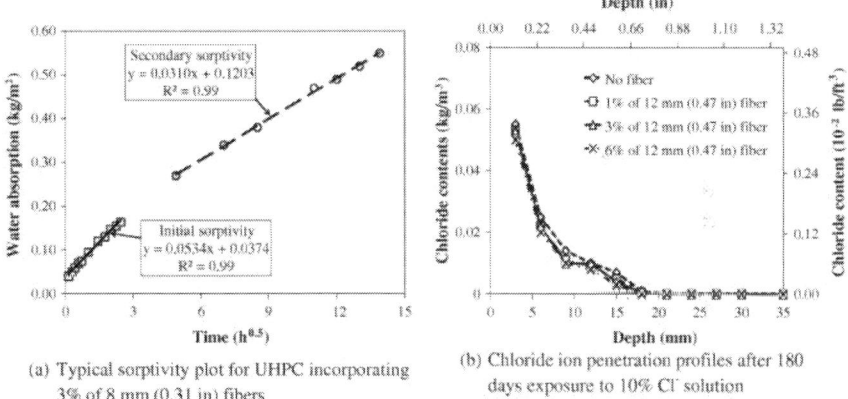

(a) Typical sorptivity plot for UHPC incorporating 3% of 8 mm (0.31 in) fibers

(b) Chloride ion penetration profiles after 180 days exposure to 10% Cl⁻ solution

Figure 8: Sorptivity plot and chloride ions penetration into various UHPC specimens.

Electrical Resistance

The short and dispersed steel fibers did not cause a short circuiting problem during the rapid chloride ion penetrability (RCPT) testing of UHPC specimens. There was also no significant rise in temperature due to electrical heating of the tested UHPC specimens. This is in agreement with previous studies [1],[33] and [64].

RCPT test results for UHPC specimens are summarized in Table 9. All UHPC specimens showed very high resistance to chloride transport and exhibited coulomb values less than 100, indicating negligible ASTM C1202 chloride ion penetrability [42]. Similar results were reported in previous studies. For instance, Graybeal [1] reported rapid chloride ions penetrability values less than 40 coulombs after 28 days for steam cured specimens. Another study conducted by Ahlborn et al. [33] showed Coulomb values less than 100 for both air-cured and heat-treated specimens, which is very low even compared to that of high-performance concrete (around 216 Coulombs) and normal strength concrete (1736 Coulombs) [18], indicating superior resistance of UHPC against chloride ions penetration.

Table 9: Rapid chloride ion penetrability of UHPC mixtures under various chloride ion exposures

Mixture	Steel fiber		Average coulombs passed					
	Length mm (in)	Dosage (%)	3.00%		3.50%		10.00%	
			28 days	56 days	28 days	56 days	28 days	56 days
1	–	–	71	70	72	71	80	78
2	8 (0.31)	1	60	60	59	60	65	63
3		3	45	43	47	44	50	49
4		6	36	35	38	33	42	40
5	12 (0.47)	1	60	58	59	60	67	65
6		3	47	48	49	48	52	53
7		6	38	39	40	40	40	40
8	16 (0.62)	1	65	54	66	65	72	70
9		3	52	53	54	52	56	54
10		6	43	40	45	42	48	46

In this study, RCPT results of UHPC at 28 and 56 days were comparable. The chloride concentration (3%, 3.5% and 10%) had minimum effect on the number of coulombs passed. Furthermore, no significant effect of the fiber length on RCPT results was observed. However, the steel fiber dosage had a significant effect on the passed coulomb values. For instance, the UHPC mixtures incorporating 3% and 6% of the 8 mm (0.62 in) steel fibers exhibited 27 and 35 lower coulomb values respectively, compared to that of the control mixture without fiber addition. This can be ascribed to the role of steel fibers, which can restrict the formation and growth of plastic and drying shrinkage cracks, resulting in decreased penetrability[62] and [64]. MIP analysis confirmed that the addition of fibers reduced the porosity in UHPC, leading to improved durability properties (Fig. 7). Furthermore, SEM analysis (Fig. 6) showed a dense fiber–matrix and matrix–aggregate interface, indicating lower penetrability properties due to reduced porosity.

Chloride Ions Penetration

Fig. 8(b) illustrates the chloride penetration in UHPC specimens versus the depth from the ponding solution. It can be observed that the chloride ions penetration in UHPC was very limited. The maximum chloride content in UHPC specimens was 0.055 kg/m^3 (0.0034 lb/ft^3), which is much lower than the corrosion threshold value of 0.60 kg/m^3 (0.037 lb/ft^3) [65]. Similar results were reported in previous studies for both air-cured and heat-treated specimens [1] and [28]. It was observed that the steel fiber length and dosage had minimal effect on the chloride ions penetration of UHPC (Fig. 8(b)).

No significant effect of the testing age (i.e. 90 and 180 days) and the chloride ion concentration (i.e. 3.5% and 10%) was observed on the chloride penetration into UHPC. It was observed that fibers on the UHPC specimen surface were corroded after chloride ions exposure (Fig. 9(a)). A thin slice at a depth of 3 mm (0.12 in) from the specimen surface was prepared in order to examine signs of corrosion on the embedded fibers. However, there was no evidence of corrosion on the embedded fibers at this depth (Fig. 9(b)). This indicates that the corrosion of fibers was limited to the surface. This was attributed to the very low porosity of UHPC specimens, which does not allow the penetration of chloride ions, moisture and oxygen required for the onset of corrosion. Fig. 10 shows optical and SEM images of a surface corroded fiber. It was observed that the embedded fibers were corroded only to a depth of 1 mm (0.04 in) (Fig. 10(b)). Energy dispersive X-ray (EDX) analysis (Fig. 10(d)) showed higher peaks of iron and oxygen, which also confirms the corrosion activity at the surface fibers.

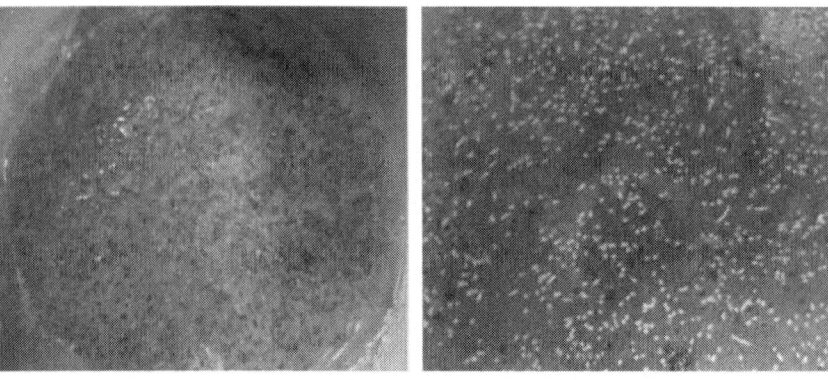

(a) Surface corrosion of steel fibers after 30 days of salt ponding

(b) No corrosion at 3 mm (0.12 in) depth

Figure 9: Corrosion of surface steel fibers (salt ponding specimen incorporating 6% by mixture volume of 8 mm (0.31 in) steel fibers).

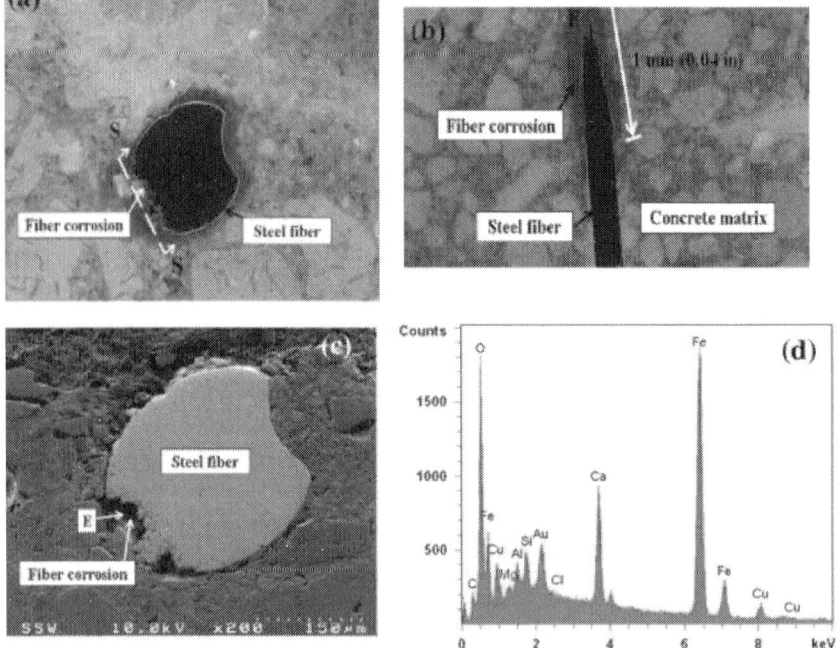

Figure 10: Optical and SEM images of surface fiber: (a) corrosion of surface fiber, (b) penetration of corrosion in the embedded fiber (Section at S–S), (c) SEM image, and (d) EDX analysis 'E'.

Effects of Chloride Ions Exposure on Mechanical Properties

A white deposit of salt material was observed on the surface of UHPC specimens after chloride ions exposure (Fig. 11). However, no surface degradation on UHPC specimens was detected (Fig. 11(b)). Table 10 shows the relative compressive, splitting tensile and flexural strengths of UHPC specimens without fibers and with 3% of the 12 mm (0.47 in) steel fibers after 1, 3, 5 and 6 months of exposure to various chloride ion solutions (i.e. 3.5% and 10%). No degradation of UHPC mechanical properties was observed due to chloride ions exposure. This indicates the superior resistance of UHPC against chloride ions exposure, which should result into more durable and sustainable concrete structures.

(a) UHPC specimen after 180 days of salt ponding

(b) UHPC specimens after chloride ions immersion test during drying cycle

Figure 11: Deposition of salt material on the surface of UHPC specimens.

Table 10: Mechanical properties of UHPC specimens after various chloride exposures

UHPFRC mixture	Exposure solution (NaCl)	Exposure duration (Months)	Relative strength factor (R)		
			Compressive strength	Splitting tensile strength	Flexural strength
No fiber	3.50%	1	1.04	1.06	1.03
		3	1.07	1.08	1.07
		5	1.1	1.11	1.08
		6	1.12	1.1	1.1
	10.00%	1	1.03	1.07	1.04
		3	1.06	1.06	1.07
		5	1.09	1.1	1.1
		6	1.11	1.09	1.1
3% by mixture volume of 12 mm (0.62 in)	3.50%	1	1.04	1.04	1.04
		3	1.09	1.1	1.12
		5	1.1	1.08	1.1
		6	1.11	1.1	1.09
	10.00%	1	1.03	1.04	1.05
		3	1.07	1.09	1.1
		5	1.1	1.08	1.11
		6	1.09	1.1	1.12

$R = (R_{Cl}/R_{28})$; R_{28} is the strength after 28 days, R_{Cl} is the strength measured after exposure period to chloride ions.

CONCLUSIONS

This study explored the effects of the steel fiber length and dosage on the mechanical and durability properties of UHPC. The following specific conclusions can be drawn from this study.

(1) The compressive strength of UHPC slightly increased with steel fiber addition, while the fiber length had insignificant effect on compressive strength. It was also observed that fiber addition improved the failure pattern from sudden explosive to ductile behavior.

(2) A significant increase in UHPC splitting tensile and flexural strengths was observed with higher dosage of steel fibers. Moreover, the fiber length considerably influenced the peak load carrying capacity and load-deflection behavior. For instance, UHPC mixtures incorporating short steel fibers achieved higher peak load capacity and exhibited enhanced strain hardening behavior compared to that of the mixture with a similar dosage of longer fibers.
(3) UHPC exhibited improved durability properties owing to its very low porosity and denser micro-structure, which was confirmed through MIP and SEM analyses. There was no significant effect of the fiber length on the durability properties of UHPC. However, at higher fiber dosage, UHPC mixtures exhibited relatively improved durability properties. No deterioration of UHPC mechanical properties was observed after various severe exposures to chloride ions.

REFERENCES

1. Graybeal B. Material property characterization of ultra-high performance concrete. FHWA-HRT-06-103, U.S. Department of Transportation; 2006. p. 176.
2. Larrard F, Sedran T. Optimization of ultra-high performance concrete by the use of a packing model. Cem Concr Res 1994;24:997–1009.
3. Richard P, Cheyrezy M. Composition of reactive powder concretes. Cem Concr Res 1995;25:1501–11.
4. Habel K, Viviani M, Denarie E, Bruehwiler E. Development of the mechanical properties of an ultra-high performance fiber reinforced concrete (UHPFRC). Cem Concr Res 2006;36:1362–70.
5. Schmidt M, Fehling E. Ultra-high-performance concrete: research, development and application in Europe. 7th International symposium on utilization of high strength high performance concrete, vol. 1; 2005. p. 51–77.
6. Vernet P. Ultra-durable concretes: structure at the micro- and nano-scale. Mater Res Soc 2004;29(5):324–7.
7. Wille K, Naaman A, Montesinos G. Ultra-high performance concrete with compressive strength exceeding 150 MPa (22 ksi): a simpler way. ACI Mater J 2011;108(1):46–54.
8. Yazici H. The effect of curing conditions on compressive strength of ultra high strength concrete with high volume mineral admixtures. Build Environ 2006;42(5):2083–9.
9. Soutsos M, Millard S, Karaiskos K. Mix design, mechanical properties, and impact resistance of reactive powder concrete (RPC). International workshop

on high performance fibre-reinforced cementitious composites in structural applications; 2005. p. 549–60.
10. Tue N, Orgass M, Ma J. Influence of addition method of superplasticizer on the properties of fresh UHPC. Proceedings of the 2nd international symposium on ultra high performance concrete, Kassel (Germany); 2008. p. 93–100.
11. Ma J, Schneider H. Properties of ultra-high performance concrete. Leipzig Annual Civil Engineering Report (LACER). 2002; 7: 25–32.
12. de Larrard F, Sedran T. Optimization of ultra-high-performance concrete by the use of a packing model. Cem Concr Res 1994;24(6):997–1009.
13. Bayard O, Ple O. Fracture mechanics of reactive powder concrete: material modeling and experimental investigations. Eng Fract Mech 2003;70(7–8):839–51.
14. Graybeal B, Davis M. Cylinder or cube: strength testing of 80–200 MPa (11.6– 29 ksi) ultra high performance fibre-reinforced concrete. ACI Mater J 2008;105(6):603–9.
15. Bonneau O, Lachemi M, Dallaire E, Dugat J, Aitcin P. Mechanical properties and durability of two industrial reactive powder concretes. ACI Mater J 1997;94(4):286–90.
16. Xing F, Huang D, Cao L, Deng L. Study on preparation technique for low-cost green reactive powder concrete. Key Eng Mater 2006;302–303:405–10.
17. Reda M, Shrive G, Gillott E. Microstructural investigation of innovative UHPC. Cem Concr Res 1999;29(3):323–9.
18. [Schmidt M, Fehling E, Teichmann T, Bunje K, Bornemann R. Ultra-high performance concrete: perspective for the precast concrete industry. Concr Pre-casting Plant Technol 2003;69(3):16–29.
19. Herold G, Muller H, Measurement of porosity of ultra-high strength fibrereinforced concrete. Proceedings of the international symposium on ultra-high performance concrete. Kassel (Germany); 2004. p. 685–94.
20. Jun P, Taek K, Tae K, Wook K. Influence of the ingredients on the compressive strength of UHPC as a fundamental study to optimize the mixing proportion. Proceedings of the 2nd international symposium on ultra high performance concrete. Kassel (Germany); 2008. p. 105–12.
21. Steil T, Karihaloo B, Fahling E. Effect of casting direction on the mechanical properties of CARDIFRC. Proceedings of the international symposium on ultra high performance concrete. Kassel (Germany); 2004. p. 481–93.
22. Lappa E, Braam C, Walraven J. Static and fatigue bending tests of UHPC. Proceedings of the international symposium on ultra high performance concrete. Kassel (Germany); 2004. p. 449–58.

23. Wille K, Parra-Montesinos G. Effect of beam size, casting method, and support conditions on flexural behaviour of ultra-high performance fibre-reinforced concrete. ACI Mater J 2012;109(3):379–88.
24. Pansuk W, Sato Y, Sato H, Shionaga R. Tensile behaviour and fibre orientation of UHPC. Proceedings of the 2nd international symposium on ultra high performance concrete. Kassel (Germany); 2008. p. 161–8.
25. Kazemi S, Lubell A. Influence of specimen size and fibre content on mechanical properties of ultra-high performance fibre-reinforced concrete. ACI Mater J 2012;109(6):675–84.
26. Orgass M, Klug Y. Fibre-reinforced ultra-high strength concretes. Proceedings of the international symposium on ultra high performance concrete. Kassel (Germany); 2004. p. 637–48.
27. Thomas M, Green B, O'Neal E, Perry V, Hayman S, Hossack A. Marine performance of UHPC at Treat Island. Proceedings of the 3rd international symposium on UHPC and nanotechnology for high performance construction materials. Kassel (Germany); 2012. p. 365–70.
28. Scheydt J, Muller H, Herold G. Long term behaviour of ultra high performance concrete under the attack of chlorides and aggressive waters. Proceedings of the 2nd international symposium on ultra high performance concrete. Kassel (Germany); 2008. p. 231–8.
29. [2Adeline R, Cheyrezy M. The Sherbrooke Footbridge: the first RPC structure. La Technique Française du Béton, AFPC-AFREM, XIII Congrès de La Fib, Amsterdam; 1998. p. 343–8.
30. Tang C. High performance concrete – past, present and future. Proceedings of the international symposium on UHPC. Kassel (Germany); 2004. p. 3–9.
31. Behloul M, Causse G, Etienne D. Ductal footbridge in Seoul. First Fib Congress. Osaka (Japan); 2002. p. 13–9.
32. Buitelaar P, Ultra high performance concrete: development and applications during 25 years. Plenary Session: Proceedings of the 1st international symposium on ultra-high performance concrete. Kassel (Germany); 2004. p. 28.
33. Ahlborn T, Peuse E, Misson D, Gilbertson C. Durability and strength characterization of ultra-high performance concrete under variable curing regimes. Proceedings of the 2nd international symposium on ultra high performance concrete. Kassel (Germany); 2008. p. 197–204.
34. Magureanu C, Sosa I, Negrutiu C, Heghes B. Mechanical properties and durability of ultra-high performance concrete. ACI Mater J 2012; 109:177–83.

35. Recchia M, Smith C, Tully J. Precast concrete tunnel lining segments-EglintonScarborough cross town light rail transit contract drives long term thinking on design and manufacture. Tunneling Association of Canada Conference. Montreal; 2012. p. 5.
36. ASTM C39/C39M. Standard test method for compressive strength of cylindrical concrete specimens. West Conshohocken (PA): American Society for Testing and Materials; 2010. p. 7.
37. Hassan A, Jones S, Mahmud G. Experimental test methods to determine the uniaxial tensile and compressive behavior of ultra-high performance fiber reinforced concrete (UHPFRC). Constr Build Mater 2012;37:874–82.
38. ASTM C496/C496M. Standard test method for splitting tensile strength of cylindrical concrete specimens. West Conshohocken (PA): American Society for Testing and Materials; 2011. p. 5.
39. ASTM C1609/C1609M. Standard test method for flexural performance of fiberreinforced concrete (Using Beam with Third Point Loading). West Conshohocken (PA): American Society for Testing and Materials; 2010. p. 9.
40. ASTM C642. Standard test method for density, absorption, and voids in hardened concrete. West Conshohocken (PA): American Society for Testing and Materials; 2006. p. 3.
41. ASTM C1585. Standard test method for measurement of rate of absorption of water by hydraulic cement concretes. West Conshohocken (PA): American Society for Testing and Materials; 2011. p. 6.
42. ASTM C1202. Standard test method for electrical indication of concrete's ability to resist chloride ion penetration. West Conshohocken (PA): American Society for Testing and Materials; 2010. p. 7.
43. Balouch S, Forth J, Granju J. Surface corrosion of steel Fiber reinforced concrete. Cem Concr Res 2010;40:410–4.
44. Matsumura T, Shirai K, Saegusa T. Verification method for durability of reinforced concrete structures subjected to salt attack under high temperature conditions. Nucl Eng Des 2008;238:1181–8.
45. NT BUILD 492. Chloride Migration Coefficient from Non-Steady State Migration Experiments. NORDTEST, Finland; 1999. p. 8.
46. Guoping L, Fangjian H, Yongxian W. Chloride ion penetration in stressed concrete. J Mater Civ Eng 2011;23:1145–53.
47. AASHTO T259. Standard test method for resistance of concrete to chloride ion penetration. Ann Arbor (MI): American Association of State and Highway Transportation Officials; 2002. p. 4.
48. McGrath P, Hooton R. Re-evaluation of the AASHTO T259 90 day salt ponding test. Cem Concr Res 1999;29:1239–48.

49. ASTM C114. Standard test methods for chemical analysis of hydraulic cement, section 19-chlorides. West Conshohocken (PA): American Society for Testing and Materials; 1997. p. 111–112.
50. FHWA-RD-72-12. Sampling and testing of chloride ion in concrete. Washington (DC): Federal Highway Administration Offices of Research and Development; 1977. p. 28.
51. ASTM D870. Standard practice for testing water resistance of coatings using water immersion. West Conshohocken, (PA USA): American Society for Testing and Materials; 2009. p. 3.
52. ASTM G31-72. Standard practice for laboratory immersion corrosion testing of metals. West Conshohocken, (PA USA): American Society for Testing and Materials; 2004. p. 8.
53. Granju J, Balouch S. Corrosion of steel fiber reinforced concrete from the cracks. Cem Concr Res 2005;35:572–7.
54. Hong K. Cyclic wetting and drying and its effects on chloride ingress in concrete. MASc Thesis, The University of Toronto, Canada; 1998. p. 123.
55. Konin A, Francois R, Arliguie G. Penetration of chlorides in relation to the micro-cracking state into reinforced ordinary and high strength concrete. Mater Struct 1998;31:310–6.
56. El-Dieb A. Mechanical, durability and microstructural characteristics of ultrahigh-strength self-compacting concrete incorporating steel fibers. Mater Des 2009;30:4286–92.
57. Lee P, Chisholm D. Reactive powder concrete. Branz, Study Report No. 146; 2005. p. 29.
58. Sheng J. Micro Fiber reinforced cement based material system: some mechanical and durability considerations. PhD thesis, Universite Laval Quebec; 1995. p. 287.
59. Park S, Kim D, Ryu G, Koh K. Tensile behavior of ultra-high performance hybrid fiber reinforced concrete. Cem Concr Compos 2012;34(2):172–84.
60. Kang S, Lee Y, Park Y, Kim J. Tensile fracture properties of an ultra-high performance fiber reinforced concrete (UHPFRC) with steel fibers. Compos Struct 2010;92(1):61–71.
61. Roque R, Kim N, Kim B, Lopp G. Durability of fiber reinforced concrete in Florida environments. Final Report, Florida Department of Transportation; 2009. p. 254.
62. Banthia N, Bhargava A. Permeability of stressed concrete and role of fiber reinforcement. ACI Mater J 2007;104:77–85.
63. Teichmann T, Schmidt M. Influence of the packing density of fine particles on structure, strength and durability of UHPC. Proceedings of the 1st international symposium on ultra-high performance concrete. Kassel (Germany); 2004. p. 313–23.

64. Vaishali G, Rao H. Strength and permeability characteristics of fiber reinforced high performance concrete with recycled aggregates. Asian J Civil Eng (Building and Housing) 2012;13:55–77.
65. Angst U, Elsener B, Larsen C, Vennesland O. Critical chloride content in reinforced concrete – A review. Cem Concr Res 2009;39(12):1122–38.

CITATION

Safeer Abbas, Ahmed M. Soliman, Moncef L. Nehdi, Exploring mechanical and durability properties of ultra-high performance concrete incorporating various steel fiber lengths and dosages, Construction and Building Materials, Volume 75, 30 January 2015, Pages 429-441, ISSN 0950-0618, http://dx.doi.org/10.1016/j.conbuildmat.2014.11.017.

CHAPTER 8

Seismic Performance Evaluation of Corroded Reinforced Concrete Structures by Using Default and User-Defined Plastic Hinge Properties

Hakan Yalçiner[1] and Khaled Marar[1]

[1]European University of Lefke, Department of Civil Engineering, Mersin,, Turkey

INTRODUCTION

There are several methods exist to define the seismic performance levels of reinforced concrete (*RC*) structures. Among these methods, the nonlinear dynamic and the static analyses in which both methods involve sophisticated computational procedures because of the non-linear behaviour of the *RC* composite materials. In order to simplify these analyses for engineers, different suggested guidelines such as *FEMA-356* (Federal emergency management agency [*FEMA-356*], 2000) and *ATC-40* (Applied Technology Council [ATC-40, 1996]) were prepared to define the plastic hinges properties for *RC* structures in the United States, and thus they have been used by many computer programs (i.e., ETABS [CSI, 2003], SAP2000 [CSI, 2008]) as a default or ready plastic hinge documents. However, there are still contradictions exist in the available literature due to the use of these ready documents in which the buildings are not designed based on the earthquake code of United States. The assessment of seismic performance of structures under future earthquakes is an important problem in earthquake engineering (Abbas, 2011). The use of methods and assumptions to define the seismic performance levels of *RC* buildings become more and more important issue with time dependent effects of corrosion. Moreover, to the knowledge of the author, no any study has been performed up to date, which studies define the possible difference in the time-dependent seismic performance levels of *RC* buildings under the impact of corrosion by using default and user-defined plastic hinge properties.

The primary objectives of this study was to investigate the effects of default hinge properties based on *FEMA-356* (FEMA-356, 2000) and user-defined hinge properties on the time-dependent seismic performance levels of corroded *RC* buildings. An assumed corrosion rate was used to predict the capacity curve of the buildings by using default and user-defined plastic hinge properties as a function of time (t: 25 years, and t: 50 years). Two, four and seven stories of *RC* buildings were considered to represent the effects of default and user-defined hinge properties on story levels. For the modelling of user-defined hinge properties, the time-dependent moment-curvature relationships of structural members were predicted as a function of corrosion rate for two different time periods in order to perform push-over analyses, while default hinge properties were used for the other case based on the ready documents by *FEMA-356* (FEMA-356, 2000). Then, the nonlinear time-history analyses for both corroded and non-corroded buildings were performed by using 20 individual earthquake motion records. Seismic performance levels of non-corroded buildings and predicted time-dependent seismic performance levels of corroded buildings were compared based on their story levels as a result of user-defined and default hinge properties. Limit–states at each performance levels (e.i. immediate occupancy, life safety, collapse prevention and collapse) were obtained. The obtained results were summarized to compare the differences in the results of seismic response of the buildings due to user-defined and default hinge properties for both corroded and non-corroded cases.

Nonlinear Material Modelling

It is vital to accurately determine the effects of corrosion on the seismic analyses of *RC* buildings. Mainly, corrosion causes loss in the cross sectional area of the reinforcement bars and reduction in bond strength between reinforcement bars and concrete. A study done by Sezen and Moehle (Sezen & Moehle, 2006) indicated that, slip deformations contributed 25% to 40% of the total lateral displacement in the case of non-corroded reinforcement bars. These displacements might be more dramatic by lowering the bond strength due to corrosion. For the user-defined plastic hinge properties, time-dependent bond-slip relationships

INTRODUCTION

can be taken into account by modifying the moment-curvature relationships. Modified target post-yield stiffness of each structural member ensures the bond-slip relationships in nonlinear analyses. However, in the case of assigning the default hinge properties, available programs are not capable to consider the bond-slip relationships as a consequence of corrosion effects. Thus, it is inevitable to obtain huge difference in the result of time-dependent seismic performance levels of analysed buildings by using the default hinge properties. Therefore, in this study, the reduction in cross sectional area of reinforcement bars only considered as a function of time which can be also obtained by using available computer software programs.

A corrosion rate of 2.79 $\mu A/cm^2$ was assumed in order to predict the loss in the cross sectional area of reinforcement bars as a function of time where 0.0116 was used as a conversion factor of $\mu A/cm^2$ into mm/year for steel. For the user-defined plastic hinge properties, the obtained time dependent loss in cross sectional area of reinforcement bars were used to predict the moment-curvature relationships of RC sections. Developed model by Kent and Park (Kent & Park 1971) was used to model the stress-strain relationships of confined columns. Fig. 1 shows the well known developed model by Kent and Park (Kent & Park 1971) which was adapted for modelling the stress-strain relationships of RC sections in this study. Basically, the developed model by Kent and Park (Kent & Park 1971), has two branches. For the ascending branch (A-B), the curve reaches to maximum stress level at a strain of 0.002. After reaching maximum stress, two other different braches occurs (B-C, B-D) where two straight lines indicate different behaviour of concrete for confined and unconfined concrete. For the descending branch of the curve assumed to be linear and its slope specified by determining the strain when the concrete stress is decreased to half of its stress value as suggested by Park et al. (Park et al, 1982).

Mander's (Mander, 1984) model was used for each time periods (i.e., t: 25 years, and t: 50 years) for modelling the stress-strain relationships of steel as can be seen from Fig. 2. The developed model by Mander (Mander, 1984) includes linear elastic region up to yield, elastic-perfect-plastic region, and strain hardening region. The Mander's model (Mander, 1984) has control on both strength and ductility where the descending branch of the curve that first branch increases linearly until yield point then the curve continues as constant. In order to model the material properties, the following required assumptions were made. The modulus of elasticity of concrete $Ec=3250fc----\sqrt{}+14000MPa$ was calculated according to TS500

(TSI, 2000). The mechanical properties of steel in the analyses were selected according to TS500 (TSI, 2000), where the minimum rupture strength (*fsu*) was equal to 500 MPa, the yield strain (εy) was equal to 0.0021, the strain hardening (εsh) was equal to 0.008, the minimum rupture extension (εsu) was equal to 0.12% and the modulus of elasticity of steel (*Es*) was taken as 200,000 MPa.

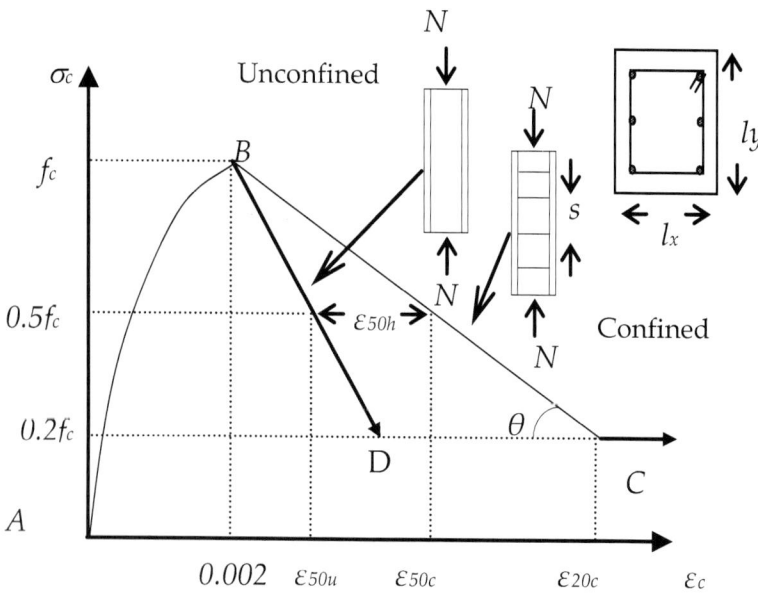

Figure 1: Used stress-strain relationship of concrete (Kent & Park 1971).

DESCRIPTION OF STRUCTURES

Three *RC* buildings having two, four and seven stories were considered in this study. The assessed three*RC* buildings were selected among the typical constructed *RC* buildings in North Cyprus where the buildings were designed according to Turkish earthquake code (TEQ, 1997). The soil classes were classified as soft clay (group D), the building importance factor was taken as 1, and the effective ground acceleration coefficient (*A0*) was taken as 0.3g (seismic zone 2) according to Turkish earthquake code (TEQ, 1997). The buildings were remodelled to select the most critical frames by using the existing plans of the buildings. Fig. 3 shows the three dimensional modelling of a two story of *RC* building. InFig. 3, the total height of the building is 6 m where the typical floor height is identical and equal to 3 m. The slab thicknesses of the building are same and equal to 0.15 m. The dead (G) and live (Q) loads of the slabs were designed to be 5.15 kN/m² and 1.96 kN/m², respectively. Additional wall load on the beams were designed to be 3.19 kN/m².

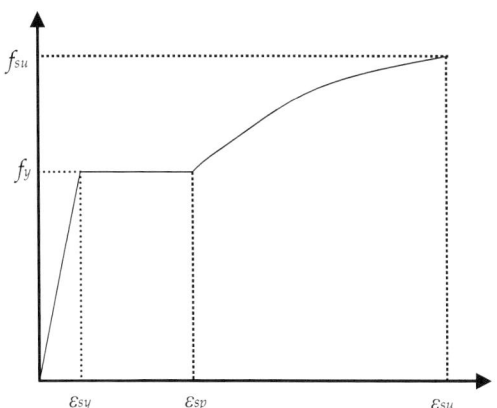

Figure 2: Stress-strain relationship of steel (Mander, 1984).

Figure 3: Three dimensional view of two story reinforced concrete building.

Fig. 4 shows two dimensional view of the selected frame from the two story of *RC* building. In Fig. 4, the member names and sectional dimensions of columns and beams with the amount of longitudinal reinforcement bars are also represented. The vertical distributed loads that were used in the analyses are also depicted in Fig. 4. The frame has a first-mode period of T_1: 0.40 seconds having a total weight of 19.69 tons. For the second case study, a four story RC building was selected to be analysed which represents a typical apartment buildings in North Cyprus. Figs. 5 and 6 show three dimensional view and the selected frame of the building, respectively. In Fig. 5, the total height of the building is equal to 12 m where the typical floor height is identical and equal to 3 m. The slab thicknesses of the building are same and equal to 0.17 m. The dead and live loads on the slabs are 5.64 kN/m² and 1.96 kN/m², respectively. Additional wall load on the beams are identical and equal to 3.19 kN/m². In Fig. 6, the sectional dimensions of all beams are identical and equal to 0.25 m by 0.60 m, with the same details of reinforcement bars. The frame has a first-mode period of T_1: 1.09 seconds having a total weight of 55.53 tons.

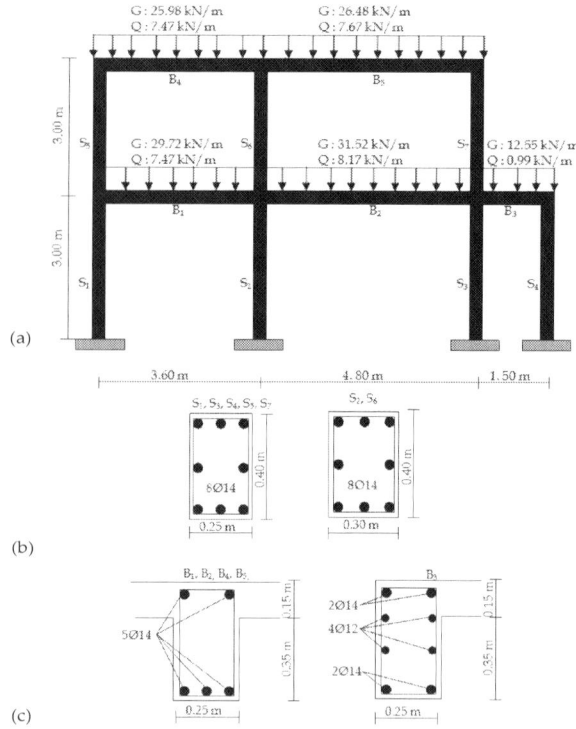

Figure 4: Dimensional and reinforcement details of a two story frame: (a) Used vertical loads in the analyses, (b) Reinforcement details of columns, (c) Reinforcement details of beams.

DESCRIPTION OF STRUCTURES   201

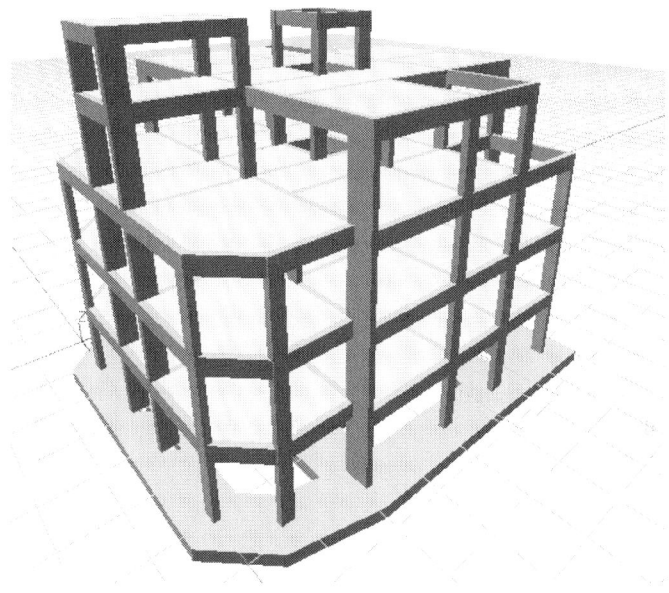

Figure 5: Three dimensional view of a four story *RC* building.

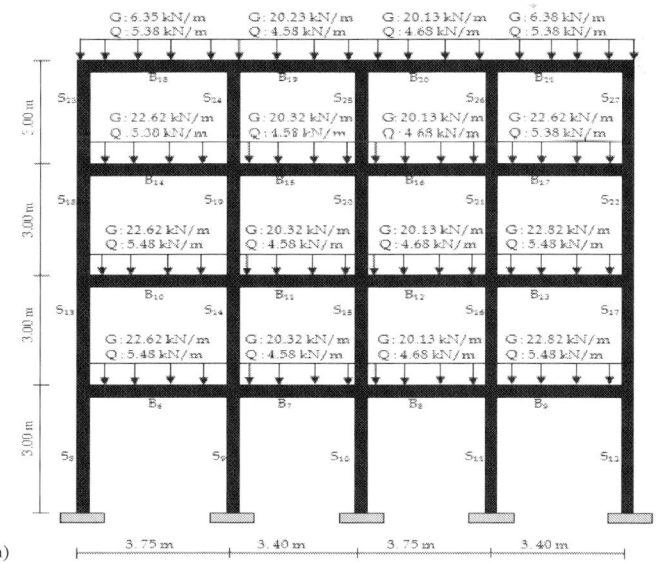

(a)

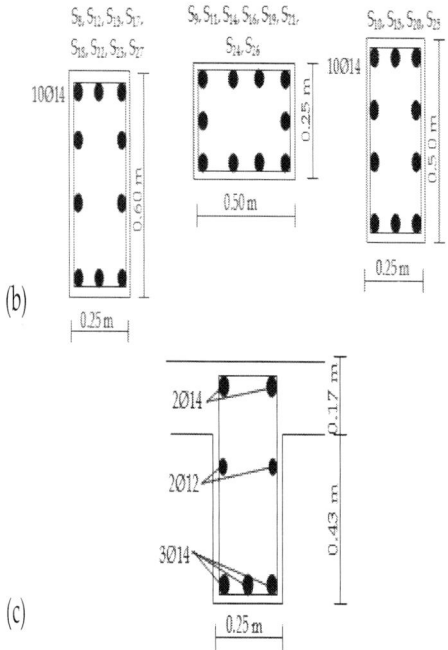

Figure 6: Dimensional and reinforcement details of a four story frame: (a) Used vertical loads in the analyses, (b) Reinforcement details of columns, (c) Reinforcement details of beams.

The third case study deals with an existing seven story of a *RC* building. Figs. 7 and 8 show three dimensional view and the selected frame of the analysed building, respectively. In Fig. 7, the total height of the building is equal to 27 m where the typical floor height is identical and equal to 4.50 m. The slab thicknesses of the building are same and equal to 0.17 m. The dead and live loads on the slabs were 6.25 kN/m² and 4.90 kN/m², respectively. Additional wall load on the beams were identical and equal to 3.19 kN/m². Because of having more than twenty different reinforcement details of beams, only reinforcement details of columns are shown in Fig. 8. In Fig. 8, the depicted vertical distributed loads of seventh floors are same for other floors. The frame has a first-mode period of T_1: 4.27 seconds having a total weight of 151.56 tons.

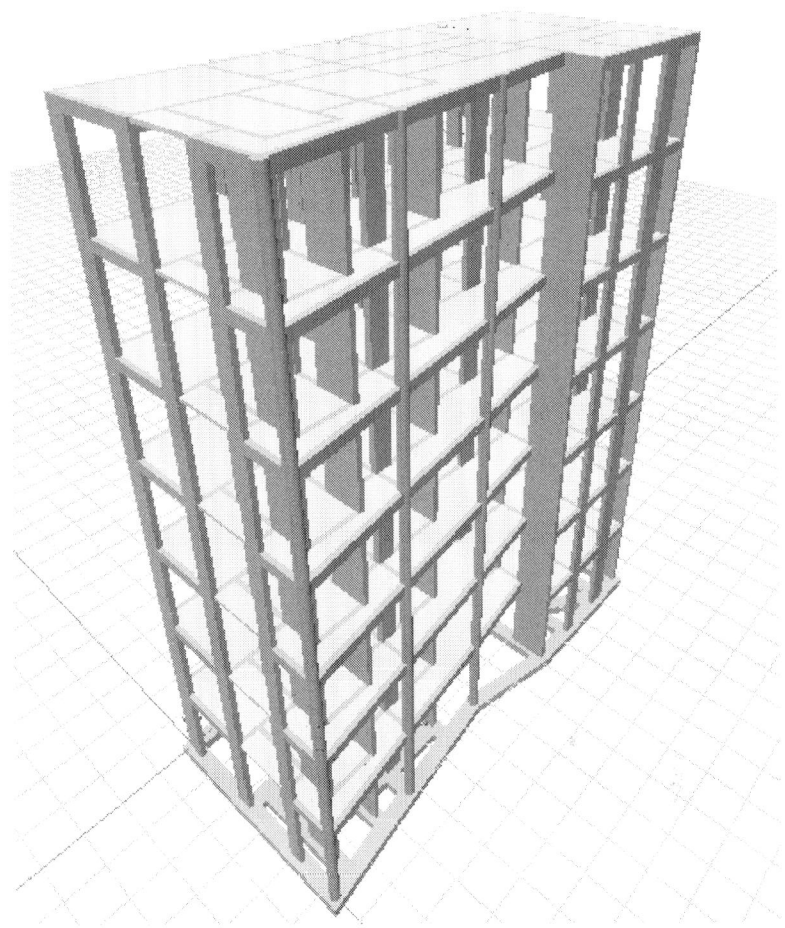

Figure 7. Three dimensional view of a seven story of *RC* building.

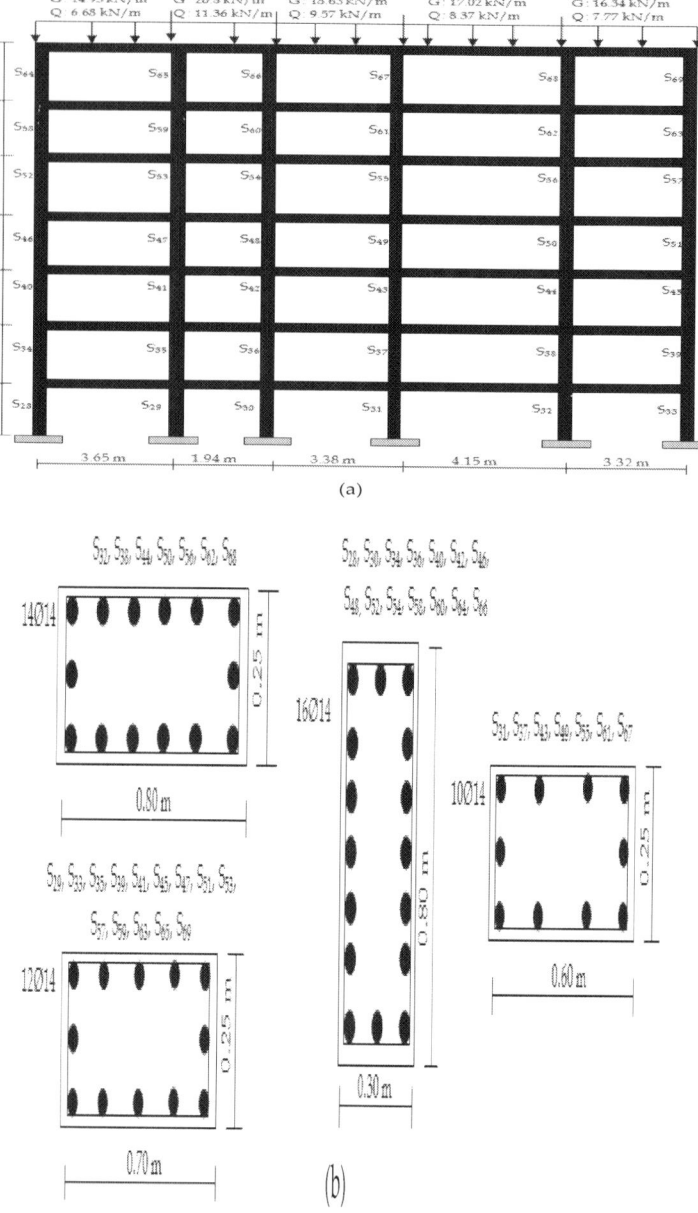

Figure 8: Dimensional and reinforcement details of seven story of frame: (a) Used vertical loads in the analyses, (b) Reinforcement details of columns.

MOMENT-CURVATURE RELATIONSHIPS

Moment-curvature relationships were predicted in order to define the user-defined plastic hinge properties as a function of time. Moment-curvature relationships of columns were carried out from the calculated section properties and constant axial forces acting on the elements. Axial loads on the beams were assumed to be zero. A total of 210 plastic hinge properties as a function of time (t: 0 (non-corroded), t: 25 years, t: 50 years) were defined to be used in the nonlinear static push-over analyses. In order to predict the moment-curvature relationships, a new developed software program *SEMAp* (Ineltal., 2009) was used. *SEMAp* (Ineltal., 2009) models the stress-strain relationships of steel and concrete by the user. Fig. 9 shows the predicted moment-curvature relationships of randomly selected *RC* columns and beams as a function of time for different story levels. In Fig. 9, time dependent moment-curvature relationships of the assessed *RC* members basically indicates three segments; the elastic region prior to cracking, the post-cracking branch between the cracking and yield points and the post-yield segment beyond yielding, respectively. As shown in Fig. 9, premature yielding occurs due to the loss in cross sectional area of the reinforcement bars. For instance, for the same story level, the premature yielding moments of the *S1* column corresponding to time periods of 25 and 50 years were 18% and 39%, respectively. As shown in Fig. 9, at the same moment values, curvature of a structural section increases as a function of time which affects the demand capacity of the frame by the defined plastic hinge properties.

In Fig. 9, the area under moment-curvature represents the storage energy capacity of a section in inelastic behaviour. As shown, in Fig. 9, the area under the curvature decreases due to premature yielding of reinforcement bars which causes cracking of concrete at early stages. The results of time period of 50 years showed that concrete crushes before the reinforcing bars exceed the strain hardening region with increased corrosion level.

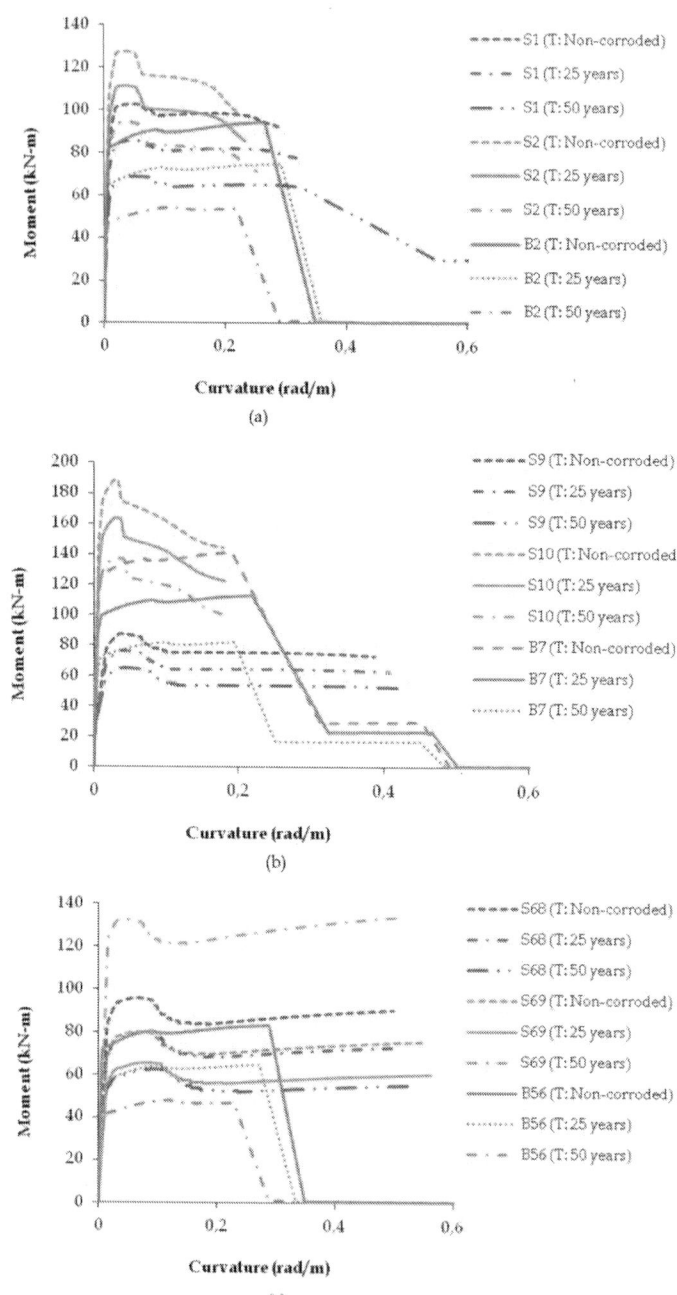

Figure 9: Moment-curvature relationships of *RC* members as a function of time: (a) Two story, (b) Four story, (c) Seven story.

NONLINEAR STATIC ANALYSIS

SAP2000 (CSI, 2008) computer program was used to analyse the selected frames as a function of time. For the user-defined plastic hinge properties, the force-deformation behaviour needs to be plotted to define the behaviour of plastic hinges. Fig. 10 shows a typical force-deformation relationship to define the behaviour of plastic hinges by *FEMA-356* (FEMA-356, 2000) and also the required acceptance criteria of immediate occupancy (*IO*), life safety (*LS*), collapse prevention (*CP*) and collapse (*C*).

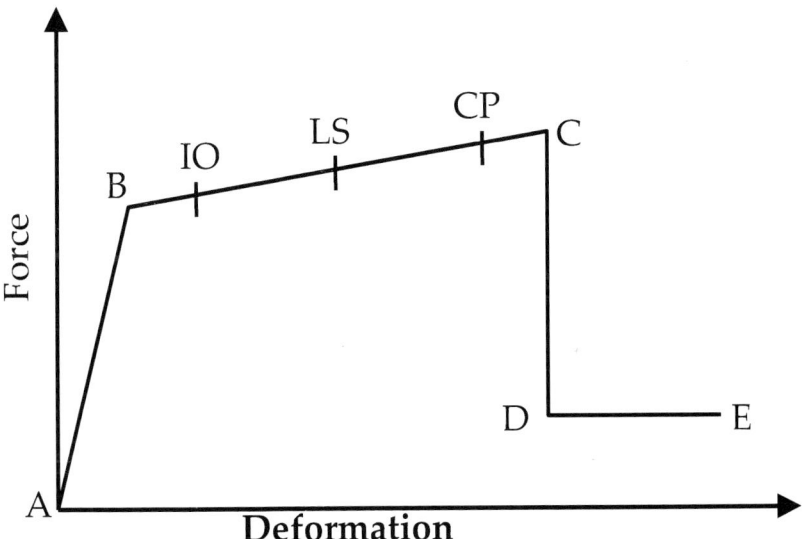

Figure 10: Force-deformation relationship of a plastic hinge.

In Fig. 10, point *A* corresponds to unloaded condition of hinge deformation. Point *B* represents yielding of structural elements that controlled by moment-curvature relationships. Hinge deformation shows strength degradation at point *D* where the structure might show sudden failure after this point. The failure of the structure can be defined by reaching the point *D* and *E*. In this study, the locations of the hinges of the selected frames were located according to the study done by Inel and Ozmen (Inel & Ozmen, 2006). The lengths of the plastic hinges were used to calculate the moment-rotation instead of moment-curvature given by Eq. 1 (Varghese, 2006):

$$\int \varphi ds : \varphi L p : \theta$$

(1)

where θ is the rotation of plastic hinge, Lp is the plastic hinge length, and ϕ is the curvature at a point.

There are different proposed models available in the literature to calculate the length of the plastic hinges. Since the mechanical properties of reinforcement bars play an important role for the user-defined plastic hinge properties, proposed model by Paulay and Priestley (Paulay & Priestley, 1992) to calculate the length of plastic hinges was used according to the given Eq. 2.

$$Lp = 0.08L + 0.022 d_b f_y$$

(2)

where L is the critical distance from the critical section of the plastic hinge to the point of contra flexure, f_y and db are the yield strength and the diameter of longitudinal reinforcement bar, respectively.

As shown in Eq. 2, the proposed model by Paulay and Priestley (Paulay & Priestley, 1992) is important to ensure the effect of corrosion on the length of plastic hinges as a function of time. Shear strength hinge properties were calculated by using Eq. 3 according to ACI 318 code (ACI 318, 2005):

$$Vc = 0.17 x f_{cx} \times b \times d \times (1 + N14Ac)$$

(3)

where Vc is the shear strengths provided by concrete, b is the section width, d is the effective depth, fc is the unconfined concrete compressive strength, N is the axial load on the section, and Ac is the concrete area.

The calculated plastic hinge properties were assigned to each floor at both ends of the beams and columns of the assessed frames according to the corresponding time periods. Triangular lateral load pattern was applied to the frames to perform nonlinear push-over analyses. There are different options are available in *SAP2000* (CSI, 2008) to define the loading of the hinge properties. In this study, unload entire structure option was selected for the method of hinge unloading. When the hinges reach point C in Fig. 10, the program continues to increase the base shear force. After point D

the lateral displacement begins to reduce with the reduced base shear force and the structural elements starts to be unloaded. Fig. 11 shows the predicted time-dependent push-over analyses of the selected frames as a function of time for both of user-defined and default hinge properties.

As can be seen from Figs. 11a-c, the collapse mechanisms of non-corroded frames were affected by corrosion as a function of time. For instance, by using the user-defined plastic hinge properties, the collapse mechanism of the non-corroded frame of the two story of *RC* building started at a top displacement of 0.2633 m when the base shear force was 206 kN (see Fig. 11a). However, for the time periods of 25 and 50 years, collapse mechanism started at top displacements of 0.2608 m and 0.2612 m when the base shear forces were 170 kN and 130 kN, respectively. Same behaviour can be also observed for other performed frames. When the results were compared for the default hinge properties, the effect of corrosion can be also observed. However, there is a huge difference for the collapse mechanism of the assessed frames by default hinge properties. For instance, the time period of 50 years of the seven story of a *RC* building (see Fig. 11c), the recorded top displacement by user-defined plastic hinge properties was 0.92 m when the base shear force was 217 kN. For the same time period of the seven story building (see Fig. 11c), the recorded top displacement by default hinge properties was 0.36 m when the base shear force was 137 kN. Thus, it is clear that the shear capacities obtained from default hinge properties gave underestimate results when compared with the user-defined hinge properties for each case and time periods. For the time period of 25 years, the hinge patterns of two, four and seven stories of frames are plotted in Fig. 12.

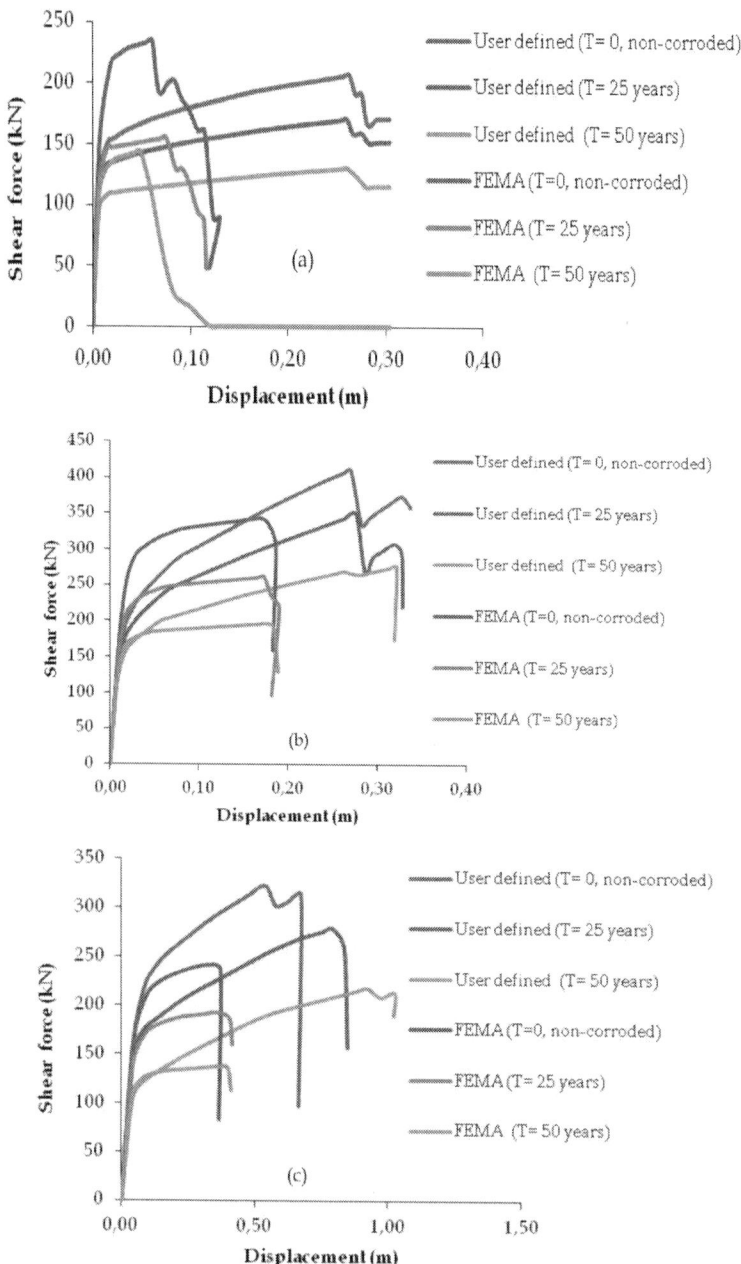

Figure 11: Time dependent load-displacement relationships by using default and user defined plastic hinge properties: (a) Two story, (b) Four story, (c) Seven story.

NONLINEAR STATIC ANALYSIS

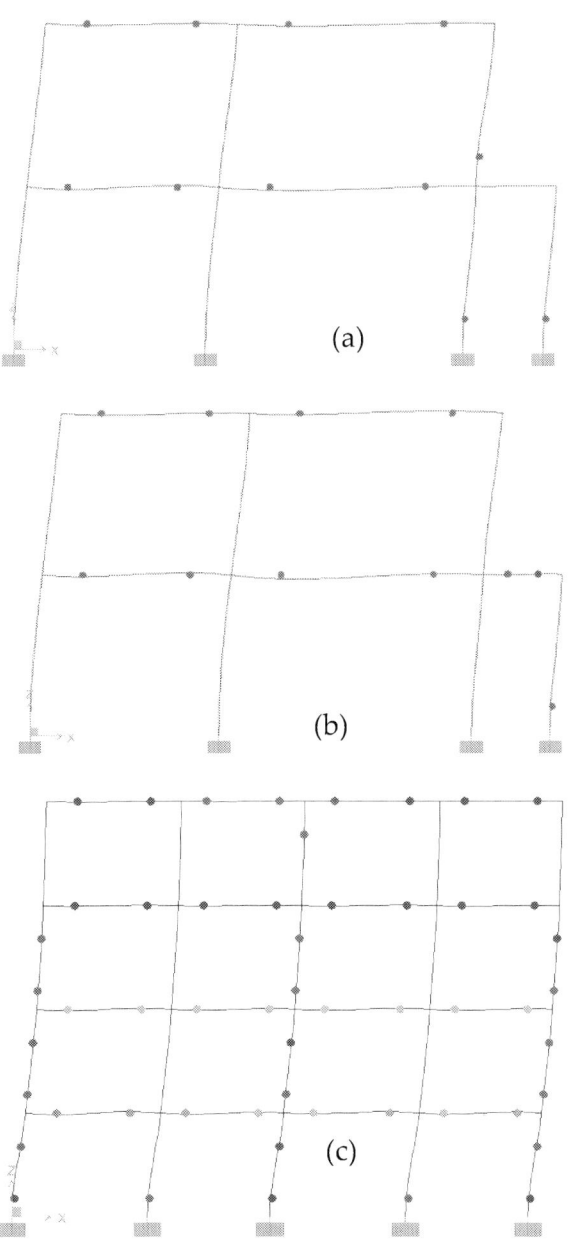

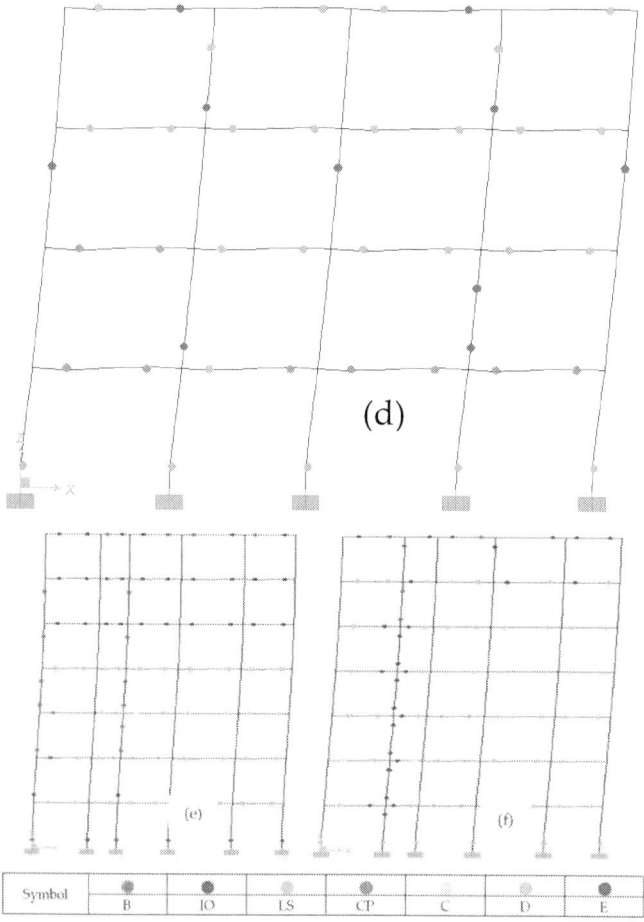

Figure 12: Plastic hinge patterns by using default and user defined plastic hinge properties: (a) Two story user-defined, (b) Two story *FEMA-356*, (c) Four story user-defined, (d) Four story *FEMA-356*, (e) Seven story user-defined, (f) Seven story *FEMA-356*.

As can be seen from Fig. 12, significant differences in hinging by the user-defined and default hinge properties. By increasing the number of stories, the differences become more significant. In both plastic hinge properties, plastic hinge formations at both columns and beams show almost similar behaviour for a two story of *RC* frame. However, for upper stories, hinge formations especially in columns show significant differences. Non-linear time history analyses were performed in the following section to define the effects of both plastic hinges modelling on performance levels.

SEISMIC PERFORMANCE ANALYSES

Incremental dynamic analyses (*IDA*) were performed to predict the performance levels of the assessed frames as a function of corrosion rate by the using user-defined and default hinge properties. For *IDA*, the 5% damped first-mode spectral acceleration (*Sa* (T_1, 5%)) was selected. Twenty ground motion records were used to predict the performance levels of the building as a function of time. For the current study, the associated roof drift ratios corresponding to performance levels, *IO*, *LS* and *CP* were adopted from the study done by Stanish et al. (Stanish et al., 1999) and reduced drift values of 0.5%, 1%, and 2% were used for *IO*, *LS*, and *CP*, respectively. In order to perform *IDA*, *NONLIN* (Charney, 1998) a software computer program was used. By using the NONLIN (Charney, 1998), the material nonlinearity could be taken into account by specifying the yield strength and initial and post yield stiffness, which were calculated from the time-dependent load-displacement relationships (see Fig. 11). Twenty ground motion records were used to predict the performance levels of the buildings as a function of time, where the randomly selected motions records of pseudo velocity versus to period in seconds are shown in Fig. 13., where earthquake moment magnitudes (*M*) ranged from 4.7 to 7.51, *PGA* varied from 0.016 to 0.875g, and peak ground velocity (*PGV*) ranged between 1.65 to 117 cm/sec.

Figs. 14a-c, 15a-c and 16a-c show fragility curves of two, four and seven stories of *RC* buildings, respectively. In Figs. 14, 15 and 16, the obtained time dependent fragility curves which were in terms of *PGA*, compare the differences in the results of performance levels of the buildings as consequences of user-defined and default hinge properties.

The obtained fragility curves indicated that the performance levels of *RC* structures obtained by the default hinge properties based on *FEMA* may under-estimate or over-estimate results. Moreover, in the case of corroded conditions, the response of the buildings obtained by the default hinge properties does not represent the actual behaviour of the structures due to ductility problems of the structural members. Although the collapse mechanism of structures were affected by corrosion; directly reduced cross sectional area of reinforcement bars to perform ready documents hinge properties based on *FEMA* might provide more ductile structural members which might also over-estimate results in the performance levels of *RC* structures. For instance in Fig. 14b, when the *PGA* is equal to 0.4g, the probability of exceeding the limit state corresponding to *LS* is 11% for user-defined plastic hinge properties while this probability is 2% based on *FEMA* ready documents plastic hinge properties. Such differences can be also observed in the case of non-corroded conditions. From Fig. 15a, it can be seen that, when the *PGA* is equal to 0.4g, the probability of exceeding the limit state corresponding to *IO* is 43% for the user-defined plastic hinge properties while this probability is 23% based on *FEMA* ready documents plastic hinge properties. It should be noted that for any story level, the maximum story displacements thus roof drift ratios occurred at different times according to user-defined or ready documents plastic hinge properties. Moreover, the results clearly showed that, the percentage of errors (i.e., *IO*, *LS*, *CP*) occurred due to use ready document plastic hinge properties were not proportional with story levels.

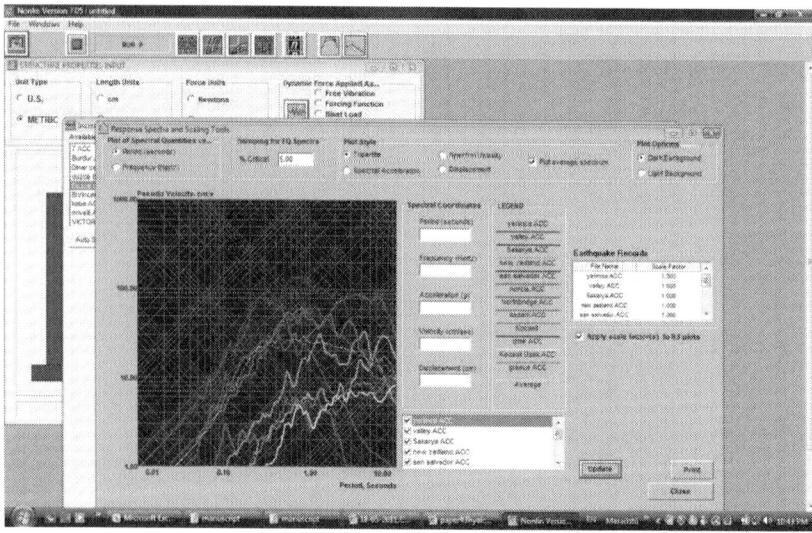

Figure 13: Pseudo velocity spectrum for used ground motion records.

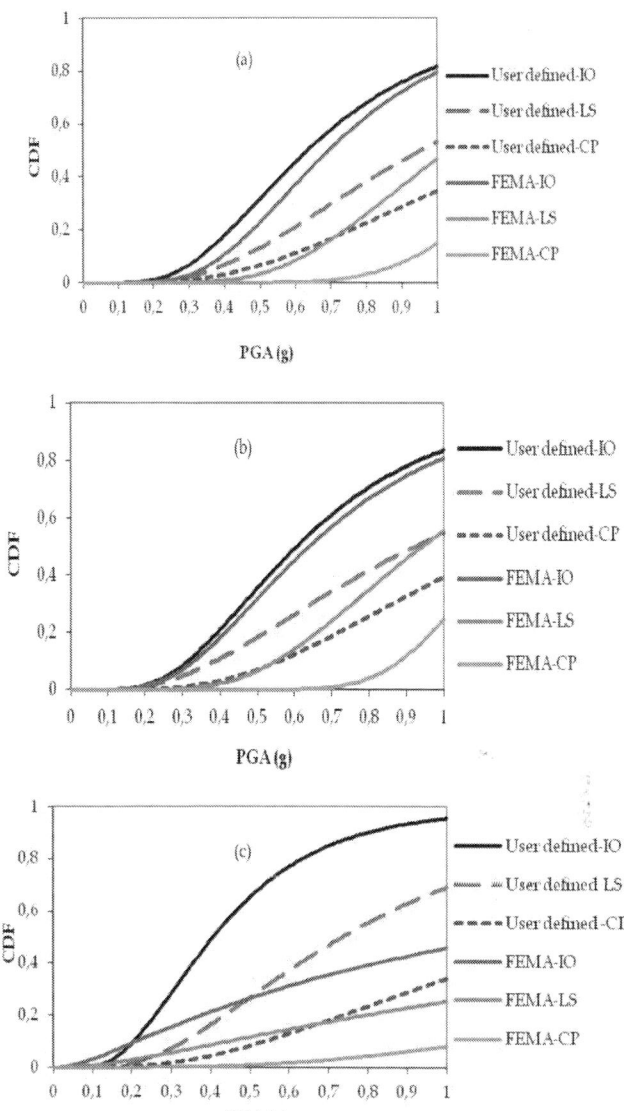

Figure 14: Fragility curves of two story *RC* building: (a) Non-corroded, (b) *T*: 25 years, (c) *T*: 50 years.

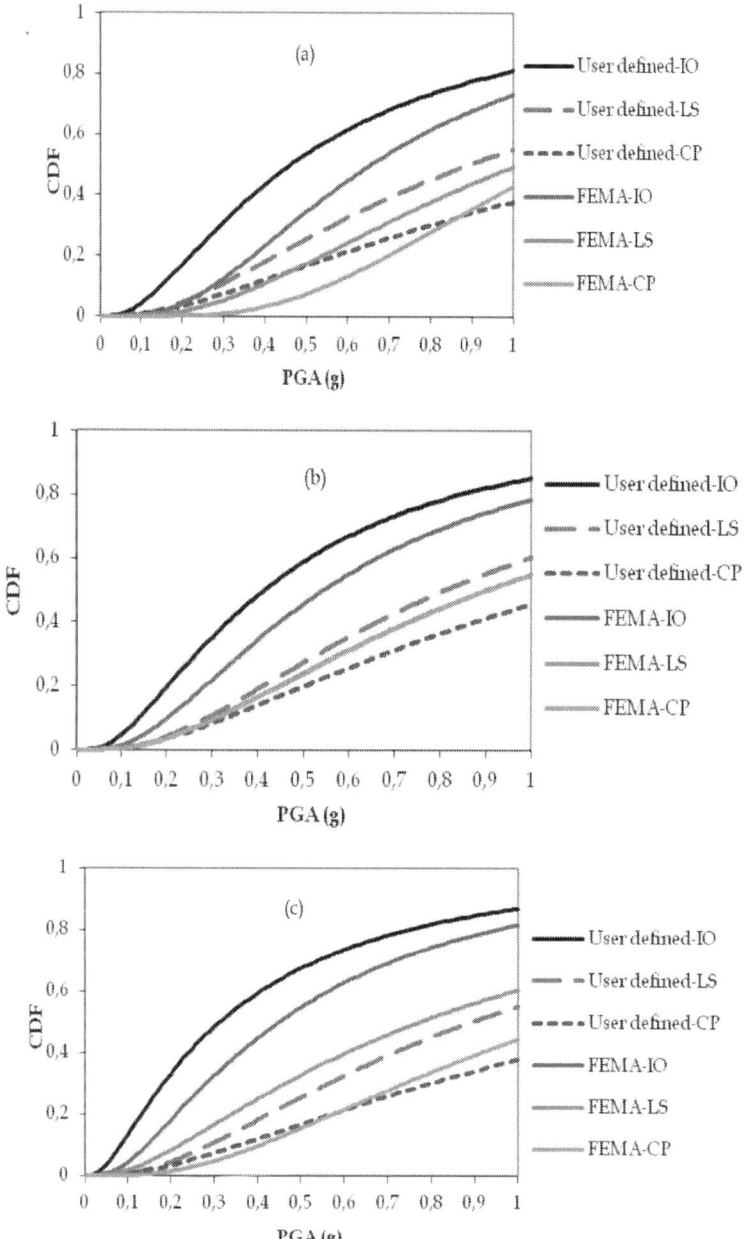

Figure 15: Fragility curves of four story *RC* building: (a) Non-corroded, (b) *T*: 25 years, (c) *T*: 50 years.

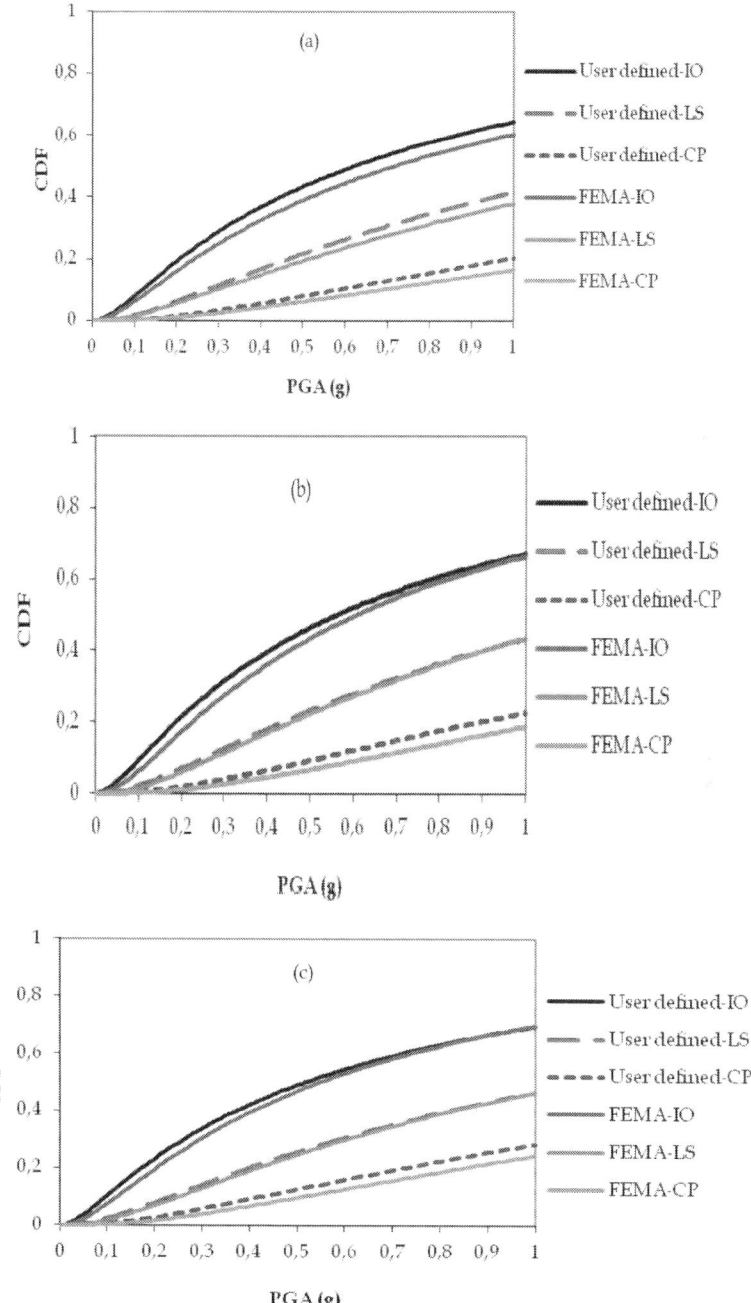

Figure 16: Fragility curves of seven story *RC* building: (a) Non-corroded, (b) *T*:25 years, (c) *T*:50 years.

CONCLUSION

Incremental dynamic analyses for three *RC* buildings having 2, 3 and 7 stories were carried out as a function of time. The performed push-over analyses and *IDA* clearly showed that there were important differences due to the use of the plastic hinge properties based on ready documents and user defined hinge properties. If the user knows the capability of the program where *SAP2000* (CSI, 2008) automatically stops the analysis when a plastic hinge reaches its curvature capacity, ready document plastic hinge properties might be used for rapid and preliminary assessment of *RC* buildings. However, the obtained time dependent results clearly showed that the user defined plastic hinge properties give better and correct results than default hinge properties. Additional studies are also required for accurate performance assessment of multi-degree-of-freedom systems. Bond-slip relationships and cover cracking of concrete due to corrosion need to be taken into account in seismic analyses where the effect of additional displacement due to slippage of reinforcement bars can be provided by the modification of plastic hinge properties. When the effects of corrosion on seismic performance levels and economical impacts in construction industry are considered, time-dependent nonlinear models rather than walk-down surveys are required for better decision making of strengthening of RC buildings to prevent serious damage under the expected seismic motions.

REFERENCE

1. Abbas, Moustafa 2011 Seismic Performance Evaluation of Corroded Reinforced Concrete Structures by Using Default and User-Defined Plastic Hinge Properties. American Society of Civil Engineers, Journal of Structural Engineering, 137 3 456467.
2. ACI Committee 318 2005 Building Code Requirements for Reinforced Concrete and Commentary. American Concrete Institute, Detroit, Michigan, 423
3. F. A. Charney, 1998 NONLIN 7 Nonlinear Dynamic Time History Analysis of Single Degree of Freedom Systems, Blacksburg, Virginia, Advanced Structural Concepts.
4. CSI, ETABS 2003 Integrated design and analysis software for building systems, California, USA, Computers and Structures Inc.

REFERENCE

5. CSI, 2000 12 2008): Integrated finite element analysis and design of structures basic analysis reference manual, Berkeley, Computers and Structures Inc.

6. M. Inel, H. B. Ozmen, 2006 Seismic Performance Evaluation of Corroded Reinforced Concrete Structures by Using Default and User-Defined Plastic Hinge Properties. Engineering Structures, 28 11 14941502 ,

7. M. Inel, H. B. Ozmen, H. Bilgin, 2009 SEMAp: Modelling and Analysing of Confined and Unconfined Concrete Sections. Scientific and Technical Research Council of Turkey, Project 105M024

8. D. C. Kent, R. Park, 1971 Flexural members with confined concrete. American Society of Civil Engineers, Journal of the Structural Division, 97 7 19691990 .

9. J. B. Mander, 1984 Seismic Performance Evaluation of Corroded Reinforced Concrete Structures by Using Default and User-Defined Plastic Hinge Properties, Ph.D. Thesis, Department of civil engineering, University of Canterbury, New Zealand.

10. R. Park, M. J. N. Priestly, W. D. Gill, 1982 Ductility of Square Confined Concrete Columns. ASCE Journal of Structural Engineering, 108 11 929951 .

11. H. Sezen, J. P. Moehle, 2006 Seismic Performance Evaluation of Corroded Reinforced Concrete Structures by Using Default and User-Defined Plastic Hinge Properties American Concrete Institute Structural Journal, 103 6 824849 , ISSN

12. K. Stanish, R. D. Hooton, S. J. Pantazopoulou, 1999 Seismic Performance Evaluation of Corroded Reinforced Concrete Structures by Using Default and User-Defined Plastic Hinge Properties. American Concrete Institute Structural Journal, 96 6 (November-December 1999), 915922 .

13. Turkish Earthquake Code (TEQ) 2007 Ministry of Public Works and Settlement Government of Republic of Turkey, Specification for Structures to be Built in Disaster Areas, Earthquake Disaster Prevention, Ankara, Turkey.

14. Turkish Standards Institute (TSI), 500 (2000). Requirements for Design and Construction of Reinforced Concrete Structures, Ankara, Turkey.

15. Varghese, PC (2006). Allowable rotation for collapse load analysis, In: Advanced reinforced concrete design 2nd edition, pp. 399-402, Prentice-Hall press, 81-203-2787-X, India.)

CITATION

Hakan Yalçiner and Khaled Marar (2012). Seismic Performance Evaluation of Corroded Reinforced Concrete Structures by Using Default and User-Defined Plastic Hinge Properties, Earthquake Engineering, Prof. Halil Sezen (Ed.), ISBN: 978-953-51-0694-4, InTech, DOI: 10.5772/47783.

CHAPTER 9

Strain Rate Dependent Properties of Ultra High Performance Fiber Reinforced Concrete (Uhp-Frc) Under Tension

Sukhoon Pyo[a], Kay Wille[b], Sherif El-Tawil[c], Antoine E. Naaman[c]

[a] Korea Railroad Research Institute, 176 Railroad Museum Road, Uiwang-si, Gyeonggi-do 437-757, South Korea
[b] Department of Civil & Environmental Engineering, University of Connecticut, 261 Glenbrook Road Unit 2037, Storrs, CT 06269-2037, USA
[c] Department of Civil & Environmental Engineering, University of Michigan, 2350 Hayward, G.G. Brown, Ann Arbor, MI 48109-2125, USA

ABSTRACT

The results of an experimental investigation of UHP-FRC tensile response under a range of low strain rates are presented. The strain rate dependent tests are conducted on dogbone specimens using a hydraulic servo-controlled testing machine. The experimental variables are strain rate, which ranges from 0.0001 1/s to 0.1 1/s, fiber type, and fiber volume fraction. Five different types of fibers are considered including straight and twisted fibers with different geometric properties. The rate sensitivity of the composite material in tension is evaluated in terms of its first cracking strength, post-cracking strength, energy absorption capacity, strain capacity, elastic modulus, fiber tensile stress and number of cracks. The test results show pronounced rate effects on post-cracking strength and energy absorption capacity. Further, post cracking strength varies linearly with the fiber reinforcing index and energy absorption capacity varies linearly with the product of the fiber length and the reinforcing index, as predicted from the theory for fiber reinforced concrete.

INTRODUCTION

The development of high performance concrete materials having significant strength and enhanced ductility, particularly in tension is a subject of continuous research. Examples of concretes with enhanced tensile properties include Fiber Reinforced Concretes (FRC) [1], Slurry Infiltrated Fiber Concrete (SIFCON)[2], Multi-Scale Cement Composite (MSCC) [3], Hybrid Fiber Concretes (HFC) [4], High Performance Fiber Reinforced Cement Composites (HPFRCC) [5], Engineered Cementitious Composite (ECC) [6] and Ultra High Performance Fiber Reinforced Concrete (UHP-FRC) [7]. As is clear from their names, all of the concretes listed above are reinforced with fibers (polymeric or/and steel).

Fiber reinforced concretes resist tensile stress through composite action between the cementitious matrix and embedded fibers. The transmission of forces between these two components occurs through interfacial bond. After cracking, fibers bridge the cracks, providing resistance to crack opening, enhancing structural behavior and durability. Many researchers have focused on characterization of the tension behavior of these cementitious materials. Examples of such research efforts can be found in Refs. [6], [8],[9], [10], [11] and [12].

A relatively recent and most promising fiber reinforced cementitious material for structural applications is UHP-FRC. Although different researchers have defined UHPC and UHP-FRC using several criteria [13],[14] and [15], American Concrete Institute (ACI) Committee 239 suggests the following definition: *"Concrete, ultra-high performance – concrete that has a minimum specified compressive strength of 150 MPa (22,000 psi) with specified durability, tensile ductility and toughness requirements; fibers are generally included to achieve specified requirements."* There appears to be consensus in the research literature that well designed UHP-FRC can be highly durable against chemical attack, freeze–thaw cycles, abrasion and chloride penetration [14], [16] and [17], and therefore there is strong interest in exploring its material properties.

The objective of this study is to obtain more detailed knowledge of the tensile behavior of UHP-FRC under loading with various strain rates. The experimental variables are fiber type, fiber volume fraction and strain rate. Five different fiber types are considered, including both straight and twisted fibers with a variety of geometric properties. The strain rates considered range from 0.0001 1/s, which represents pseudo static loading, to 0.1 1/s, which is commonly considered as representative of seismic loading rates [18] and [19]. Dogbone shaped UHP-FRC specimens are used in the test program and their performance is evaluated in terms of their first cracking strength, post-cracking strength, energy absorption capacity, strain capacity, elastic modulus, fiber tensile stress and number of cracks within the gage length. The fiber tensile stress is the average stress resisted by the fibers at the maximum post-cracking strength of the composite. Dogbone shaped specimens are selected for a tensile testing program following earlier research reported in Wille et al. [20]. Such specimens fulfill AASHTO T 132-87 [21] requirements.

STRAIN RATE EFFECT ON FRC UNDER TENSION

Numerous tests have been carried out to investigate the response of FRC under tension loading applied at various strain rates. Körmeling and Reinhardt [22] carried out experimental studies to characterize the tensile behavior of steel fiber reinforced concrete at strain rates of 1.25×10^{-6}–20 1/s. They noted a considerable increase in tensile strength, strain at maximum stress and fracture energy associated with the rate effect. Zhu et al. [23] conducted a series of dynamic tensile tests (loading speed: 1000 mm/s) for three types of fabric (carbon, alkali resistant glass, and polyethylene) reinforced cement composites using a high speed servo-hydraulic testing machine, and compared their mechanical properties with the cases under quasi-static loading. They reported a significant decrease in strength of the carbon composite with increased loading rate. They concluded that the strength decrease under high speed loading is mainly due to sliding friction of filaments against their neighboring filaments. Other experimental research on the impact behavior of FRC can also be found in [24], [25] and [26].

HPFRCC materials were investigated for their use in seismic applications by several researchers[27] and [28]. Maalej et al. [29] investigated the tensile behavior of ECC containing a combination of high-modulus steel fibers and relatively low-modulus polyethylene fibers at strain rate ranging from 2×10^{-6} to 0.2 1/s. They concluded that the tensile strength doubled but there was no obvious change in strain capacity with increasing strain rate. Yang and Li [30] investigated the tensile strength of ECC reinforced with polyvinyl alcohol (PVA) fibers at strain rate increasing from 10^{-5} to 10^{-1} 1/s. They found that the tensile strength of ECC doubled whereas the strain capacity decreased from 3% to 0.5%. They attributed these observations to the sensitivity of fiber interfacial chemical bond strength to loading speed. Kim et al. [31] investigated the tensile behavior of HPFRCC using two types of deformed, high-strength steel fibers, namely twisted and hooked fibers at strain rates of 10^{-4} to 10^{-1} 1/s. They observed that the dynamic increase factors (DIF) increased up to 2.0 and 1.7 for the first-cracking strength and post-cracking strength, respectively, depending on fiber type, fiber volume fraction, and matrix strength. In contrast to the findings of Yang and Li [30] for ECC, strain capacity remained largely unaffected by strain rate. Douglas and Billington [19] investigated the strain rate sensitivity of two HPFRCC materials (PVA fiber reinforced ECC and twisted high strength steel fiber reinforced HPFRCC) in cylindrical specimens under monotonic tension at strain rates ranging from 2×10^{-5} to 0.2 1/s. Both HPFRCC materials showed enhancements in strengths (25–120% for ECC and 77–165% for steel fiber reinforced HPFRCC) as strain rate increased. Both types of materials experienced a 50–55% decrease in strain capacity as the strain rate increased. They noted that unlike thin coupon specimens which tend to align fibers along the loading direction, cylindrical specimens allow for random, three-dimensional fiber alignment, leading to less efficient fiber bridging.

Fujikake et al. [32] investigated the tensile behavior of UHP-FRC using the commercial mix of Ductal Premix under various strain rates ranging from 10^{-6} to 0.5 1/s. Based on the obtained stress-elongation relationships in uniaxial tension, they proposed a rate-dependent bridging law expressing the relation between tensile stress and crack opening. Wille et al. [33] reported on the strain rate dependent tensile behavior of UHP-FRC with different fiber volume fraction at strain rates ranging from 0.0001 to 0.1 1/s. They observed that strength and energy absorption capacity both increased with an increase in fiber volume fraction for a given strain rate.

As previously noted in the introduction, and as is clear from the limited references listed, there is little information on the strain rate-dependent response of UHP-FRC, especially under tension. This study is geared towards addressing this gap by providing information on key parameters that influence UHP-FRC behavior under tension.

EXPERIMENTAL PROGRAM

A targeted experimental program was carried out to investigate the rate sensitivity of UHP-FRC under uniaxial tension. The mix proportions of UHP-FRC with three different volume fractions of fiber are given in Table 1. Further details about the mix constituents can be found in Wille et al. [11]. Five different fibers were used with properties described in Table 2. The twisted fibers were made in the lab out of round wire stock. They had square cross section with a pitch of 5 mm, where the fiber Pitch is the length of one full (360-degree) twist around the fiber axis [34]. The different fibers are designated by their type, diameter and length as shown in Table 2, e.g. S-0.2-25 is a straight smooth fiber with 0.2 mm diameter and 25 mm length. Each UHP-FRC test series is designated by the fiber used to make it, appended with a number that reflects the volume fraction of fiber, e.g. S-0.2-25-2% represents a mix with 2% fiber volume fraction. Nine series of UHP-FRC tensile test specimens were prepared and tested as shown in Table 3.

Table 1: Mixtures proportions by weight (based on Wille et al. [11])

Fiber volume fraction	1%	2%	3%
Cement	1	1	1
Silica fume	0.25	0.25	0.25
Glass powder	0.25	0.25	0.25
Water	0.22	0.22	0.22
Superplasticizer	0.005	0.005	0.005
Sand A[a]	0.27	0.26	0.26
Sand B[b]	1.07	1.05	1.02

[a]Maximum grain size = 0.2 mm (1/128 in.).
[b]Maximum grain size = 0.8 mm (1/32 in.).

Table 2: Properties of steel fibers used in this study

Notation	Form	d_f (mm)	l_f (mm)	l_f/d_f	Tensile strength MPa (ksi)
T-0.3-25	Twisted[a]	0.3	25	83	2670 (387)[b]
S-0.2-25	Straight	0.2	25	125	2860 (415)
S-0.4-25	Straight	0.4	25	62.5	1850 (268)
S-0.3-18	Straight	0.3	18	60	2330 (338)
T-0.3-18	Twisted[a]	0.3	18	60	2670 (387)[b]

[a]Manufactured out of round wire with d_f = 0.30 mm at the University of Michigan Structural Laboratories.
[b]Tensile strength of fiber after twisting.

Table 3: UHP-FRC test series investigated in this study

Test series	Fiber volume fraction (V_f)			$V_f(l_f/d_F)$	Strain rate
	1%	2%	3%		
T-0.3-25-1%	X			0.83	0.1
T-0.3-25-2%		X		1.67	0.01
T-0.3-25-3%			X	2.5	0.001
					0.0001
S-0.2-25-1%	X			1.25	0.1
					0.01
S-0.2-25-2%		X		2.5	0.001
					0.0001
S-0.4-25-2%		X		1.25	0.1
					0.01
S-0.4-25-3%			X	1.88	0.001
					0.0001
S-0.3-18-2%		X		1.2	0.1
					0.01
					0.001
					0.0001
T-0.3-18-2%		X		1.2	0.1
					0.01
					0.001
					0.0001

Four strain rates were considered in the tests, ranging from quasi-static ($\dot{\varepsilon} = 0.0001$ 1/s) to seismic ($\dot{\varepsilon} = 0.1$ 1/s), and applied using a hydraulic servo-controlled testing machine (MTS-810). At least three specimens for each loading rate in each test series were tested, resulting in 108 specimens in total (9 series × 4 loading rates × 3 specimens) as summarized in Table 3.

Materials and Fabrication

A Hobart type laboratory mixer was used to prepare the UHP-FRC mixture. The mixing procedure followed Wille et al. [11]. First, silica fume was mixed with all the sand for approximately 5 min. Then, cement and glass powder were added and mixed together for at least another 5 min before the water and superplasticizer were added. The cementitious mixture became fluid after approximately 5 min of adding the liquid ingredients. When the cementitious mixture started to show adequate flowability and viscosity, fibers were dispersed by hand in the mix. The cementitious mixture with uniformly distributed fibers was poured into dogbone shaped molds without any vibration. After casting, the specimens were covered with plastic sheets and stored at room temperature for 24 h prior to demolding. The specimens were then placed in a water tank for an additional 26 days without any special curing such as heat or pressure treatment. All specimens were tested in a dry condition at the age of 28 days, after 24 h of drying in the laboratory environment.

Test Setup and Procedure

Wille et al. [20] surveyed the various techniques used for tensile testing of cementitious composites and selected a specific method with the best ability to capture strain hardening behavior. Fig. 1 shows the geometry of the specimen and test set up that were also adopted in this research. The gage length of the dogbone type tensile test specimen is 76 mm (=3 inch). Tests were conducted using displacement control, where the displacement of the actuator in the hydraulic servo-controlled testing machine was used as the control variable. Two optical markers for noncontact displacement measurement were attached to the surface of the specimen. Three cameras were used to measure the movement of the markers, from which the measurement system calculated the deformation of the specimen. The applicability of the measurement system for this type of application was confirmed in a similar study done by Kim et al. [31] for HPFRCC. The tensile load history was recorded via the load cell attached to the testing machine and synchronized to the deformation history recorded by the measurement system.

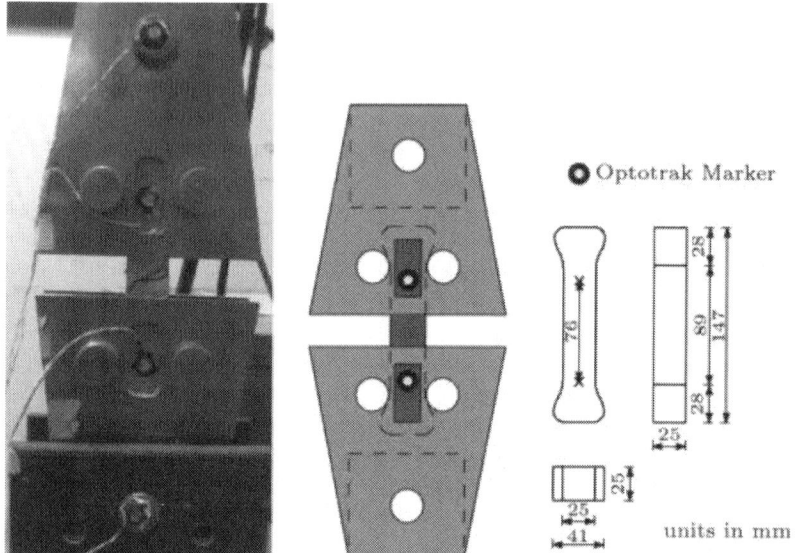

Figure 1: Direct tensile test setup for this experimental research according to Wille et al. [20].

Test Results

Referring to Fig. 2a and b, the parameters considered in this investigation are: (1) the first cracking strength (σ_{cc}), (2) the post cracking strength (σ_{pc}), (3) the energy absorption capacity (g), (4) the strain capacity, which is the strain value at post cracking strength (ε_{pc}), (5) the elastic modulus (E_{cc}) up to the first cracking strength, (6) the fiber tensile stress (σ_{fpc}) and (7) number of cracks within the gage length. It is important to note that σ_{cc} is the turning point between elastic and strain hardening parts and does not necessarily represent the stress when the first crack developed. It also should be noted that g is the area under stress–strain curve up to an arbitrarily selected value: $\sigma_u = 0.95$. It was noted from experience that consistent softening behavior generally occurred beyond this point (compare the softening regimes in Fig. 2a and b).

EXPERIMENTAL PROGRAM

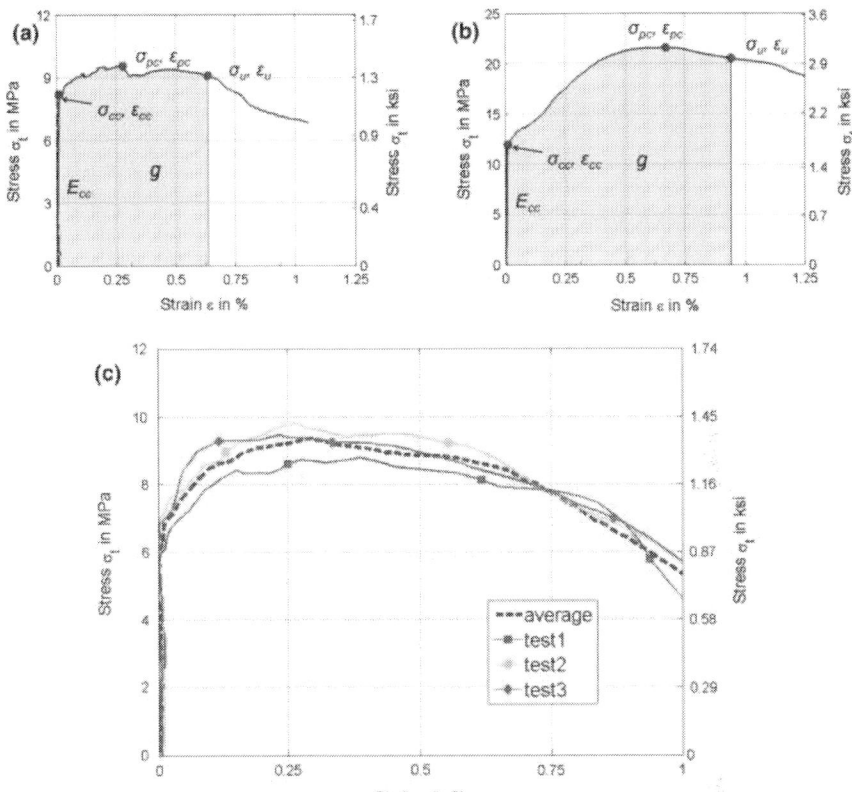

Figure 2: Typical interpretation and measurement of test results: (a and b) typical variables; (c) typical averaging of stress–strain curve (T-0.3-18-2% & $\dot{\varepsilon} = 0.01$ 1/s) (*Note:* strain valid up to peak stress only.)

The variable σ_{fpc} represents the average tensile stress in the fiber at maximum stress in the composite. It is calculated by Eq. (1) as follows [20]:

$$\sigma_{fpc} = \frac{\sigma_{pc}}{\phi \cdot V_f} \tag{1}$$

where ϕ is a fiber orientation factor, taken as 0.9 for all series following Wille et al. [20]. The number is quite high because most of the fibers are aligned in the load direction due to the casting method. V_f is the fiber volume fraction. Table 4 summarizes all the test results. All the values given in Table 4 are averaged values from at least three specimens. A dynamic increase factor (DIF), which is the ratio between dynamic response and static response, is computed for four parameters (σ_{cc}, σ_{pc}, g, ε_{pc}) to effectively illustrate the effect of strain rate on strength and other material parameters. In this study the values measured at the lowest strain rate of 0.0001 1/s were considered to be the reference, i.e. pseudostatic loading rate. The DIF values are also listed in Table 4.

Table 4: Rate effect on the average tensile properties of UHP-FRC

Test series	Strain rate (1/s)	First cracking strength (σ_{cc})			Post cracking strength (σ_{pc})			Energy absorption capacity (g)		Strain capacity (ε_{pc})		Elastic modulus (E_{cc})	Fiber tensile stress (σ_{fpc})	Average number of cracks
		Avg. MPa	Std. dev.	DIF	Avg. MPa	Std. dev.	DIF	kJ/m³	DIF	%	DIF	GPa	MPa	EA
T-0.3-25-1%	0.0001	6.89	0.84	1	8.31	0.28	1	33.5	1	0.19	1	54.5	923	7
	0.001	7.54	0.29	1.09	8.91	0.65	1.07	37.9	1.1	0.23	1.2	54.6	990	7
	0.01	7.04	0.83	1.02	9.42	0.71	1.13	47.6	1.4	0.26	1.38	57.1	1050	8
	0.1	8.13	1.1	1.18	10.22	0.88	1.31	60.7	1.8	0.34	1.8	58.9	1200	9
T-0.3-25-2%	0.0001	9.03	0.31	1	11.6	1.2	1	63.5	1	0.42	1	60.1	644	13
	0.001	9.14	0.92	1.01	12.6	0.85	1.08	83.4	1.3	0.5	1.2	60.4	698	14
	0.01	8.7	1.3	0.9	13	0.59	1.1	84.1	1.3	0.55	1.3	54.3	764	11

EXPERIMENTAL PROGRAM

Group														
		3		7	8		9							
	0.1	8.87	1.9	0.998	14	0.14	1.21	103	1.61	0.7	1.66	60.8	779	11
T-0.3-25-3%	0.0001	11.8	0.28	1	20.9	1.8	1	108	1	0.48	1	71.3	775	13
	0.001	11.5	0.5	0.97	21.7	0.94	1.04	143	1.32	0.62	1.28	70	805	13
	0.01	12	0.61	1.02	23.5	2.1	1.12	177	1.64	0.68	1.41	70.3	869	12
	0.1	12.1	0.68	1.03	24.1	1.9	1.15	186	1.72	0.79	1.64	76.1	892	12
S-0.2-25-1%	0.0001	6.22	0.79	1	8.11	1.4	1	34.5	1	0.33	1	47.7	901	9
	0.001	7.86	0.19	1.26	9.49	0.6	1.17	51.9	1.5	0.49	1.47	56.1	1050	13
	0.01	6.92	2	1.11	8.29	2	1.02	48.1	1.39	0.39	1.17	51.1	921	10
	0.1	6.7	0.65	1.08	9.01	1.4	1.11	47.6	1.38	0.43	1.3	54	1000	13
S-0.2-25-2%	0.0001	9.73	0.77	1	14.9	0.44	1	84.4	1	0.55	1	66.6	829	13
	0.001	9.19	1.7	0.94	14.6	2.3	0.98	83.1	0.98	0.59	1.09	63.2	813	12
	0.0	1	0.	1	1	2.	1	87.	1.0	0.	1.	64.8	858	14

		1	1.2	98	.15	5.55	7	.04	2	3	61	11			
		0.1	9.65	1.7	0.99	15.5	0.26	1.04	89.3	1.06	0.55	1.02	67.8	862	17
S-0.4-25-2%	0.0001	6.86	0.83	1	8.32	1.9	1	35.9	1	0.27	1	54.5	462	8	
	0.001	6.78	0.55	0.99	8.65	0.72	1.04	36.8	1.02	0.3	1.1	51.2	481	10	
	0.01	8.13	0.88	1.18	9.96	0.43	1.2	48.3	1.34	0.35	1.28	57.7	553	9	
	0.1	8.3	1.5	1.21	9.72	1	1.17	56.1	1.56	0.42	1.55	55	540	9	
S-0.4-25-3%	0.0001	7.67	0.35	1	9.41	0.62	1	41.3	1	0.24	1	61.4	349	10	
	0.001	8.2	1	1.07	9.63	0.47	1.02	48.5	1.17	0.28	1.15	61	357	9	
	0.01	8.69	1.2	1.13	12.5	1.4	1.33	74.8	1.81	0.47	1.96	62.8	463	9	
	0.1	8.71	1.4	1.14	12.8	0.48	1.36	78.8	1.91	0.47	1.94	59.9	474	7	
S-0.3-18-2%	0.0001	7.05	0.8	1	8.44	0.26	1	20	1	0.17	1	56.5	469	7	
	0.001	8.45	1.3	1.2	9.8	0.98	1.16	28.1	1.4	0.21	1.22	61.1	544	7	

	0.001	7.05	0.063	1	9.32	1.4	1.1	34.2	1.7	0.23	1.35	53.1	518	6
	0.1	7.66	0.9	1.09	9.7	0.6	1.15	38.3	1.91	0.31	1.8	53.4	539	6
T-0.3-18-2%	0.0001	7.73	0.24	1	9.04	0.29	1	27.8	1	0.27	1	56.1	502	7
	0.001	8.53	0.89	1.1	10.4	0.52	1.15	34.6	1.25	0.22	0.86	55	578	8
	0.01	7.32	0.12	0.95	9.63	0.18	1.07	47.3	1.7	0.29	1.15	50.4	535	7
	0.1	7.37	1.5	0.95	10.8	1.2	1.2	54.5	1.96	0.44	1.7	56.9	600	7

Fig. 2c shows an example averaged stress–strain curve, plotted along with raw data to give a sense of the spread in the test data. In order to get an averaged curve, the raw data set for each individual plot is divided into two parts, one before and one after the peak stress point. Each point on either part is assigned a fraction of peak stress strain. The data at each 'fraction point' is then averaged across all curves for both portions of each curve. It is clear from Fig. 2c, and this was generally observed throughout the test program, that specimen-to-specimen variability within each strain rate was relatively low for all parameters. The average tensile stress–strain curves for each series under four different strain rates are plotted in Fig. 3, Fig. 4 and Fig. 5. All test series maintained performance Level 3 as classified in Naaman and Reinhardt[35] for all loading rates, i.e., strain hardening was observed for all loading rates. Furthermore, T-0.3-25-2%, T-0.3-25-3%, and S-0.2-25-2% series maintained performance Level 4 as classified by Naaman and Reinhardt [35], which is defined as high energy absorbing with $g \geqslant 50$ kJ/m^3 as suggested by Wille et al.[20] for all loading rates.

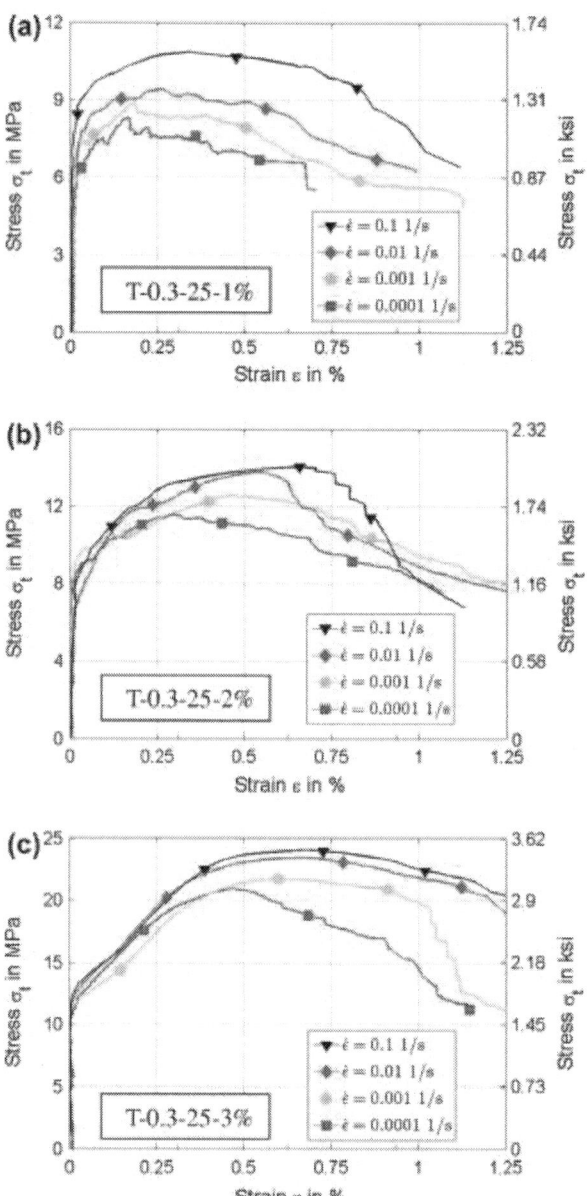

Figure 3: Observed rate effect on the tensile behavior of UHP-FRC using twisted steel fibers with increasing volume content and same fiber aspect ratio (*Note:* strain valid up to peak stress only.)

EXPERIMENTAL PROGRAM 235

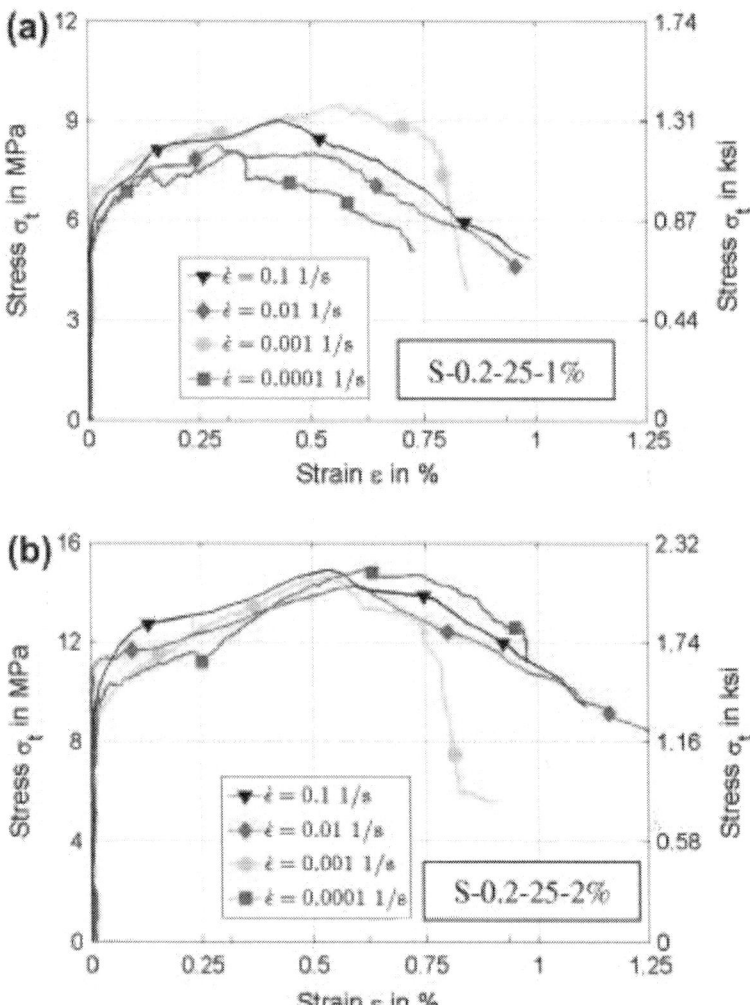

Figure 4: Observed rate effect on the tensile behavior of UHP-FRC using smooth steel fibers having a high aspect ratio with increasing volume content (*Note:* strain valid up to peak stress only.)

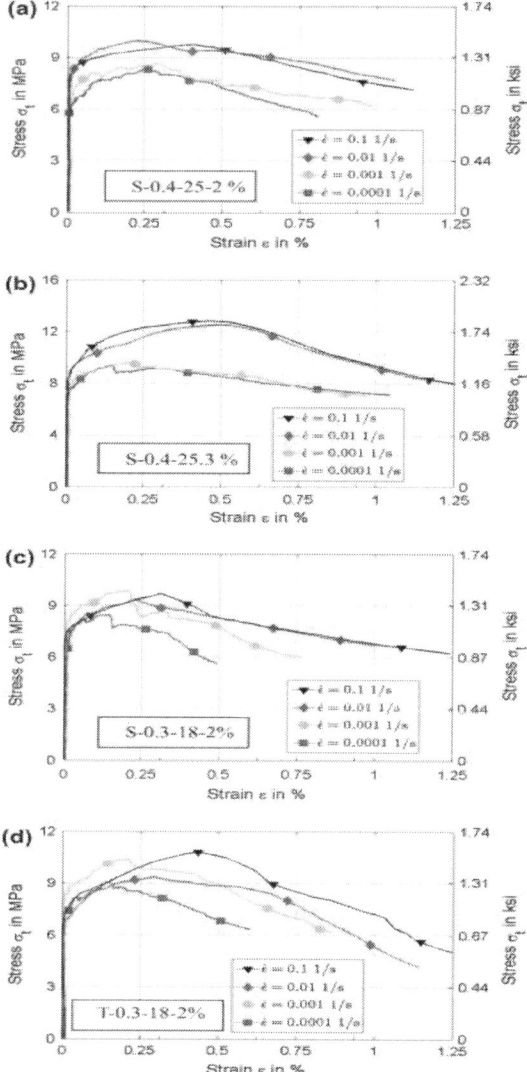

Figure 5: Observed rate effect on the tensile behavior of UHP-FRC using fibers with about same aspect ratio (*Note:* strain valid up to peak stress only.)

According to Naaman [36], the post cracking strength of composites reinforced with short discontinuous fibers is proportional to fiber aspect ratio and their volume fraction as follows:

$$\sigma_{pc} = \lambda_{\tau eq} V_f (l_f / d_f) \qquad (2)$$

where λ is a factor equal to the product of several coefficients that account for average pullout length, group reduction effect, and fiber orientation effect. τ_{eq} is the equivalent bond strength. The product $V(l_f/d_f)$ is termed the fiber reinforcing index. Table 5 shows the effect of the fiber aspect ratio on post cracking strength based on the results of Eq. (2) for the pseudostatic and seismic loading rates and 2% fiber content. Moreover, the surface energy of pull-out, which gives a good estimate of the fracture energy in the type of tensile tests carried out here can be put in the following form [10]:

$$\gamma_g = \zeta \tau_{eq} V_f (l_f^2/d_f) \qquad (3)$$

where ζ is the product of λ in Eq. (2) and the ratio of average bridging stress to the maximum post-cracking stress over the expected maximum pull-out length. It can be thus observed that the expected fracture energy varies with the square of the fiber length or with the product of the fiber length by the reinforcing index.

Table 5: Derivation of equivalent bond strength for UHP-FRC with 2% fiber content

Test series	l_f/d_f	$\dot{\varepsilon} = 0.0001$ 1/s		$\dot{\varepsilon} = 0.1$ 1/s	
		σ_{pc} (MPa)	$\lambda \tau_{eq}$ (MPa)	σ_{pc} (MPa)	$\lambda \tau_{eq}$ (MPa)
T-0.3-25-2%	83	11.6	6.98	14.0	8.44
S-0.2-25-2%	125	14.9	5.97	15.5	6.20
S-0.4-25-2%	62.5	8.32	6.65	9.72	7.78
S-0.3-18-2%	60	8.44	7.04	9.70	8.08
T-0.3-18-2%	60	9.04	7.53	10.8	9.00

EVALUATION OF EXPERIMENTAL RESULTS

The average numerical results are summarized in Table 4. Fig. 3, Fig. 4, Fig. 5, Fig. 6 and Fig. 7 and Fig. 9 illustrate the effects of various parameters on the response of the composite at various strain rates.

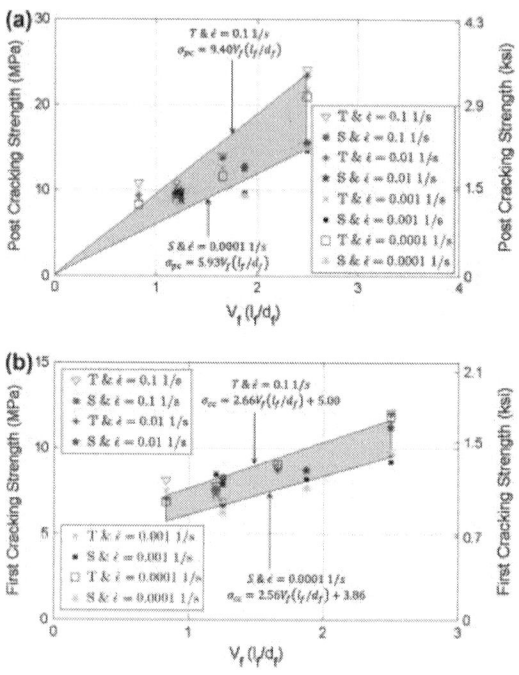

Figure 6: Effects of the fiber reinforcing index on mechanical properties of UHP-FRC: (a) Post cracking strength; (b) first cracking strength.

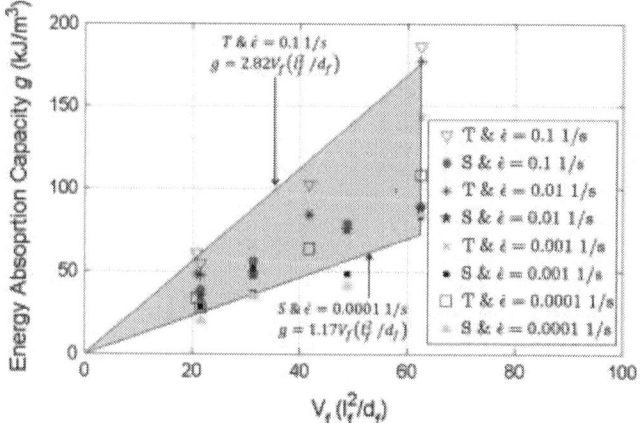

Figure 7: Effects of $V_f(l_f^2/d_f)$ on energy absorption capacity of UHP-FRC.

Separate Effects of Fiber Volume Fraction, Aspect Ratio, Length, and Type

Fig. 3 illustrates the effects of volume fraction of fiber and strain rate on the tensile response of the composite reinforced with twisted steel fibers. It can be observed that the post-cracking strength and the corresponding strain generally increase with an increase in both volume fraction of fiber and strain rate. Typically the post-cracking strength more than doubles when the volume fraction of fiber goes from 1% to 3%. Fig. 4 illustrates similar trends when using smooth steel fibers with 1% and 2% volume content. Fig. 5 shows the response of four series of tests using either smooth or twisted steel fibers having about the same aspect ratio; here also the effect of strain rate is clear, that is, the post-cracking strength increases with an increase in strain rate.

The effect of fiber aspect ratio by itself, l_f/d_f, can be evaluated by comparing series with 2% volume fraction from the three figures (Fig. 3, Fig. 4 and Fig. 5 and Table 4). Not only does the post cracking strength generally increases with fiber aspect ratio, but also the first cracking strength. For example, S-0.2-25-2% shows better performance than S-0.4-25-2% by 33% in first cracking strength, 66% in post cracking strength, 78% in strain capacity and 100% in energy absorption averaged across all strain rates, respectively.

Although the elastic modulus was measured for all specimens tested, the large scatter in the data makes it difficult to draw a firm conclusion regarding the effect of aspect ratio or volume fraction of fiber or strain rate on elastic modulus (see Table 4). Overall, the twisted fiber series showed better performance than straight fiber series because of the additional anchorage effect associated with the untwisting action that occurs during pullout. Kim et al. [37] discuss the untwisting mechanism. However, for a given volume fraction, the T-0.3-25 and the S-0.2-25 fibers show similar mechanical performances. For example, The T-0.3-18-2% series shows marginally better mechanical properties than S-0.3-18-2%. The increase is most prominent in energy absorption capacity, where it is 36% higher on average across all strain rates. This is attributed to the fact that the number of S-0.2-25 fibers, which are thinner than the T-0.3-25 fibers, is 225% more in a unit volume than T-0.3-25 fibers, making up for the additional anchorage mechanism of the twisted fibers. The good performance of S-0.2-25 fibers, however, comes at a price because the larger number of the fibers makes it more difficult to mix. In general, it was not possible to make mixes with more than 2% volume fraction of S-0.2-25 fibers. In contrast, mixes with up to 3% volume fraction of T-0.3-25 fibers were feasible. The opposite argument can be made for the T-0.3-25 fibers versus the S-0.4-25 fibers.

Equivalent Bond

The results of all test series with 2% volume fraction of fibers are summarized in Table 5. Looking across all entries in the 4th and 6th columns suggests that $\lambda_{\tau eq}$ is almost independent of fiber type, diameter and length. This unexpected result may be attributed to the use of a UHPC matrix where the bond for smooth straight fiber is reported to be excellent due to surface abrasion (see recent paper by Wille and Naaman[34]), and is likely due to the very high packing density of the cementitious matrix around the fiber. It is also noted that the equivalent bond is an average value estimated over a small crack opening (related to ε_{pc}) and is different from a similar value obtained from a complete fiber pull-out curve.

Effect Of Reinforcing Index $V_f(L_f/F)$

Eq. (2) suggests that, for the same $\lambda_{\tau eq}$, the post-cracking strength of the composite is directly proportional to the fiber reinforcing index, that is, the product of the volume fraction times the aspect ratio of fiber. Fig. 6a provides a summary of the data observed for the post-cracking strength versus the reinforcing index at different strain rates. The trend predicted by Eq. (2) is clearly confirmed by the data, that is, the post-cracking strength increases with the fiber reinforcing index. To best quantify the data, the least square fit lines for the four loading rates ranging from quasi-static (0.0001 1/s) to seismic (0.1 1/s) are listed in Table 6 and the extreme ones shown in Fig. 6. It can be further observed that the post-cracking strength increases with strain rate.

Table 6: The least square fit approximations of results in Fig. 7 and Fig. 8

Index	Strain rate (1/s)	T-fiber	S-fiber
First cracking strength	$\dot{\varepsilon} = 0.0001$	$\sigma_{cc} = 2.98 V_f(l_f/d_f) + 4.25$	$\sigma_{cc} = 2.56 V_f(l_f/d_f) + 3.86$
	$\dot{\varepsilon} = 0.001$	$\sigma_{cc} = 2.31 V_f(l_f/d_f) + 5.59$	$\sigma_{cc} = 1.10 V_f(l_f/d_f) + 6.33$
	$\dot{\varepsilon} = 0.01$	$\sigma_{cc} = 3.11 V_f(l_f/d_f) + 3.96$	$\sigma_{cc} = 2.93 V_f(l_f/d_f) + 3.67$
	$\dot{\varepsilon} = 0.1$	$\sigma_{cc} = 2.66 V_f(l_f/d_f) + 5.00$	$\sigma_{cc} = 1.67 V_f(l_f/d_f) + 5.52$
Post cracking strength	$\dot{\varepsilon} = 0.0001$	$\sigma_{pc} = 7.98 V_f(l_f/d_f)$	$\sigma_{pc} = 5.93 V_f(l_f/d_f)$
	$\dot{\varepsilon} = 0.001$	$\sigma_{pc} = 8.41 V_f(l_f/d_f)$	$\sigma_{pc} = 6.12 V_f(l_f/d_f)$
	$\dot{\varepsilon} = 0.01$	$\sigma_{pc} = 9.04 V_f(l_f/d_f)$	$\sigma_{pc} = 6.65 V_f(l_f/d_f)$
	$\dot{\varepsilon} = 0.1$	$\sigma_{pc} = 9.40 V_f(l_f/d_f)$	$\sigma_{pc} = 6.76 V_f(l_f/d_f)$
Energy absorption capacity	$\dot{\varepsilon} = 0.0001$	$g = 1.66 V_f(l_f^2/d_f)$	$g = 1.17 V_f(l_f^2/d_f)$
	$\dot{\varepsilon} = 0.001$	$g = 2.16 V_f(l_f^2/d_f)$	$g = 1.27 V_f(l_f^2/d_f)$
	$\dot{\varepsilon} = 0.01$	$g = 2.56 V_f(l_f^2/d_f)$	$g = 1.48 V_f(l_f^2/d_f)$
	$\dot{\varepsilon} = 0.1$	$g = 2.82 V_f(l_f^2/d_f)$	$g = 1.55 V_f(l_f^2/d_f)$

Fig. 6b illustrates the variation of the first cracking strength of the composite versus the fiber reinforcing index at different strain rates. Here also the least square fit lines of the data are plotted and suggest that the first cracking strength increases with both the reinforcing index and the strain rate.

Energy Absorption Capacity

Following the format of Eq. (3), the observed energy absorption capacity (see Fig. 2) is plotted in Fig. 7versus the quantity $V_f(l_f^2/d_f)$ for different strain rates. The trend observed confirms theoretical predictions; that is, the energy increases with both $V_f(l_f^2/d_f)$ and the strain rate. The least square fit lines provided in the figures and Table 6 offer a good mean to quantify the data.

Cracking

Fig. 8 shows observed cracking patterns of UHP-FRC, which indicates UHP-FRC materials exhibit multiple cracking after first cracking along with hardening behavior. The number of cracks within the gage length is also an important indicator of energy absorption capacity and strain at peak load. It appears from the results (see Table 4) that both the number of cracks and the energy capacity increase as fiber volume fraction increases. However, it is difficult to draw a clear conclusion about the effect of strain rate on the number of cracks, due to the variability in the test data. It should be noted that the crack counting process itself is somewhat subjective because of the difficulty of ascertaining the presence of a crack after unloading.

EXPERIMENTAL PROGRAM 243

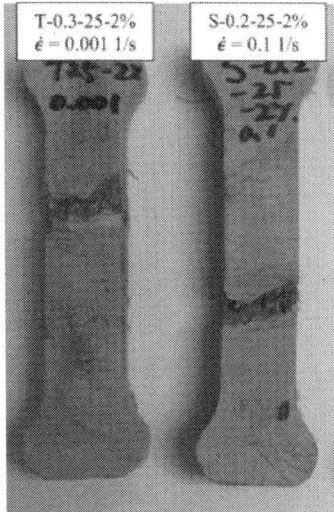

Figure 8: Representative multiple cracking patterns in UHP-FRC specimens.

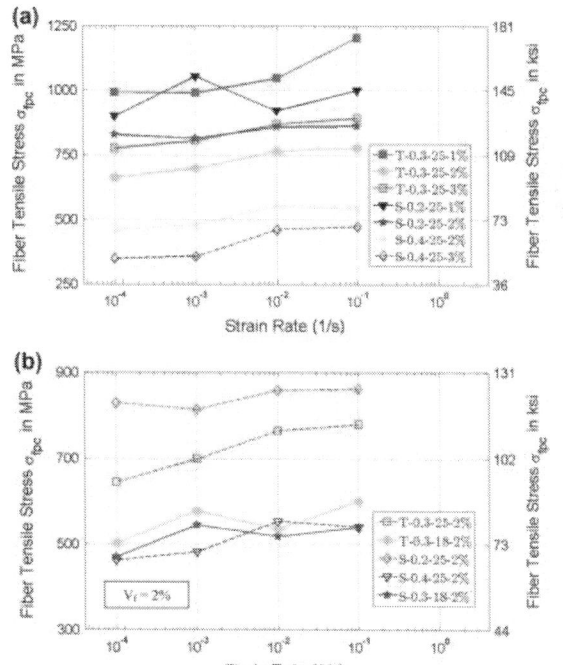

Figure 9: Rate effect on the fiber tensile stress of UHP-FRC at maximum load: (a) with different fiber volume fractions; (b) with 2% fiber volume fraction.

Fiber Tensile Stress

Fig. 9 shows rate effects on the fiber tensile stress of UHP-FRC. The fiber tensile stress (Eq. (1)), which represents the effectiveness of fiber usage, increases as strain rate increases. Even though higher fiber volume fraction led to better mechanical properties of UHP-FRC, it generally decreases, from a qualitative perspective, as fiber volume fraction increases, in contrast to other mechanical parameters (see Table 4), likely due to the fiber-group effect. Increases in performance attributed to volume fraction are also limited by another practical limitation, i.e. difficulty of mixing with a large quantity of fibers.

DIF

DIF was evaluated for four key parameters, first cracking strength, post cracking strength, energy absorption capacity and strain capacity. Plots of DIF versus strain rate for first cracking strength, post cracking strength, energy absorption capacity and strain capacity are compared in Fig. 10. The increases in DIF for the four parameters can be reasonably simulated by a log-linear trend with the increase in strain rate. Fig. 10a shows that the highest rate sensitivity of first cracking strength occurs in series with S-0.4-25 fibers, while the lowest occurs in series with T-0.3-18 fibers. Similarly, S-0.4-25 series shows the highest rate sensitivity in post cracking strength, energy absorption capacity and strain capacity, but S-0.2-25 series shows less rate sensitivity in those parameters as shown in Fig. 10b–d, respectively.

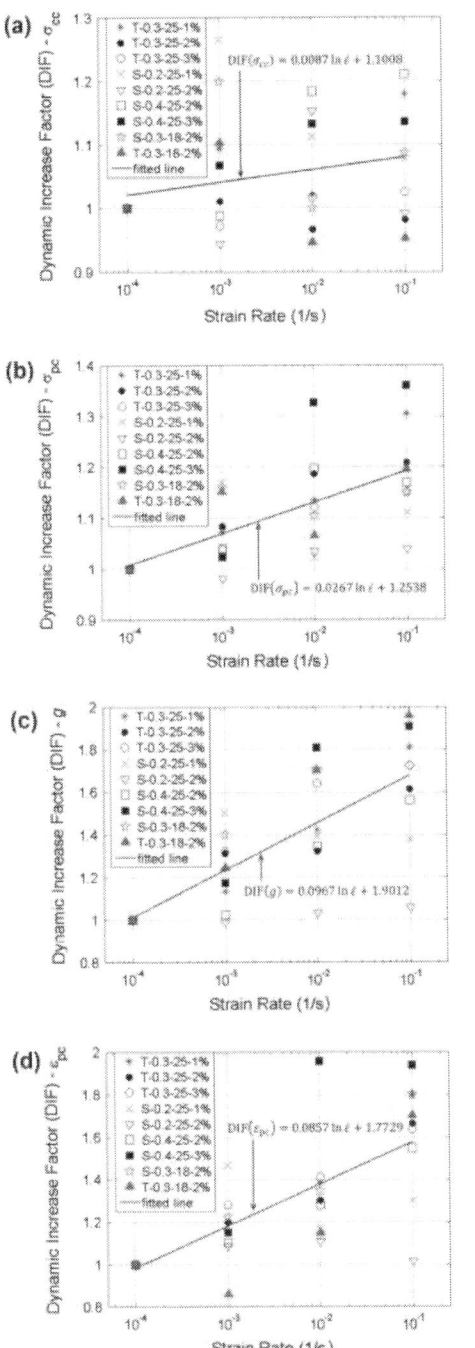

Figure 10: Dynamic increase factor (DIF) of UHP-FRC: (a) First cracking strength; (b) post cracking strength; (c) energy absorption capacity; (d) strain capacity.

Fitted log-linear relationships for first cracking strength, post cracking strength, energy absorption capacity and strain capacity with strain rate are shown in Fig. 10. The figure indicates that the increase in DIF is moderate and nearly linear in log-linear space at strain rates up to 0.1 1/s. This general tendency has been observed in other cement-based materials such as ordinary concrete, high-performance concrete or UHPC [31], [33] and [38].

General Trends in the Test Data

Several general trends can be seen in the tension test results: (1) Tensile strength, energy absorption capacity and strain capacity all increase as fiber volume fraction increases for all fiber series under all strain rates; (2) while twisted fibers provided somewhat better performance than equivalent straight fibers, the mechanical anchorage advantage of twisted fibers over smooth fibers for post-cracking strength could be overcome by increasing fiber aspect ratio; (3) the fiber aspect ratio influences post cracking strength and fiber tensile stress, while fiber shape and length play important roles for strain capacity and energy absorption capacity; (4) similar to other concretes, log-linear relationships of DIF of UHP-FRC are moderate and nearly linear at strain rates up to 0.1 1/s.; (5) even though there is no rate sensitivity in the number of cracks (see Table 4), multiple cracks developed in all UHP-FRC specimens (see Fig. 8) promoting strain hardening behavior of UHP-FRC under tension.

Twisted fibers lead to better overall mechanical performance than equivalent straight fibers, primarily because their mechanical anchorage mechanism is beneficial. However, the fact that the S-0.2-25 series shows similar mechanical performance to the T-0.3-25 series at the 2% fiber content suggests that the mechanical anchorage advantage can be accounted for by changing fiber aspect ratio. This has practical significance because the price of straight fiber is generally cheaper than that of twisted fiber.

Unlike results reported for HPFRCC (e.g. in Yang and Li [30] and Douglas and Billington [19]), the strain capacity does not decrease as strain rate increases. In fact, it almost doubles depending on fiber type and other mechanical properties such as tensile strengths and energy absorption capacity also increase substantially as strain rate increases. These results indicate that UHP-FRC is particularly promising for applications that involve seismic, impact or blast.

CONCLUSIONS

This experimental study investigated the direct tensile behavior of UHP-FRC with five different steel fibers at strain rates ranging from quasi-static (0.0001 1/s) to seismic (0.1 1/s). The tests were conducted using a hydraulic servo-controlled testing machine and results were evaluated in terms of first cracking strength, post-cracking strength, energy absorption capacity, strain capacity, elastic modulus, fiber tensile stress and number of cracks within the gage length. Log-linear relationships of DIF for first cracking strength, post-cracking strength, energy absorption capacity and strain capacity were presented based on the test data. The key observations and findings of this study can be summarized as follows:

1. An increase in the fiber volume fraction led to increases in the composite tensile strength, energy absorption capacity, strain capacity and elastic modulus for all fiber tested under all strain rates. In contrast, the fiber tensile stress did not show a clear trend, likely because it was influenced by the fiber group effect.
2. For the UHP-FRC tested in this study, the equivalent bond strength for the straight steel fibers seems to be of the same order as that of the twisted fibers.
3. For similar equivalent bond strength, the observed post cracking strength of the composite varies linearly with the fiber reinforcing index ($V_f(l_f/d_f)$) as predicted theoretically from Eq. (2).
4. For similar equivalent bond, the observed energy absorption capacity up to peak load varies linearly with the product $V_f(l_f^2/d_f)$, as predicted from theory.
5. For the range of strain rates used in this study (0.0001 1/s to 0.1 1/s) both the post-cracking peak strength and the fracture energy up to peak load increase with an increase in strain rate. It was difficult to draw a firm conclusion regarding the rate sensitivity of the elastic modulus and the number of cracks because of observed large scatter in the test data.

It is hoped that the results from this experimental research will provide some basic information for developing rate dependent constitutive models for UHP-FRC. Such models are needed to simulate the response of UHP-FRC structures subjected to extreme loading. Furthermore, the fact that UHP-FRC shows substantial increases in energy absorption capacity as strain rate increases implies that the material is especially promising for blast and impact applications. However, additional research is needed to investigate and characterize the response of UHP-FRC at strain rates higher than those selected in this study and for a much broader range of parameters.

ACKNOWLEDGMENTS

The research described herein was sponsored by the National Science Foundation - United States under Grant No. CMS 0928193 and the University of Michigan, Ann Arbor. The opinions expressed in this paper are those of the authors and do not necessarily reflect the views of the sponsors.

REFERENCES

1. AASHTO T 132-87. Standard method of test for tensile strength of hydraulic cement mortars. American Association of State and Highway Transportation Officials; 2009. 8 pages.
2. Bindiganavile V, Banthia N, Aarup B. Impact response of ultra-high-strength fiber-reinforced cement composite. ACI Mater J 2002;99:543–8.
3. Brandt AM. Fibre reinforced cement-based (FRC) composites after over 40 years of development in building and civil engineering. Compos Struct 2008;86:3–9.
4. Douglas KS, Billington SL. Strain rate dependence of HPFRCC cylinders in monotonic tension. Mater. Struct. 2011;44:391–404.
5. Dugat J, Roux N, Bernier G. Mechanical properties of reactive powder concretes. Mater Struct 1996;29:233–40.
6. Fischer G, Fukuyama H, Li VC. Influence of matrix ductility on tensionstiffening behavior of steel reinforced engineered cementitious composites (ECC). ACI Struct J 2002;99:104–11.
7. Fujikake K, Senga T, Ueda N, Ohno T, Katagiri M. Effects of strain rate on tensile behavior of reactive powder concrete. J Adv Concr Technol 2006;4:79–84.
8. Graybeal BA. Material Property Characterization of Ultra-High Performance Concrete Report No. FHWA-HRT-06-103. Washington, DC: Federal Highway Administration; 2006.
9. Graybeal BA. Ultra-high performance concrete Report No. FHWA-HRT-11-038. Washington, DC: Federal Highway Administration; 2011.

10. Kim DJ, El-Tawil S, Naaman AE. Loading rate effect on pullout behavior of deformed steel fiber. ACI Mater J 2008;105:576–84.
11. Kim DJ, El-Tawil S, Naaman AE. Rate-dependent tensile behavior of high performance fiber reinforced cementitious composites. Mater Struct 2009;42:399–414.
12. Körmeling HA, Reinhardt HW. Strain rate effects on steel fibre concrete in uniaxial tension. Int J Cem Compos Lightweight Concr 2006;36:1371–8.
13. Li VC, Wang S, Wu C. Tensile strain-hardening behavior of polyvinyl alcohol engineered cementitious composite (PVA-ECC). ACI Mater J 2001;98:483–92.
14. Li VC. On engineered cementitious composites (ECC). J Adv Concr Technol 2003;1:215–30.
15. Lok TS, Zhao PJ. Impact response of steel fiber-reinforced concrete using a split Hopkinson pressure bar. J Mater Civ Eng 2004;16:54–9.
16. Maalej M, Quek ST, Zhang J. Behavior of hybrid-fiber engineered cementitious composites subjected to dynamic tensile loading and projectile impact. J Mater Civ Eng 2005;17:143–52.
17. Markovic I. High-performance hybrid-fibre concrete: development and utilisation, PhD thesis. Delft University of Technology; 2006.
18. Mechtcherine V, Millon O, Butler M, Thoma K. Mechanical behaviour of strain hardening cement-based composites under impact loading. Cem Concr Compos 2011;33:1–11.
19. Naaman AE, Homrich JR. Tensile stress-strain properties of SIFCON. ACI Mater J 1989;86:244–51.
20. Naaman AE, Reinhardt HW. Proposed classification of HPFRC composites based on their tensile response. Mater Struct 2006;39:547–55.
21. Naaman AE, Reinhardt HW. Setting the stage: toward performance based classification of FRC composites. In: Proceedings of 4th international RILEM workshop on high performance fiber reinforced cement composites (HPFRCC 4); 2003. p. 1–4.
22. Naaman AE, Wille K. The path to ultra-high performance fiber reinforced concrete (UHP-FRC): five decades of progress. In: 3rd International symposium on UHPC and nanotechnology for high performance construction materials. Kassel: Kassel University Press; 2012. p. 3–16.
23. Naaman AE. Engineered steel fibers with optimal properties for reinforcement of cement composites. J Adv Concr Technol 2003;1:241–52.
24. Naaman AE. Toughness, ductility, surface energy and deflection-hardening FRC composites. In: Proceedings of the JCI international workshop on ductile fiber reinforced cementitious composites (DFRCC) – application and evaluation (DFRCC-02), Takayama, Japan; 2002. p. 33–57.

25. Parra-Montesinos G. HPFRCC in earthquake-resistant structures: current knowledge and future trends. In: Proceedings of 4th international RILEM workshop on high performance fiber reinforced cement composites (HPFRCC 4). RILEM Publications; 2003. p. 453–72.
26. Pfeifer CG, Moeser B, Giebson C, Stark J. Durability of ultra-high-performance concrete. In: Tenth ACI international conference on recent advances in concrete technology and sustainability issues. No. SP-261-1; 2009.
27. Pyo S, El-Tawil S. Crack velocity-dependent dynamic tensile behavior of concrete. Int J Impact Eng 2013;55:63–70.
28. Rossi P. High performance multimodal fiber reinforced cement composites (HPMFRCC): the LCPC experience. ACI Mater J 1997;94:478–83.
29. Rossi P. Ultra high performance concretes. Concr Int 2008;30:31–4.
30. Scott BD, Park R, Priestley MJN. Stress–strain behavior of concrete confined by overlapping hoops at low and high strain rates. ACI J Proc 1982;79:13–27.
31. Wille K, El-Tawil S, Naaman AE. Properties of strain hardening ultra high performance fiber reinforced concrete (UHP-FRC) under direct tensile loading. Cem Concr Compos 2014;48:53–66.
32. Wille K, El-Tawil S, Naaman AE. Strain rate dependent tensile behavior of ultra-high performance fiber reinforced concrete. In: High Performance Fiber Reinforced Cement Composites 6. Netherlands: Springer; 2012. p. 381–87.
33. Wille K, Kim DJ, Naaman AE. Strain-hardening UHP-FRC with low fiber contents. Mater Struct 2011;44:583–98.
34. Wille K, Naaman AE, Parra-Montesinos GJ. Ultra-high performance concrete with compressive strength exceeding 150 MPa (22 ksi): a simpler way. ACI Mater J 2011;108:46–54.
35. Wille K, Naaman AE. Pullout behavior of high-strength steel fibers embedded in ultra-high-performance concrete. ACI Mater J 2012;109:479–87.
36. Wille K, Parra-Montesinos GJ. Effect of beam size, casting method, and support conditions on flexural behavior of ultra-high-performance fiber-reinforced concrete. ACI Mater J 2012;109:379–88.
37. Yang E, Li VC. Rate dependence in engineered cementitious composites. In: Proceedings HPFRCC-2005 international workshop. Honolulu, Hawaii, USA; 2005.
38. Zhu D, Peled A, Mobasher B. Dynamic tensile testing of fabric–cement composites. Constr Build Mater 2011;25:385–95.

CITATION

Sukhoon Pyo, Kay Wille, Sherif El-Tawil, Antoine E. Naaman, Strain rate dependent properties of ultra high performance fiber reinforced concrete (UHP-FRC) under tension, Cement and Concrete Composites, Volume 56, February 2015, Pages 15-24, ISSN 0958-9465, http://dx.doi.org/10.1016/j.cemconcomp.2014.10.002.

CHAPTER 10

Development of an Eco-Friendly Ultra-High Performance Concrete (Uhpc) with Efficient Cement and Mineral Admixtures Uses

R. Yu[a], P. Spiesz[a,b], H.J.H. Brouwers[a]

[a] Department of the Built Environment, Eindhoven University of Technology, P.O. Box 513, 5600 MB Eindhoven, The Netherlands
[b] HeidelbergCement Benelux, The Netherlands

ABSTRACT

This paper addresses the development of an eco-friendly Ultra-High Performance Concrete (UHPC) with efficient cement and mineral admixtures uses are investigated. The modified Andreasen & Andersen particle packing model is utilized to achieve a densely compacted cementitious matrix. Fly ash (FA), ground granulated blast-furnace slag (GGBS) and limestone powder (LP) are used to replace cement, and their effects on the properties of the designed UHPC are analyzed. The results show that the influence of FA, GGBS or LP on the early hydration kinetics of the UHPC is very similar during the initial five days, while the hydration rate of the blends with GGBS is mostly accelerated afterwards. Moreover, the mechanical properties of the mixture with GGBS are superior, compared to that with FA or LP at both 28 and 91 days. Due to the very low water amount and relatively large superplasticizer dosage in UHPC, the pozzolanic reaction of FA is significantly retarded. Additionally, the calculations of the embedded CO_2 emission demonstrate that the cement and mineral admixtures are efficiently used in the developed UHPC, which reduce its environmental impact compared to other UHPCs found in the literature.

INTRODUCTION

Since 1980s, High Strength Concrete (HSC) has attracted a lot of attention, which later triggered the development of Reactive Powder Concrete (RPC) [1], [2] and [3]. In the components of RPC, coarse aggregates are normally eliminated with active powders (e.g. cement, ground granulated blast-furnace, silica fume) as the main ingredients. Due to the relatively dense and homogenous microstructure of RPC, its maximum compressive strength can even exceed 200 MPa [4] and [5]. However, with the quickly developing construction industry, concrete expect the compressive strength is also required to have high flexural strength, workability and durability, which resulted the development of Ultra-High Performance Concrete (UHPC) and Ultra-High Performance Fibre Reinforced Concrete (UHPFRC) [6], [7] and [8]. Nevertheless, as the sustainable development is currently a pressing global issue and various industries have strived to achieve energy savings, the high material cost, high energy consumption and CO_2 emission for UHPC are the typical disadvantages that restrict its wider application [9], [10] and [11]. Hence, how to efficiently produce UHPC, based on materials point of view, still needs further investigation.

By far, the measures pursued to reduce the economic and environmental disadvantages of UHPC are limited in most cases to the application of industrial by-products or waste materials without sacrificing the UHPC performance [7], [8], [12], [13], [14] and [15]. Nevertheless, in most cases in the literature, for the mix design of UHPC, the amounts of mineral admixtures (e.g. fly ash (FA), ground granulated blast-furnace (GGBS), limestone powder (LP) and silica fume (SF)) are given directly, without any detailed explanations or theoretical support. Moreover, due to the complex cementitious system of UHPC (extremely low water amount and relatively high SP content), the influence of different mineral admixtures on the hydration kinetics and properties of UHPC still needs further clarification [6], [7], [8], [11], [12], [13], [14] and [15]. As commonly known, GGBS has hydraulic properties although the rate of the reaction with water is low [16]. The reaction can be activated by several methods, but the hydration product is always C–S–H. In blended cements, GGBS is chemically activated by $Ca(OH)_2$ and gypsum [17] and [18]. In most cases, GGBS reacts very fast, which causes that the enhancement of mechanical properties of mortar or concrete with GGBS can be observed already during the early age [19], [20] and [21]. On the contrary, the pozzolanic reaction of FA is relatively slow, and the addition of FA can retard the hydration of cement [22],[23] and [24]. The retardation phenomenon is related to the presence and properties of FA. It is suggested that the FA surface acts somewhat like a calcium-sink, and calcium in solution is

INTRODUCTION

removed by the abundant aluminum associated with FA, as AFt phases preferentially forms on the surface of FA [22] and [23]. This depresses the Ca^{2+} concentration in solution during the first 6 h of hydration, and the formation of a Ca-rich surface layer on the clinker minerals is also postponed [22] and [23]. Therefore, the $Ca(OH)_2$ and C–S–H nucleation and crystallization are delayed and the cement hydration is simultaneously retarded [23]. Nevertheless, with a slow increase of the $Ca(OH)_2$ concentration in normal strength concrete (NSC), the pozzolanic reaction of FA can be further proceeded and the mechanical properties of NSC at 91 days can be further enhanced [25], [26] and [27]. Additionally, the activity of LP in the cementitious system is still under debate. Many researchers treat LP as a filler and have experimentally demonstrated that the principal properties of cement are not negatively affected if small quantities of LP (5–6%) are added during the cement grinding [28], [29], [30] and [31]. On the other hand, some investigations [32], [33] and [34] showed that, during the hydration process of cement with LP, tricalcium aluminate (C_3A) can react with calcium carbonate to form both high- and low carbonate forms of calcium carboaluminate (CCA) in much the same manner as C_3A reacts with calcium sulfate to form high- and low-sulfate forms of calcium sulpoaluminate (CSA). Furthermore, the reaction of LP largely depends on its fineness, which can be demonstrated by the phenomenon that the LP with d_{50} of about 0.7 μm could effectively enhance the heat flow of cement during the hydration process [35]. Although a significant amount of investigations regarding the effect of mineral admixtures on the physical and chemical characteristics of mortar or concrete can be easily found, they all focus only on NSC, in which the water to binder ratio is relatively high and very limited SP dosage is utilized. However, the cementitious system of UHPC is very different from that of NSC, which cause that it is difficult to evaluate the influence of mineral admixtures on the cement hydration and properties development of UHPC, based on the knowledge obtained from NSC. Therefore, to efficiently develop UHPC, it is important to understand the effect of different mineral admixtures on the properties and hydration process of UHPC.

For the design of mortars and concretes, several mix design tools are in use. Based on the properties of multimodal, discretely sized particles, De Larrard and Sedran [36] and [37] postulated different approaches to design concrete: the Linear Packing Density Model (LPDM), Solid Suspension Model (SSM) and Compressive Packing Model (CPM). Furthermore, Fennis et al. [38] developed a concrete mix design method based on the concepts of De Larrard and Sedran [36] and [37]. However, all these design methods are based on the packing fraction of individual solid components (cement, sand, etc.) and their combinations, and therefore it is complicated to include very fine particles in these mix design tools, as it is difficult to determine the packing fraction of such fine materials or their combinations. Another possibility for mix design is offered by an integral particle size distribution approach of continuously graded mixes (modified Andreasen & Andersen particle packing model), in which very fine particles can be integrated with considerably lower effort, as detailed in [39]. Additionally, based on the previous experiences and investigations of the authors [40], [41], [42] and [73], by applying this modified Andreasen & Andersen particle packing model, it is possible to produce a dense and homogeneous skeleton of UHPC or UHPFRC with a relatively low binder amount (about 650 kg/m^3). Consequently, it can be shortly concluded that such an optimized design of concrete with appropriate amount of mineral admixtures can be a promising approach to produce Ultra-High Performance Concrete (UHPC) in an efficient way.

In general, based on these premises, the objective of this study is to develop UHPC and evaluate the influence of different mineral admixtures on the fresh and hardened behavior, hydration kinetics and thermal properties of the developed UHPC. Techniques such as isothermal calorimetry, thermal analysis and scanning electron microscopy are employed to investigate the hydration mechanism and microstructure development of concrete. Additionally, to evaluate the environmental impacts of the designed UHPC, its embedded CO_2 emission is calculated and compared with that of UHPCs found in the literature.

MATERIALS AND EXPERIMENTAL METHODOLOGY

Materials

The cement used in this study is Ordinary Portland Cement (OPC) CEM I 52.5 R, provided by ENCI (the Netherlands). A polycarboxylic ether based superplasticizer is used to adjust the workability of UHPC. The FA, GGBS and LP are used to replace cement. Two types of sand are used, one is a normal sand with the fraction 0–2 mm and the other one is a micro-sand with the fraction 0–1 mm (Graniet-Import Benelux, the Netherlands). One type of nano-silica slurry is selected as an high active pozzolanic material in this study. More detailed information and characteristics of the used materials are shown in Table 1, Table 2, Table 3 and Table 4 and Fig. 1 and Fig. 2. It can be noticed that the particle size distribution of the used

FA, GGBS and LP is comparable to that of cement. Therefore, when the cement is replaced by FA, GGBS or LP, the particle packing of the whole solid skeleton is only slightly affected.

Table 1: Materials types and densities

Materials	Type	Specific density (kg/m³)	Pozzolanic activity index (28 days)
Cement	CEM I 52.5 R	3150	–
FA	–	2293	83
GGBS	–	2893	96
LP	–	2710	–
Fine sand	Micro-sand	2720	–
Coarse sand	Sand 0–2	2640	–
Superplasticizer	Polycarboxylate ether	1050	–
Pozzolanic material	Nano-silica (nS)	2200	113

Table 2: Characterization of the used nano-silica

Type	Slurry
Stabilizing agent	Ammonia
Specific density (g/cm³)	2.2
pH (at 20 °C)	9.0–10.0
Solid content (% w/w)	20
Viscosity (mPa s)	⩽100
BET (m²/g)	22.7
PSD by LLS (μm)	0.05–0.3
Mean particle size (μm)	0.12

Table 3: Characteristics of the powder materials

Materials	Specific density, $\rho_s(g/cm^3)$	Water demand (Puntke test), amw-p (g)	Computed void fraction, φ (%)	Particle shape factor, $\zeta_{Reschke}$ (–)
CEM I 52.5 R	3.15	13.2	45.4	1.68
LP	2.72	10.8	37	1.26
FA	2.29	11.2	33.9	1.2
GGBS	2.89	13.2	43.3	1.58

Table 4: Oxide composition of cement, FA, GGBS, LP and nS

Substance	Cement (mass%)	FA (mass%)	GGBS (mass%)	LP (mass%)	nS (mass%)
CaO	64.6	4.46	38.89	89.56	0.08
SiO_2	20.08	55.32	34.18	4.36	98.68
Al_2O_3	4.98	22.45	13.63	1	0.37
Fe_2O_3	3.24	8.52	0.51	1.6	–
K_2O	0.53	2.26	0.43	0.34	0.35
Na_2O	0.27	1.65	0.33	0.21	0.32
SO_3	3.13	1.39	1.41	–	–
MgO	1.98	1.89	10.62	1.01	–
TiO_2	0.3	1.17	–	0.06	0.01
Mn_3O_4	0.1	0.11	–	1.605	–
P_2O_5	0.74	0.76	–	0.241	0.15
Cl^-	0.05	0.02	–	–	0.04

MATERIALS AND EXPERIMENTAL METHODOLOGY

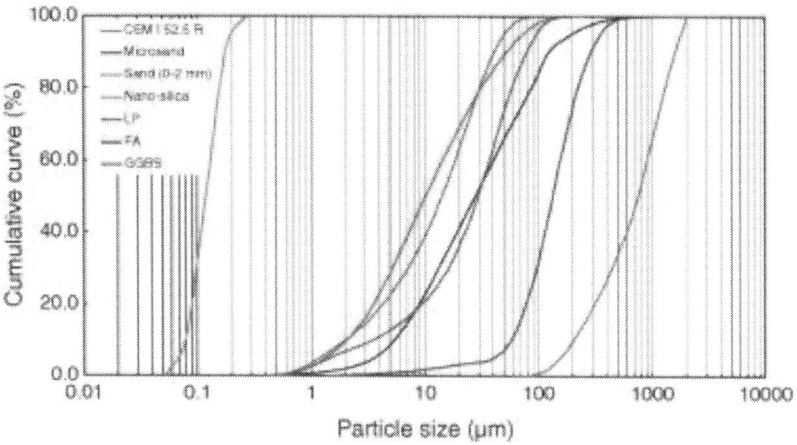

Figure 1: Particle size distributions of the used materials.

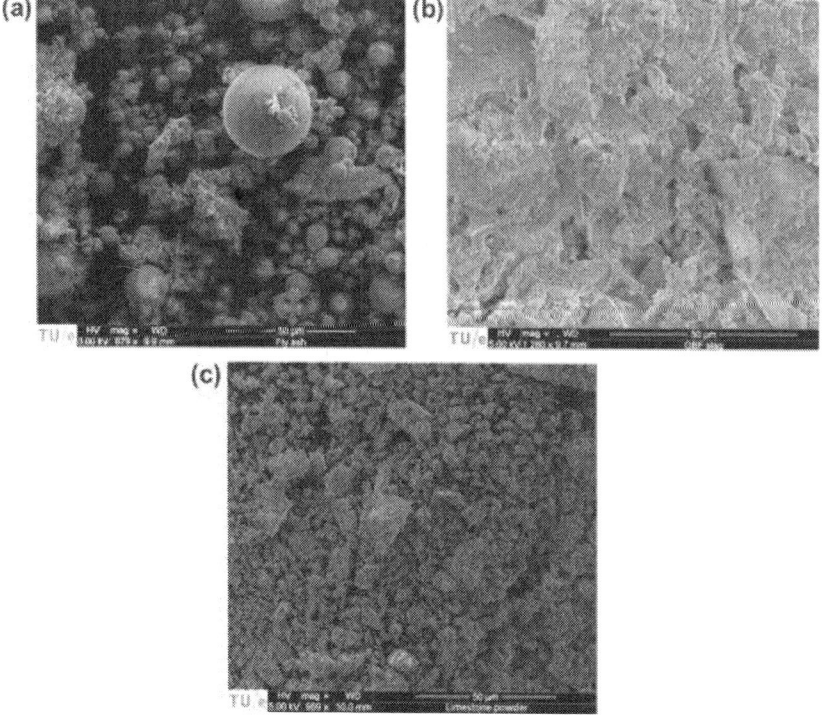

Figure 2: SEM pictures of used FA (a), GGBS (b) and LP (c).

EXPERIMENTAL METHODOLOGY

Mix Design of UHPC

In this study, the modified Andreasen and Andersen model is utilized to design all the concrete mixtures, which reads as follows [43], [86], [87] and [88]:

$$P(D) = \frac{D^q - D_{min}^q}{D_{max}^q - D_{min}^q} \qquad (1)$$

where D is the particle size (μm), $P(D)$ is a fraction of the total solids being smaller than size D, D_{max} is the maximum particle size (μm), D_{min} is the minimum particle size (μm) and q is the distribution modulus.

As presented in the literature [44], [45], [46] and [47], different types of concrete can be designed using Eq.(1) by applying different values of the distribution modulus q, as it determines the proportion between the fine and coarse particles in the mixture. As recommended in [44], considering that a high amount of fine particles is utilized to produce the UHPC, the value of q is fixed at 0.23 in this study. The modified Andreasen and Andersen model (Eq. (1)) acts as a target function for the optimization of the composition of mixture of granular materials. The proportions of each individual material in the mix are adjusted until an optimum fit between the composed mix and the target curve is reached, using an optimization algorithm based on the Least Squares Method (LSM), as presented in Eq. (2). When the deviation between the target curve and the composed mix, expressed by the sum of the squares of the residuals (RSS) at defined particle sizes, is minimized, the composition of the concrete is considered the best one (optimized packing)[47].

$$RSS = \sum_{i=1}^{n} (P_{mix}(D_i^{i+1}) - P_{tar}(D_i^{i+1}))^2 \qquad (2)$$

where P_{mix} is the composed mix, and the P_{tar} is the target grading calculated from Eq. (1).

The developed UHPC mixtures are listed in Table 5. In total, three different types of UHPC and one reference are designed, and three different water to binder ratios are chosen. Compared to the reference sample, about 30% of Portland cement (by mass) is replaced by FA, GGBS or LP in the UHPC mixtures. It can be noticed from Fig. 3, that the resulting integral grading curves of all the designed concretes are comparable to each other. The deviation between the target curves and composed mixes (RSS) are 101, 79, 85 and 74 for the reference mixture and the mixtures with FA, GGBS and LP, respectively.

Table 5: Mix recipes of the designed concrete

No.	C (kg/m³)	FA (kg/m³)	GGBS (kg/m³)	LP (kg/m³)	S (kg/m³)	MS (kg/m³)	nS (kg/m³)	W (kg/m³)	SP (kg/m³)	W/B	SP/C
1	582.1	259.9	0	0	1039.5	216.6	24.3	173.2	43.3	0.2	0.07
2	591.9	264.3	0	0	1057	220.2	24.7	159.3	44	0.18	0.07
3	600	267.9	0	0	1071.4	223.2	25	147.8	44.6	0.165	0.07
4	596.1	0	266.1	0	1064.5	221.8	24.8	177.4	44.4	0.2	0.07
5	606.4	0	270.7	0	1082.9	225.6	25.3	163.2	45.1	0.18	0.07
6	614.9	0	274.5	0	1098	228.8	25.6	151.5	45.8	0.165	0.07
7	592.6	0	0	264.6	1058.3	220.5	24.7	176.4	44.1	0.2#	0.07
8	602.8	0	0	269.1	1076.5	224.3	25.1	162.2	44.9	0.18#	0.07
9	611.2	0	0	272.9	1091.4	227.4	25.5	150.6	45.5	0.165#	0.07
Ref.[1]	868.8	0	0	0	1072.5	223.4	25	178.8	44.7	0.2	0.05
Ref.[2]	883.9	0	0	0	1091.2	227.3	25.5	164.4	45.5	0.18	0.05
Ref.[3]	896.3	0	0	0	1106.6	230.5	25.8	152.7	46.1	0.165	0.05

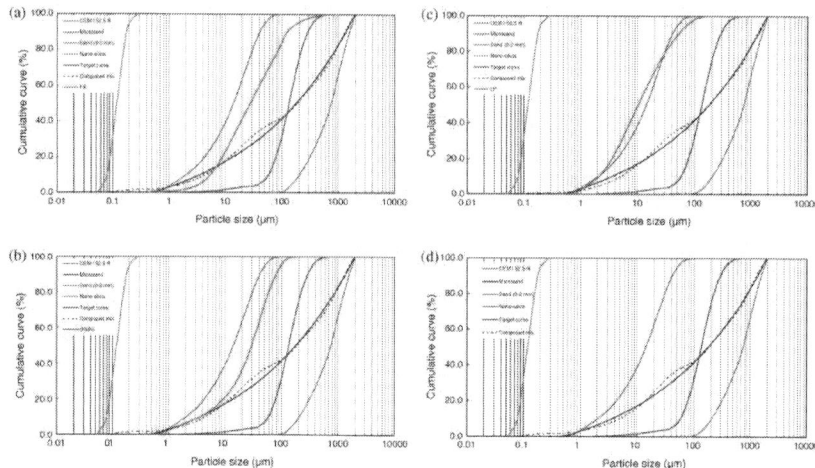

Figure 3: PSDs of the ingredients, target and optimized grading curves of the designed concretes: (a) with FA, (b) with GGBS, (c) with LP and (d) reference mixture.

Determination of Water Demand

In this study, the Puntke test is employed to evaluate the water demand of the powder materials (cement, FA, GGBS and LP). The water demand from Puntke test shows the water absorption capacity of the tested powder at the point of saturation, which depicts the transition from a coherent packing to a suspension [48]. Therefore, a fine, cohesion-free granular skeleton cannot be self-compacted to a specific packing density until the water content is sufficient for the saturation of the dense grain structure [44]. The first sign of bleeding is a glimmering surface of the water–powder mixture, which also is the evaluation target criterion of the addressed test. Additionally, Puntke test assumes that for the point of saturation the granular blend becomes free of air (the void fraction is completely filled with water), which derives a relation between the void fraction and the involved volumes of water and powder represented by their masses. Hence, void faction of the saturated powder material can be computed as follows [48]:

$$\psi = \frac{V_w}{V_p + V_w} \qquad (3)$$

where ψ is the void faction of the saturated powder material, V_w is the volumetric water demand of the powder material for saturation, V_p is the volume of the tested powder material.

Mixing Procedure
In this study, the mixing procedure follows the method shown in [40]:

1. All powders and sand fractions are added into the mixer for dry mixing (30 s at low speed).
2. Then, around 75% of water is added into the mixer. After mixing for 90 s (low speed), the mixer is stopped for 30 s.
3. Afterwards, the remaining water and SP are added, and the mixture is mixed at low speed for 180 s.
4. Finally, the mixture is mixed at high speed for 120 s.

The mixing is always executed under laboratory conditions with dried and tempered aggregates and powder materials. The room temperature while mixing and testing is constant at around 21 °C.

Flowability of UHPC
To evaluate the flowability of UHPC, the flow table tests are performed following EN 1015-3 [49]. During the test, the cone is lifted straight upwards in order to allow free flow of the mixture without any jolting (flowing suggestions from [44]). In the test, two diameters perpendicular to each other (d_1 (mm) and d_2 (mm)) are determined. Their mean is employed to compute the relative slump (Γ) via:

$$\Gamma = \left(\frac{d_1 + d_2}{2d_0}\right)^2 - 1 \tag{4}$$

where d_0 represents the base diameter of the used cone (mm), i.e. 100 mm in the case of the Hägermann cone. The relative slump Γ is a measure for the deformability of the mixture, which originally was introduced by Okamura and Ozawa [50] as the relative flow area R.

Mechanical Properties of UHPC

After preforming the flowability tests, the fresh concrete is cast in molds with the dimensions of 40 mm × 40 mm × 160 mm. The prisms are demolded approximately 24 h after casting and then cured in water at about 21 °C. After curing for 28 and 91 days, the flexural and compressive strengths of the specimens are tested according to EN 196-1 [51]. At least three specimens are tested at each age to compute the average strength.

Water-Permeable Porosity of UHPC

The water-permeable porosity of the designed UHPC is measured applying the vacuum-saturation technique, which is referred to as the most efficient saturation method [52]. The saturation is carried out on at least 3 samples (100 mm × 100 mm × 20 mm) for each mix, following the description given in NT Build 492 [53] and ASTM C1202 [54].

The water permeable porosity is calculated from the following equation:

$$\varphi_{v,water} = \frac{m_s - m_d}{m_s - m_w} \cdot 100 \tag{5}$$

where $\varphi_{v,water}$ is the water permeable porosity (%), m_s is the mass of the saturated sample in surface-dry condition measured in air (g), m_w is the hydrostatic mass of water-saturated sample (g) and m_d is the mass of oven-dried sample (g).

Calorimetry Analysis of UHPC

Following the recipes shown in Table 5, the pastes (without aggregates) are produced for the calorimetry analysis. The water to binder ratio of the prepared mixtures is fixed at 0.18 (based on the results of mechanical properties that will be shown later). All the pastes are mixed for two minutes and then injected into a sealed glass ampoule, which is then placed into the isothermal calorimeter (TAM Air, Thermometric). The instrument is set to a temperature of 20 °C. After 7 days, the measurement is stopped and the obtained data is analyzed. All results are ensured by double measurements (two-fold samples).

Thermal Test and Analysis of UHPC

A Netzsch simultaneous analyzer, model STA 449 C, is used to obtain the Thermo-gravimetric (TG) and Differential Scanning Calorimetry (DSC) curves of UHPC paste. The water to binder ratio of the tested sample is fixed at 0.18 (based on the results of mechanical properties that will be shown later). Analyses are conducted at the heating rate of 5 °C/min from 20 °C to 1000 °C in flowing nitrogen environment.

EXPERIMENTAL RESULTS AND DISCUSSION

Fresh Behavior of the Designed UHPC

The relative slump of fresh UHPC mixtures versus the volumetric water to powder (particle size < 125 μm) ratio is presented in Fig. 4. As can be seen, with an increase of the water amount, the relative slump of all the concrete mixtures increases linearly. The intersection of these linear functions with the axis of ordinates at $\Gamma = 0$ depicts the retained water ratio where no slump takes place [50]. In other words, this denotes the maximum amount of water which can be retained by the particles. Exceeding this water content will turn the coherent bulk into a concentrated suspension [44]. In this study, it can be noticed that the water demand of each mixture follow the order: FA (0.306) < LP (0.315) < GGBS (0.359) < reference sample (0.384). Nevertheless, these results are not in accordance with the results obtained from Puntke test (as shown inTable 3). This should be attributed to the following two reasons: (1) the used mineral admixtures are different from each other, which can also affect the workability of the concrete mixture. As presented in Fig. 2, a large amount of angular particles can be observed in GGBS, while that the FA particles are more spherical. The particle shape factors (shown in Table 3) of the used mineral admixtures are 1.20, 1.58 and 1.28 for FA, GGBS and LP, respectively [44]. When the shape factor is close to 1, the shape of the particle is spherical, which can further help to improve the flowability of the concrete mixture; (2) the utilized superplasticizer has different effect on the slump flow value of various powders. As described in [85], the efficiency of superplasticizer largely depends on the zeta potential along the entire surface of the tested powder particles. The experiments shown in [85] demonstrate that, in most cases, cement needs more superplasticizer to reach a certain slump flow value compared to that of FA, GGBS and LP. Hence, based on the two reasons mentioned above, the mixture with FA has the lowest demand water amount among all the analyzed concrete mixtures.

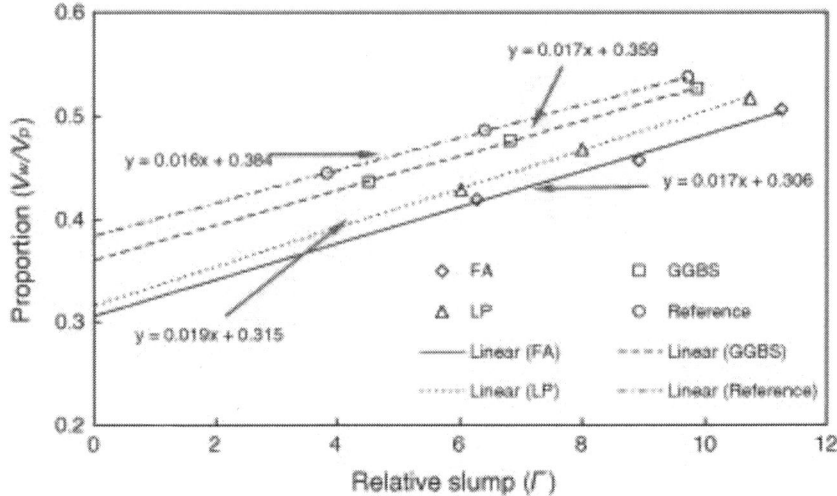

Figure 4: Relative slump (Γ) versus volumetric water /powder ratio (V_w/V_p).

The slopes of the lines shown in Fig. 4, called the deformation coefficient, represent the sensitivity of the mixture to the water amount needed to attain a certain flowability [50]. When the value of deformation coefficient is relatively small, a big change in deformability can be observed (to a certain change in water dosage), which means the mixture tends to bleed or segregate sooner than the mixtures with larger deformation coefficients [44] and [55]. In this study, the obtained deformation coefficient values are small and similar to each other, which implies that all the designed mixtures are sensitive to the water amount. This should be attributed to the specific characteristics of UHPC, which has a large amount of superplasticizer and low water content. Hence, to achieve a well flowable UHPC mixture, the added water amount should be precisely controlled.

Mechanical Properties of the Designed UHPC

The flexural and compressive strengths of UHPC at 28 and 91 days are shown in Fig. 5. A very slight variation of the strengths can be observed when the water/binder ratio increases from 0.165 to 0.18. Nevertheless, with a further increase of the water/binder ratio (from 0.18 to 0.20), the mechanical properties of the produced UHPC decrease. This phenomenon is different from that shown in [27]. In most cases, due to the fact that the excessive water can enhance the porosity of concrete, the strengths of concrete gradually decrease with an increase of the water amount. The difference between the obtained results and the results presented in the literature should be attributed to the fact that a large amount of powder and limited water are utilized to produce the UHPC. When the water to binder ratio is relatively small, the added water is more significantly absorbed by the powders (cement, FA, GGBS or LP in this study), and cannot react with cement, which causes that the amount of cement hydration products is limited and the strength development of UHPC is restricted. Hence, in this study, the strengths difference between the mixtures with lowest and medium water amount is not significant. There is an optimal value of water/binder ratio, at which the strengths of the UHPC can be highest.

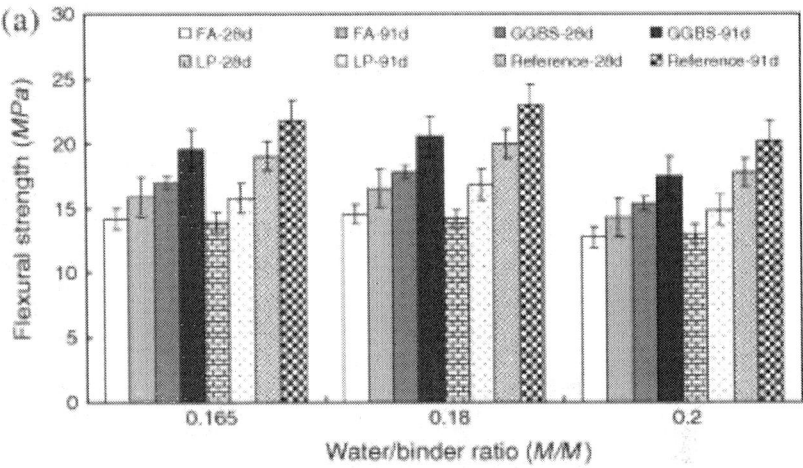

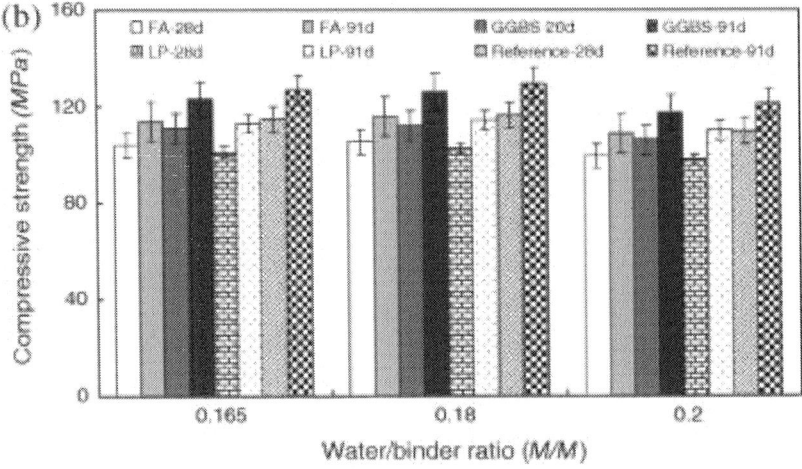

Figure 5: Mechanical properties of the developed UHPC mixtures with different mineral admixtures and water amount: (a) flexural strength and (b) compressive strength.

Furthermore, it can be found here the mixture with GGBS has superior mechanical properties at both 28 and 91 days, while that the strengths of the mixtures with FA or LP are similar to each other. The observed trend is conflicting with the results obtained for normal strength concrete [16], [17], [18], [19], [20], [21], [22],[23], [24], [25], [26] and [27]. Normally, the pozzolanic reaction of FA begins at the age of 3 days after blending with cement and water [56], [57] and [58]. Nevertheless, this pozzolanic reaction is much slower than the Portland cement hydration [56], [57] and [58]. The main hydrate of cement and fly-ash, calcium silicate hydrate (C–S–H), adopts two distinct morphologies: a low density C–S–H at the surface of cement and FA particles and a high density C–S–H deeper into the cement and FA particles [59] and [60]. After curing for 28 days, a limited amount of C–S–H gel can be generated, and the microstructure of the concrete is less dense than the one with GGBS. With an ongoing cement hydration, more portlandite can be generated and the pozzolanic reaction of FA can be accelerated, which causes that the already formed pore structure in concrete is filled by the newly generated C–S–H and the mechanical properties of concrete are significantly improved after curing for 91 days [22], [23], [24], [25], [26] and [27]. Nevertheless, in this study, the strengths of the mixture with FA are similar to that of the mixture with LP after curing for 91 days, which implies that the pozzolanic reaction of FA cannot proceed well in the cementitious system of UHPC (assuming limestone is a non-reactive material).

From the results obtained in this study, it can be summarized that the specific system of UHPC (very low water amount and high SP content) can significantly influence the pozzolanic reaction of FA and mechanical properties of the hardened UHPC. As already mentioned, the strengths of the mixture with GGBS are superior, and comparable to the reference sample (with 50% more cement). To further investigate the pozzolanic reaction of FA/GGBS or their effect on cement hydration at early age, some other techniques (isothermal calorimetry, thermal analysis) are employed and presented later.

Water-Permeable Porosity of the Designed UHPC
Fig. 6 illustrates the variation of the total water-permeable porosity (after curing for 28 or 91 days) of UHPC at different water to binder ratios. In accordance with the mechanical properties results, the water-permeable porosity of UHPC firstly remains stable and then increases with an increase of the water to binder ratio. This should also be attributed to the fact that a large amount of powder and limited water are utilized to

produce the UHPC. When the water amount is relatively low, the added water is more significantly absorbed by the powders (cement, FA, GGBS or LP in this study), and cannot react with cement, which cause that the amount of cement hydration products is limited and the water-permeable porosity is relatively high. On the other hand, when the water content is higher, the excessive water can obviously enhance the porosity of concrete, as described in [27]. Hence, there is an optimal water to binder ratio, at which the water-permeable porosity of UHPC can be minimized. Moreover, it can also be found that the water-permeable porosity of the mixtures with FA or LP is relatively higher than that with GGBS and the reference mixture, which implies that the mechanical properties of the mixtures with FA and LP are lower than that of the mixture with GGBS.

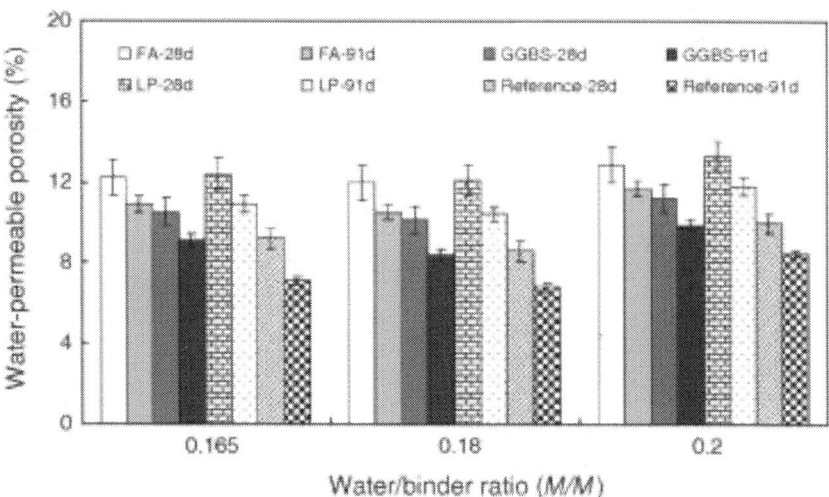

Figure 6: Total water-permeable porosity of the designed UHPC with different mineral admixtures and water amount.

To clearly determine the relationships between the water-permeable porosity and mechanical properties of UHPC, the results obtained here are compared with the existing models (as shown in Fig. 7). Historically, several general types of models have been developed for cement-based materials. From a study of the compressive strength of Al_2O_3 and ZrO_2, Ryshkewitch [61] proposed the following relationship:

$$\sigma = \sigma_0 \cdot \exp(-k \cdot p) \qquad (6)$$

where σ is the strength, σ_0 is the strength at zero porosity, p is the porosity of the tested material and k is an empirical constant.

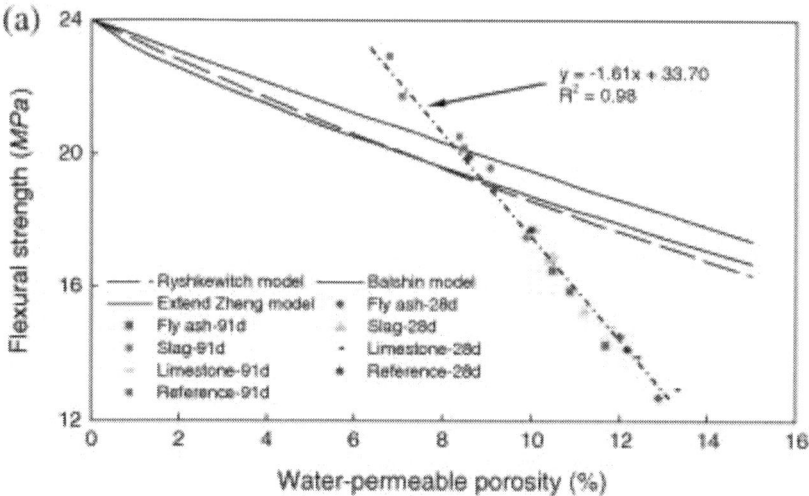
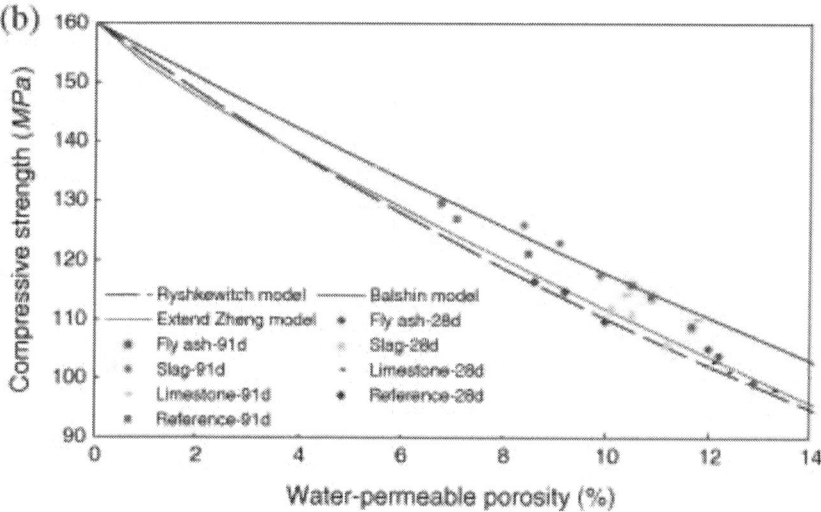

Figure 7: Relationship between water-permeable porosity and flexural strength (a) and compressive strength (b) of the developed UHPC.

Balshin [62] suggested the following relationship:

$$\sigma = \sigma_0 \cdot (1-p)^b \quad (7)$$

where b is the empirical constant.

According to the investigation shown in [63], Chen et al. proposed the extended Zheng's model:

$$\sigma = \sigma_0 \cdot \left[\left(\frac{p_c - p}{p_c}\right)^{1.85} \cdot \left(1 - p^{2/3}\right)\right]^{1/2} \tag{8}$$

where p_c is the percolation porosity at failure threshold. In the present study, all the empirical constants for the models mentioned above are chosen as recommended in [63]. Based on the empirical fitting of the experimental data to the presented models, the maximum flexural and compressive strengths (σ_0) of the UHPC (when porosity is zero) are equal to 24 and 160 MPa, respectively.

From Fig. 7, it can be found that all the presented models can well represent the relationships between the water-permeable porosity and the compressive strength of the developed UHPC. However, these models are inaccurate in predicting the obtained relationships between the water-permeable porosity and the flexural strength. The existing models obviously underestimate the flexural strength of UHPC when its water-permeable porosity is less than about 8%. Additionally, these models also overestimate the flexural strength of UHPC when its water-permeable porosity is larger than 10%. These phenomena may be attributed to the relatively low water-permeable porosity and high strengths of UHPC. For normal concrete, the water-permeable porosity is relatively high, which is the reason for lower mechanical properties (especially the flexural strength). For instance, Safiuddin and Hearn [52] reported a porosity of 20.5% of concrete produced with a water/cement ratio of 0.60, employing the same porosity measurement method as used in the present study (vacuum-saturation technique). Many of these empirical formulas are derived for normal strength concrete (NSC). However, for UHPC, its porosity is very low and its flexural strength is around 3–4 times of that of NSC. Therefore, these empirical equations are less precise to represent the relationships between the water-permeable porosity and the flexural strength of the developed UHPC.

Based on the obtained results, a new relationship between the water-permeable porosity and the flexural strength of the UHPC is shown as follows:

$$\sigma = \left(\frac{p_c - p}{p_c}\right)\sigma_0 \tag{9}$$

in which the σ_0 is about 33.7 MPa, and the p_c is around 0.21. It can be noticed that the derived p_c value is much smaller than that recommended in [63] (0.78), which could be the reason that the existing models cannot well represent the relationships between the water-permeable porosity and the flexural strength of the developed UHPC. As mentioned before, compared to NSC, UHPC has much lower porosity and higher flexural strength. Therefore, to precisely establish the relationships between the water-permeable porosity and the flexural strength of UHPC, the crucial parameters should be reasonably adjusted.

Hydration Kinetics of the Designed UHPC

Based on the calorimetry test results, the influence of the different mineral admixtures on the cement hydration of UHPC is investigated and presented in Fig. 8. It is apparent that the influence of FA, GGBS or LP on the early hydration kinetics of the designed UHPC is very similar, which can be demonstrated by the relatively small difference between the observed dormant period (calculated as the time between the lower point of the heat flow curve and the first inflection point in the main peak), relative setting time (calculated as the time between the first and the second inflection points in the heat flow curve), as well as the time to reach the maximum hydration peak. This phenomenon is not in accordance with the results shown in [19],[20], [22], [23] and [64]. In most cases, GGBS can quickly react with $Ca(OH)_2$ and generate the C–S–H gel, while the reaction between FA and portlandite is relatively slower. It is suggested that the fly ash surface acts as a Ca^{2+} sink, which is caused by the reaction of the aluminate in the fly ash with the Ca from the solution and/or chemisorption of Ca^{2+} ions on the fly ash surface [23] and [65]. This would retard the formation of C–S–H nuclei and thereby delay the end of the induction period. Hence, when the particle size distributions of GGBA, FA and LP are similar to each other, the activity of GGBS should be much higher than that of FA and LP in concrete at early age.

EXPERIMENTAL RESULTS AND DISCUSSION

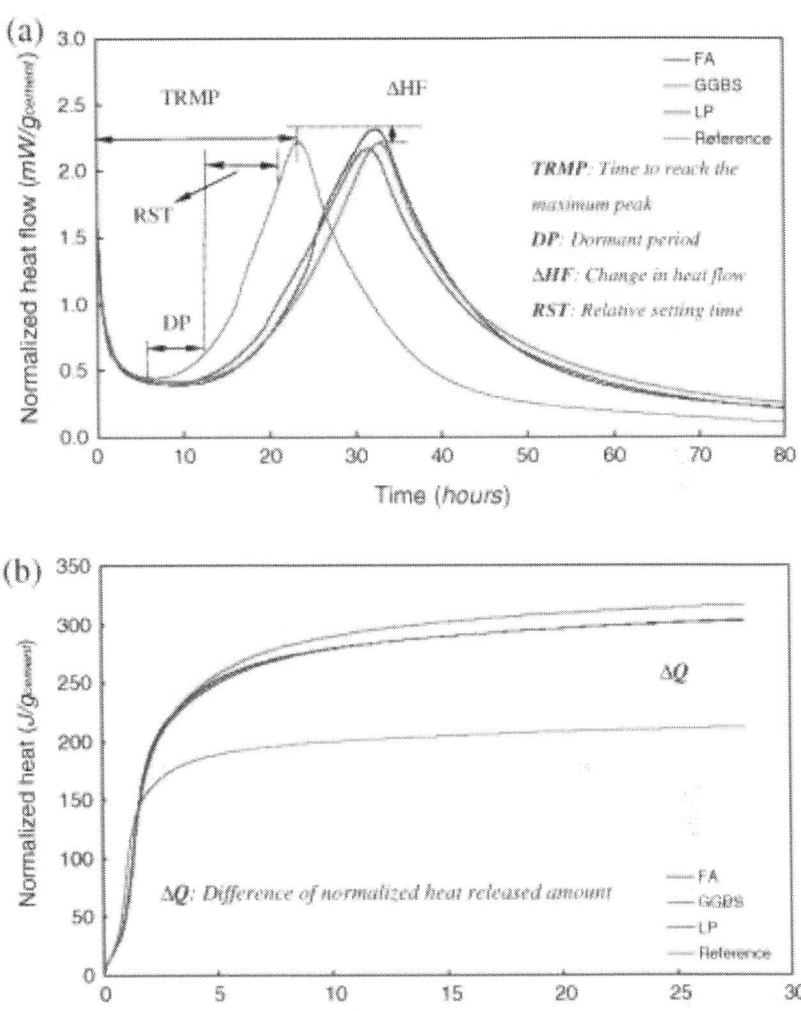

Figure 8: Calorimetry test results of UHPC pastes with different mineral admixtures: (a) normalized heat flow and (b) normalized total heat.

To better explain these phenomena, the following reasons should be considered: (1) a large amount of superplasticizer is utilized in the production of the UHPC. According to the investigation of Jansen [66], complex Ca^{2+} ions from pore solution by the superplasticizer can touch the polymer absorbed on the nuclei or the anhydrous grain surfaces, which in turn might lead to prevention of the nuclei growth or to the dissolution of the anhydrous grains. Hence, the early hydration of the cement is significantly retarded and the generation of Ca(OH)$_2$ is restrained. Due to the insufficient amount of portlandite in the mixtures, the pozzolanic reaction cannot well progress, which causes that the difference of the pozzolanic activity between FA and GGBS is not easy to be observed in the calorimetry tests; (2) low water content is used in the UHPC mixtures. To achieve good mechanical properties, high powder amount and low water content are normally used to produce UHPC, which causes that much water is absorbed by the powder materials and there is litter free water in the cementitious system. Hence, the diffusion of Ca^{2+} and OH$^-$ is restricted, and pozzolanic reaction of FA or GGBS is simultaneously postponed.

The normalized (by 1 g of cement) total heat of the designed UHPC mixtures is illustrated in Fig. 8b. The total heat is the contribution of heat produced by the cement particles themselves and by the pozzolanic reaction between the active mineral admixtures and the precipitated Ca(OH)$_2$[67]. The total heat can be related to the hydration degree of the paste, and this hydration degree is related to the compressive strength of the mixture, if the parameters of the microstructure are similar. Thus, a higher compressive strength is expected with the progressive increase of the total heat released. In this study, after 28 days it can be noticed that the normalized heat of the mixture with GGBS is the largest, which is followed by the one with FA and LP. As described before, due to effect of the large amount of superplasticizer and low water content in UHPC, the pozzolanic reaction of GGBS cannot well progress during the initial 5 days. However, afterwards, with an increasing concentration of Ca(OH)$_2$, the pozzolanic reaction of GGBS is promoted, which simultaneously causes that more heat can be released and the mechanical properties of the concrete can be enhanced. Additionally, the normalized heat of the mixture with FA is similar to that with LP, which implies that the FA and LP have similar contributions to the cement hydration after 28 days. Additionally, it can be found that the normalized heat of reference sample is significantly lower than the mixtures with mineral admixtures. This should be attributed to the fact that the calculation of normalized heat is based on the released heat per gram cement, and in the mixtures with mineral admixtures the utilized cement amount is obviously lower than that of reference sample.

Consequently, according to the results obtained in this study, it can be found that the hydration kinetics of UHPC is different from that of normal concrete. Due to the effects related to the superplasticizer and water dosages, the cement hydration and pozzolanic reaction of mineral admixtures are significantly retarded.

Thermal Analysis of the Hardened UHPC

The DSC and TG curves of the UHPC pastes after hydrating for 28 and 91 days are presented in Fig. 9 and Fig. 10. From the DSC curves, it is apparent that there main peaks exist in the vicinity of 105 °C, 450 °C and 800 °C for all the samples, which can be attributed to the evaporation of free water, decomposition of $Ca(OH)_2$ and decomposition of $CaCO_3$, respectively [68], [69], [70], [71] and [72]. Based on the test results shown in Figs. 9a and 10a, the samples for TG analysis were subjected to isothermal treatment during the test, which was set at 105 °C, 450 °C and 800 °C for 2 h. From the obtained TG curves, it can be noticed that all the tested samples show a similar tendency of losing their weight. However, their weight loss rates in each temperature range are different, which means that the amounts of the substances reacting at each treatment stage are different. It is important to note that the mass loss of portlandite of the mixture with GGBS is the smallest at 28 days, which implies that the pozzolanic activity of GGBS is relatively higher so that more portlandite has already been consumed. Fig. 8 confirms this phenomenon. However, after curing for 91 days, the mass loss of portlandite still follows the order: GGBS < FA < LP < reference concrete, while the differences between the mixtures with FA and LP is relatively small. Hence, it can be concluded that the specific cementitious system of UHPC significantly restricts the pozzolanic reaction of FA, which causes that a very limited amount FA can react with $Ca(OH)_2$ even after 91 days. Hence, it explains why the mechanical properties of the mixture with FA are lower than that with GGBS at both 28 and 91 days. The observed phenomenon is not in accordance with the results obtained in normal concrete system. As mentioned before, with the increase of the portlandite amount, the pozzolanic reaction of FA can be promoted, and the already-formed pore structure in concrete is filled by the newly generated C–S–H [16], [17], [18], [19], [20] and [21]. Consequently, it is not reliable to predict the effect of FA on the properties of UHPC, based on the results obtained on traditional concrete. Additionally, it can be noticed that the difference of the $Ca(OH)_2$ amount between the mixtures with mineral admixtures is relatively small. This phenomenon may be attributed to the reaction between nano-silica and $Ca(OH)_2$, which cause that very limited $Ca(OH)_2$ is available to react with FA or GGBS.

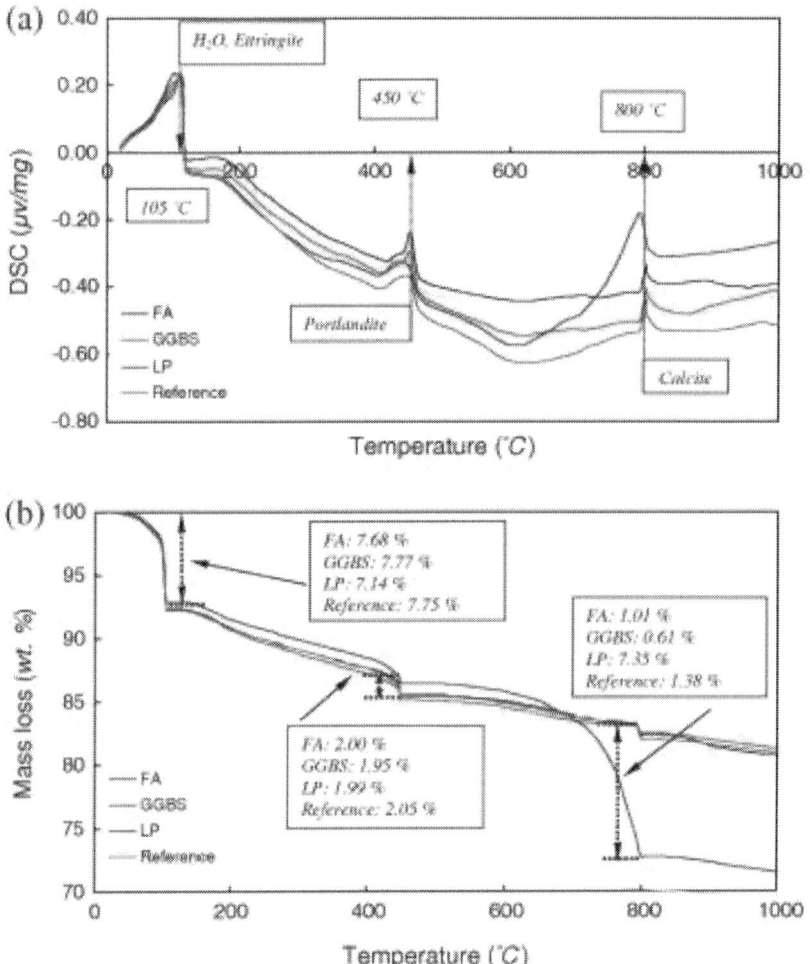

Figure 9: Thermal analysis results of UHPC pastes with different mineral admixtures (after hydrating for 28 days): (a) DSC curves and (b) TG curves.

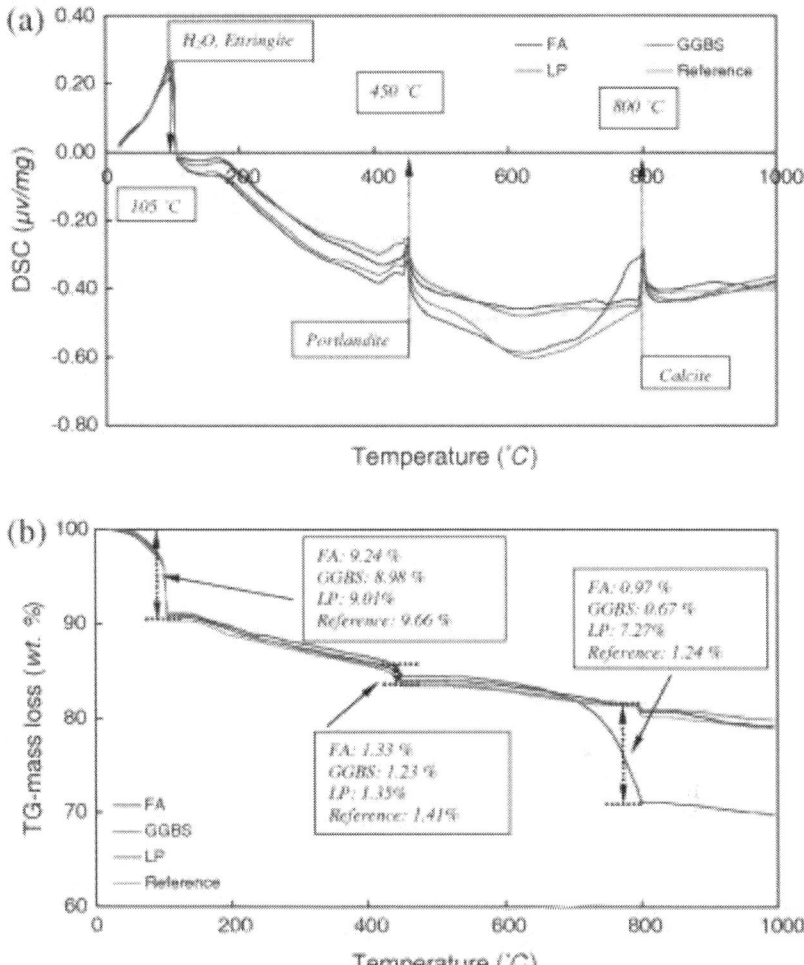

Figure 10: Thermal analysis results of UHPC pastes with different mineral admixtures (after hydrating for 91 days): (a) DSC curves and (b) TG curves.

According to the thermal analyses results, it is clear that there is more portlandite in the concrete with larger amount of cement (e.g. the reference system in this study) than the mixture with mineral admixtures, which does not play a positive role in improving the mechanical properties of concrete, especially when the portlandite hexagonal plates form distribute around the ITZ. When cement is appropriately replaced by GGBS, portlandite amount can be reduced and the already-formed pore structure in concrete can be filled by the newly generated C–S–H. Consequently, the UHPC with good mechanical properties can be produced with relatively low cement amount.

Ecological Evaluation of the Designed UHPC

To demonstrate that the designed UHPC is materials efficient and eco-friendly, its embedded CO_2 emission is evaluated in this study, focusing on the amount of materials required for 1 m^3 of compacted concrete. Based on the embodied CO_2 values for each components of concrete [74] and [89], the relationships between the CO_2 emission and the compressive strength of UHPCs are illustrated in Fig. 11. It can be noticed that the enhancement of compressive strength of all the analyzed UHPCs corresponds to an increase of the embedded CO_2 emission and environmental impact. Some of the presented UHPCs have superior mechanical properties (compressive strength is more than 200 MPa), but simultaneously, their embedded CO_2 emissions are also high (more than 1200 kg/m^3 concrete). However, it is important to notice that the data points representing UHPC developed in this study are all below of the trend line, which means the designed UHPC has a lower environmental impact than the other UHPCs. This is significant especially for the mixture with GGBS, as its compressive strength is larger than that with FA and LP, with a comparable embedded CO_2 emission at the same age. Additionally, it can be also found that the data points representing the reference concrete developed in this study are on the trend line, which implies that the relatively high cement amount is not helpful for producing UHPC with small environmental impact. This should be attributed to the fact that when the cement amount is relatively high, the cement hydration degree is smaller and the cement efficiency is lower, compared to the concrete with low cement amount[40]. Hence, to efficiently produce an eco-friendly UHPC with a reduced environmental impact, the mineral admixtures should be utilized to replace cement and the concrete design should be based on the optimized particle packing model.

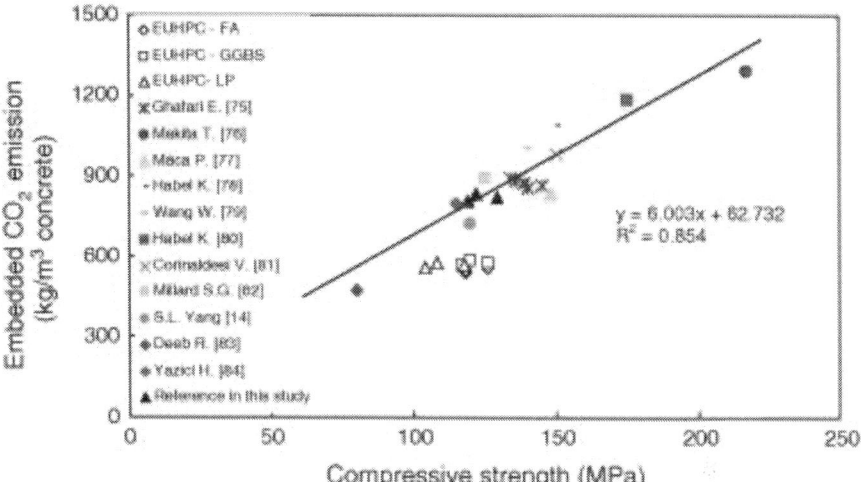

Figure 11: Comparison of embedded CO_2 emission of the developed eco-friendly UHPC (EUHPC) and other UHPCs or UHPFRCs [14], [75],[76], [77], [78], [79], [80], [81], [82], [83] and [84].

CONCLUSIONS

This paper presents the mix design and properties evaluation of an eco-friendly Ultra-High Performance Concrete (UHPC). From the results presented in this paper the following conclusions are drawn:

- In this study, based on the modified Andreasen & Andersen particle packing model, UHPC with different mineral admixtures (FA, GGBA, and LP) is produced. After comparing the embedded CO_2 emissions of the designed UHPC and other UHPCs, it is demonstrated that the proposed methodology allows production of an eco-friendly concrete with a relatively low environmental impact.
- The fresh behavior of the developed UHPC is evaluated. It is found that the water demand of each UHPC mixtures with FA, GGBS, LP and reference concrete follows the order: FA < LP < GGBS < reference. Moreover, the deformation coefficient values of UHPCs are small and close to each other, which implies that all the designed mixtures are sensitive to the water amount.
- The mechanical properties of UHPC with GGBS are obviously higher than that with FA or LP at both 28 and 91 days. Furthermore, a slight increase of the strengths can be observed when the water/binder ratio increases from 0.165 to 0.18. Nevertheless, with a further increase of

the water/binder ratio (from 0.18 to 0.20), the mechanical properties of the produced UHPC decrease.
- The existing models used to correlate the porosity and mechanical properties of concrete obviously underestimate the flexural strength of UHPC when its water-permeable porosity is less than about 8%, and overestimate the flexural strength of UHPC when its water-permeable porosity is larger than 10%. At the same time, all the presented models can well represent the relationships between the water-permeable porosity and the compressive strength of the designed UHPC.
- The hydration heat development curves of the UHPC mixtures with FA, GGBS and LP are similar to each other during the initial five days. Afterwards, the hydration rate of the mixture with GGBS is obviously accelerated. Due to the specific cementitious system of UHPC (very small water/binder ratio and relatively high SP amount), it is observed that the pozzolanic reaction of FA is significantly retarded, which causes that a very limited amount of FA can react with $Ca(OH)_2$ after curing for 91 days.

ACKNOWLEDGEMENTS

The authors wish to express their gratitude to Ir. G.C.H. Doudart de la Grée for assisting with the XRF and SEM testing, experimental work and valuable discussions. Moreover, the appreciation also goes to the following sponsors of the Building Materials research group at TU Eindhoven: Graniet-Import Benelux, Kijlstra Betonmortel, Struyk Verwo, Attero, ENCI, Provincie Overijssel, Rijkswaterstaat Zee en Delta – District Noord, Van Gansewinkel Minerals, BTE, V.d. Bosch Beton, Selor, Twee "R" Recycling, GMB, Schenk Concrete Consultancy, Geochem Research, Icopal, BN International, Eltomation, Knauf Gips, Hess ACC Systems, Kronos, Joma, CRH Europe Sustainable Concrete Centre, Cement&BetonCentrum and Heros (in chronological order of joining).

REFERENCES

1. Alarcon-Ruiz L, Platret G, Massieu E, Ehrlacher A. The use of thermal analysis in assessing the effect of temperature on a cement paste. Cem Concr Res 2005;35(3):609–13.

2. Alonso C, Fernandez L. Dehydration and rehydration processes of cement paste exposed to high temperature environments. J Mater Sci 2004;39(9):3015–24.
3. Andreasen AHM, Andersen J. Über die Beziehungen zwischen Kornabstufungen und Zwischenraum in Produkten aus losen Körnern (mit einigen Experimenten). Kolloid-Zeitschrift 1930;50:217–28 [in German..
4. ASTM C1202. Standard test method for electrical indication of concrete's ability to resist chloride ion penetration. In: Annual book of ASTM standards, vol. 04.02. American Society for Testing and Materials, Philadelphia, July 2005.
5. Balshin MY. Relation of mechanical properties of powder metals and their porosity and the ultimate properties of porous-metal ceramic materials. Dokl Askd SSSR 1949;67(5):831–4.
6. Bentz DP. Modeling the influence of LP filler on cement hydration using CEMHYD3D. Cem Concr Comp 2006;28(2):124–9.
7. Berry EE, Hemmings RT, Cornelius BJ. Mechanisms of hydration reactions in high volume FA pastes and mortars. Cem Concr Comp 1990;12(4): 253–61.
8. Bessey GE. Proceedings of the symposium on the chemistry of cements, Stockholm, lngeniarsveterskapsakaderrier, Stockholm; 1938. p. 233–4.
9. Brouwers HJH, Radix HJ. Self compacting concrete: theoretical and experimental study. Cem Concr Res 2005;35:2116–36.
10. BS-EN-1015-3. Methods of test for mortar for masonry - Part 3: Determination of consistence of fresh mortar (by flow table). British Standards Institution-BSI and CEN European Committee for Standardization; 2007.
11. BS-EN-196-1. Methods of testing cement - Part 1: Determination of strength. British Standards Institution-BSI and CEN European Committee for Standardization; 2005.
12. Carlson ET, Berman HA. Some observations on the calcium aluminate carbonate hydrates. J Res Nat Bur Stand 1960;64A(4):333–41.
13. Castellote M, Alonso C, Andrade C, Turrillas X, Campo J. Composition and microstructural changes of cement pastes upon heating, as studied by neutron diffraction. Cem Concr Res 2004;34(9):1633–44.
14. Chan Y, Chu S. Effect of silica fume on steel fiber bond characteristics in reactive powder concrete. Cem Concr Res 2004;34(7):1167–72.
15. Chen X, Wu S, Zhou J. Influence of porosity on compressive and tensile strength of cement mortar. Constr Build Mater 2013;40:869–74.
16. Constantinides G, Ulm FJ. The nanogranular nature of C–S–H. J Mech Phys Solids 2007;55:64–90.

17. Corinaldesi V, Moriconi G. Mechanical and thermal evaluation of Ultra High Performance Fiber Reinforced Concretes for engineering applications. Constr Build Mater 2012;26(1):289–94.
18. De Larrard F, Sedran T. Mixture-proportioning of high-performance concrete. Cem Concr Res 2002;32:1699–704.
19. De Larrard F, Sedran T. Optimization of ultra-high-performance concrete by the use of a packing model. Cem Concr Res 1994;24:997–1009.
20. Deeb R, Ghanbari A, Karihaloo BL. Development of self-compacting high and ultra high performance concretes with and without steel fibres. Cem Concr Comp 2012;34(2):185–90.
21. El-Dieb AS. Mechanical, durability and microstructural characteristics of ultrahigh-strength self-compacting concrete incorporating steel fibres. Mater Des 2009;30:4286–92.
22. Feldman RF, Carette GG, Malhotra VM. Studies on mechanics of development of physical and mechanical properties of high-volume FA-cement pastes. Cem Concr Comp 1990;12(4):245–51.
23. Fennis SAAM, Walraven JC, den Uijl JA. The use of particle packing models to design ecological concrete. Heron 2009;54:185–204.
24. Friedlingstein P, Houghton RA, Marland G, Hackler J, Boden TA, Conway TJ, et al. Uptake on CO_2 emissions. Nat Geosci 2010;3:811–2.
25. Funk JE, Dinger DR. Predictive process control of crowded particulate suspensions, applied to ceramic manufacturing. Boston, United States: Kluwer Academic Publishers; 1994.
26. Ghafari E, Costa H, Júlio E, Portugal A, Durães L. The effect of nanosilica addition on flowability, strength and transport properties of ultra high performance concrete. Mater Des 2014;59:1–9.
27. Grattan-Bellew PE. Microstructural investigation of deteriorated Portland cement concretes. Constr Build Mater 1996;10(1):3–16.
28. Habel K, Gauvreau P. Response of ultra-high performance fiber reinforced concrete (UHPFRC) to impact and static loading. Cem Concr Comp 2008;30(10):938–46.
29. Habel K, Viviani M, Denarié E, Brühwiler E. Development of the mechanical properties of an Ultra-High Performance Fiber Reinforced Concrete (UHPFRC). Cem Concr Res 2006;36(7):1362–70.
30. Habert G, Denarié E, Šajna A, Rossi P. Lowering the global warming impact of bridge rehabilitations by using Ultra High Performance Fibre Reinforced Concretes. Cem Concr Comp 2013;38:1–11.
31. Handoo SK, Agarwal S, Agarwal SK. Physicochemical, mineralogical, and morphological characteristics of concrete exposed to elevated temperatures. Cem Concr Res 2002;32(7):1009–18.

REFERENCES

32. Hanehara S, Tomosawa F, Kobayakawaa M, Hwang K. Effects of water/powder ratio, mixing ratio of FA, and curing temperature on pozzolanic reaction of FA in cement paste. Cem Concr Res 2001;31:31–9.
33. Hassan AMT, Jones SW, Mahmud GH. Experimental test methods to determine the uniaxial tensile and compressive behaviour of ultra-high performance fibre reinforced concrete (UHPFRC). Constr Build Mater 2012;37:874–82.
34. He J, Barry ES, Della MR. Hydration of FA-portland cements. Cem Concr Res 1984;14(4):505–12.
35. Hewlett PC. Lea's chemistry of cement and concrete. 3th ed. New York: John Wiley & Son Inc.; 1988. pp. 1–1053.
36. Hunger M. An integral design concept for ecological self-compacting concrete. PhD thesis. Eindhoven University of Technology, Eindhoven, the Netherlands; 2010.
37. Hüsken G, Brouwers HJH. A new mix design concept for each-moist concrete: a theoretical and experimental study. Cem Concr Res 2008;38:1249–59.
38. Hüsken G. A multifunctional design approach for sustainable concrete with application to concrete mass products. PhD thesis. Eindhoven University of Technology, Eindhoven, the Netherlands; 2010.
39. Hwang CL, Shen DH. The effects of blast-furnace GGBS and FA on the hydration of portland cement. Cem Concr Res 1991;21(4):410–25.
40. Jansen D, Neubauer J, Goetz-Neunhoeffer F, Haerzschel R, Hergeth WD. Change in reaction kinetics of a Portland cement caused by a superplasticizer – calculation of heat flow curves from XRD data. Cem Concr Res 2012;42(2):327–32.
41. Kevin DI, Kenneth ED. A review of LP additions to Portland cement and concrete. Cem Concr Comp 1991;13(3):165–70.
42. King D. The effect of silica fume on the properties of concrete as defined in concrete society report 74, cementitious materials. In: 37th Conference on our world in concrete and structures, Singapore, 29–31 August 2012.
43. Kovács R. Effect of the hydration products on the properties of fly-ash cements. Cem Concr Res 1975;5(1):73–82.
44. Kumar A, Oey T, Kim S, Thomas D, Badran S, Li J, et al. Simple methods to estimate the influence of LP fillers on reaction and property evolution in cementitious materials. Cem Concr Comp 2013;42:20–9.
45. Lea FM. The chemistry of cement and concrete. New York: Chemical publishing company; 1971. p. 202–33.
46. Li C, Sun H, Li L. A review: the comparison between alkali-activated GGBS (Si + Ca) and metakaolin (Si + Al) cements. Cem Concr Res 2010;40:1341–9.

47. Máca P, Sovják R, Konvalinka P. Mix design of UHPFRC and its response to projectile impact. Int J Impact Eng 2014;63:158–63.
48. Makita T, Brühwiler E. Tensile fatigue behaviour of Ultra-High Performance Fibre Reinforced Concrete combined with steel rebars (R-UHPFRC). Int J Fatigue 2014;59:145–52.
49. Millard SG, Molyneaux TCK, Barnett SJ, Gao X. Dynamic enhancement of blastresistant ultra high performance fibre-reinforced concrete under flexural and shear loading. Int J Impact Eng 2010;37(4):405–13.
50. Neville AM. Properties of concrete. Burnt Mill, Harlow, England: Longman House; 1995.
51. NT Build 492. Concrete, mortar and cement-based repair materials: Chloride migration coefficient from non-steady-state migration experiments. Nordtest method, Finland; 1999.
52. Ogawa K, Uchikawa H, Takemoto K, Yasui I. The mechanism of the hydration in the system C3S-pozzolana. Cem Concr Res 1980;10:683–96.
53. Okamura H, Ozawa K. Mix-design for self-compacting concrete. Concr Library JSCE 1995;25:107–20.
54. Olivier B, Christian V, Micheline M, Pierre-Claude A. Characterization of the granular packing and percolation threshold of reactive powder concrete. Cem Concr Res 2000;30:1861–7.
55. Papadakis VG. Effect of FA on Portland cement systems Part I: low-calcium FA. Cem Concr Res 1999;29:1727–36.
56. Plum NM. The predetermination of water requirement and optimum grading of concrete. Copenhagen: The Danish National Institute of Building Research; 1950.
57. Puntke W. Wasseranspruch von feinen Kornhaufwerken. Beton 2002;52(5): 242–8 [in German..
58. Quercia BG, Hüsken G, Brouwers HJH. Water demand of amorphous nano silica and its impact on the workability of cement paste. Cem Concr Res 2012;42:344–57.
59. Randl N, Steiner T, Ofner S, Baumgartner E, Mészöly T. Development of UHPC mixtures from an ecological point of view. Constr Build Mater 2014;67:373–8.
60. Regourd M, Thomassin JH, Baillif P, Touray JC. Blast-furnace GGBS hydration surface analysis. Cem Concr Res 1983;13:549–56.
61. Regourd M. Structure and behavior of GGBS cement hydrates. Principal Paper III, 2, VII th Int congress chemistry of cement, Paris. 1980; I(III.2): 9–26.
62. Richard P, Cheyrezy M. Composition of reactive powder concretes. Cem Concr Res 1995;25(7):1501–11.

REFERENCES

63. Rossi P. Influence of fibre geometry and matrix maturity on the mechanical performance of ultra-high-performance cement-based composites. Cem Concr Comp 2013;37:246–8.
64. Ryshkevitch R. Compression strength of porous sintered alumina and zirconia. J Am Ceram Soc 1953;36(2):65–8.
65. Safiuddin Md, Hearn N. Comparison of ASTM saturation techniques for measuring the permeable porosity of concrete. Cem Concr Res 2005;35:1008–13.
66. Schmidt W. Design concepts for the robustness improvement of selfcompacting concrete. PhD thesis. Eindhoven University of Technology, Eindhoven, the Netherlands; 2014.
67. Schutter GD. Hydration and temperature development of concrete made with blast-furnace GGBS cement. Cem Concr Res 1999;29:143–9.
68. Soroka I, Setter N. The effect of fillers on strength of cement mortars. Cem Concr Res 1977;7(4):449–56.
69. Spiesz P, Yu QL, Brouwers HJH. Development of cement-based lightweight composites – Part 2: durability related properties. Cem Concr Comp 2013;44:30–40.
70. Tayeh BA, Abu Bakar BH, Megat Johari MA, Voo YL. Mechanical and permeability properties of the interface between normal concrete substrate and ultra-high performance fibre concrete overlay. Constr Build Mater 2012;36:538–48.
71. Thomassin JH, Goni J, Baillif P, Touray JC, Jaurand NC. An XPS study of the dissolution kinetics of chrysotile in 0.1n oxalic acid at different temperatures. Phys Chem Miner 1977;1:385–98. 0 300 600 900 1200 1500 0 50 100 150 200 250 Compressive strength (MPa) Embedded CO_2 emission (kg/m3 concrete) EUHPC - FA EUHPC - GGBS EUHPC- LP Ghafari E.
72. Tuan NV, Ye G, Breugel K, Copuroglu O. Hydration and microstructure of ultrahigh performance concrete incorporating rice husk ash. Cem Concr Res 2011;41:1104–11.
73. Tuan NV, Ye G, Breugel K, Fraaij ALA, Dai BD. The study of using rice husk ash to produce ultra-high performance concrete. Constr Build Mater 2011;25:2030–5.
74. UNSTATS. Greenhouse gas emissions by sector (absolute values). United Nation Statistical Division: Springer; 2010.
75. Valcuende M, Parra C, Marco E, Garrido A, Martínez E, Cánoves J. Influence of LP filler and viscosity-modifying admixture on the porous structure of selfcompacting concrete. Constr Build Mater 2012;28(1):122–8.
76. Wang W, Liu J, Agostini F, Davy CA, Skoczylas F, Corvez D. Durability of an Ultra High Performance Fiber Reinforced Concrete (UHPFRC) under progressive aging. Cem Concr Res 2014;55:1–13.

77. Wei F, Michael WG, Della MR. The retarding effects of FA upon the hydration of cement pastes: the first 24 hours. Cem Concr Res 1985;15(1): 174–84.
78. Wu X, Jiang W, Roy DM. Early activation and properties of GGBS cement. Cem Concr Res 1990;20(6):961–74.
79. Wu X, Roy DM, Langton CA. Early stage hydration of GGBS-cement. Cem Concr Res 1983;13(2):277–86.
80. Yang SL, Millard SG, Soutsos MN, Barnett SJ, Le TT. Influence of aggregate and curing regime on the mechanical properties of ultra-high performance fibre reinforced concrete (UHPFRC). Constr Build Mater 2009;23:2291–8.
81. Yazici H, Deniz E, Baradan B. The effect of autoclave pressure, temperature and duration time on mechanical properties of reactive powder concrete. Constr Build Mater 2013;42:53–63.
82. Yazici H. The effect of curing conditions on compressive strength of ultra high strength concrete with high volume mineral admixtures. Build Environ 2007;42(5):2083–9. R. Yu et al. / Cement & Concrete Composites 55 (2015) 383–394 393
83. Yu QL, Spiesz P, Brouwers HJH. Development of cement-based lightweight composites – Part 1: mix design methodology and hardened properties. Cem Concr Comp 2013;44:17–29.
84. Yu R, Spiesz P, Brouwers HJH. Effect of nano-silica on the hydration and microstructure development of Ultra-High Performance Concrete (UHPC) with a low binder amount. Constr Build Mater 2014;65:140–50.
85. Yu R, Spiesz P, Brouwers HJH. Mix design and properties assessment of UltraHigh Performance Fibre Reinforced Concrete (UHPFRC). Cem Concr Res 2014;56:29–39.
86. Yu R, Spiesz P, Brouwers HJH. Static properties and impact resistance of a green Ultra-High Performance Hybrid Fibre Reinforced Concrete (UHPHFRC): experiments and modeling. Constr Build Mater 2014;68:158–71.
87. Yu R, Tang P, Spiesz P, Brouwers HJH. A study of multiple effects of nano-silica and hybrid fibres on the properties of Ultra-High Performance Fibre Reinforced Concrete (UHPFRC) incorporating waste bottom ash (WBA). Constr Build Mater 2014;60:98–110.
88. Zeng Q, Li K, Fen-chong T, Dangla P. Determination of cement hydration and pozzolanic reaction extents for fly-ash cement pastes. Constr Build Mater 2012;27:560–9.
89. Zhang Y, Sun W, Liu S, Jiao C, Lai J. Preparation of C200 green reactive powder concrete and its static–dynamic behaviors. Cem Concr Comp 2008;30(9): 831–8.

CITATION

R. Yu, P. Spiesz, H.J.H. Brouwers, Development of an eco-friendly Ultra-High Performance Concrete (UHPC) with efficient cement and mineral admixtures uses, Cement and Concrete Composites, Volume 55, January 2015, Pages 383-394, ISSN 0958-9465, Doi.org/10.1016/j.cemconcomp.2014.09.024.

CHAPTER 11

Concrete Mix Design for Service Life of RC Structures under Carbonation Using Genetic Algorithm

Seung-Jun Kwon[1], Byung Jae Lee[2], and Yun Yong Kim[2]

[1]Department of Civil and Environmental Engineering, Hannam University, 133 Ojeong-dong, Daedeok-gu, Daejeon 306-791, Republic of Korea
[2]Department of Civil Engineering, Chungnam National University, 99 Daehak-ro, Yuseong-gu, Daejeon 305-764, Republic of Korea

ABSTRACT

Steel corrosion in reinforced concrete (RC) structure is such a critical problem to structural safety that many researches have been performed for maintaining required performance during intended service life. This paper is for a numerical technique for obtaining optimum concrete mix proportions through genetic algorithm (GA) for RC structures under carbonation which is considered as a serious deterioration in underground sites and big cities. For this study, mix proportions and CO_2 diffusion coefficients are analyzed through the previous studies, and then the fitness function of CO_2 diffusion coefficient is derived through regression analysis. The fitness function from 69 test results includes 5 variables of mix proportions such as w/c (water to cement) ratio, cement content, sand content percentage, coarse aggregate content, and R.H. (relative humidity). Through GA technique, simulated mix proportions are obtained for 12 cases of verification and they show reasonable results with average relative error of 4.6%. Assuming intended service life and design parameters, intended CO_2 diffusion coefficients and cement contents are determined and then related mix proportions are simulated. The proposed technique can provide initial concrete mix proportions which satisfy service life under carbonation.

INTRODUCTION

CO_2 concentration is increasing due to fossil energy consumption and this causes more carbonation damage to RC structures [1, 2]. Carbonation means that pH in pore water drops to about 10.5 due to intrusion of exterior CO_2 [3] and consumption of $CaOH_2$. In carbonated concrete, embedded steel is easily corroded. It is so critical deterioration phenomenon that it should be considered in durability design for underground RC structures or those in metropolitan cities which have high CO_2 concentration.

With higher CO_2 concentration, carbonation depth increases but this can be comparatively controlled by a design of concrete mix proportions. The influencing parameters on carbonation are reported to be type of cement, unit content of cement, type of aggregate, and so on [2]. Semiempirical prediction techniques, so-called mesolevel, have been proposed and they are still utilized for the sake of simple and practical application [4, 5]. Carbonation mechanism can be explained as diffusion of CO_2 and carbonatable materials like calcium hydroxide ($Ca(OH)_2$) and calcium silicate hydrates (C–H–S). CO_2 diffusion represents how fast CO_2 gas (or liquid) intrudes into concrete, so that concrete with high CO_2 diffusion coefficient allows rapid carbonation. From the defensive point of view for carbonation, concrete with larger carbonatable materials can keep high alkali so long as they are not fully consumed due to carbonation reaction. From 1980, several physico-chemo carbonation models have been proposed. They are all constructed by both modeling on diffusion coefficient based on pore structure and modeling on carbonic reaction based on dissociation of carbonatable materials [3,6–8]. Carbonation modeling for cracked and joint concrete is similarly performed considering the larger CO_2 intrusion due to crack effect and cold joint effect [9–12]. Recently, carbonation prediction techniques are proposed through experimentally measuring CO_2 diffusion coefficient [13–15] and numerically obtaining CO_2 diffusion coefficient through neural network algorithm [16].

If environmental conditions like CO_2 concentration, temperature, and R.H. are quantitatively evaluated, intended carbonation depth in design stage can be determined considering design cover depth and intended service life. Provided that various mix proportions and the related CO_2 diffusion coefficients are experimentally given, intended CO_2 diffusion coefficient satisfying the intended service life can be obtained. Then mix proportions satisfying the intended CO_2 diffusion coefficient can be obtained through optimization technique as well.

GA (generic algorithm) technique is a representative optimization technique and widely utilized in civil engineering. Through reverse analysis, the parameters which satisfy the fitness function can be derived so that application of GA has been extended. For the application of GA to concrete researches, mix proportion optimizations are performed only for strength prediction in HPC (high performance concrete) [17, 18]. With regard to durability design for service life, very limited research has been performed for chloride attack [19]. For carbonation, optimization of mix proportions has not been carried out so far.

In this paper, CO_2 diffusion coefficients and the related mix proportions are investigated. Based on 69 mix proportions and diffusion coefficients; the fitness function for CO_2 is derived through MATLAB with parameters of mixing variables (w/c ratio, unit content of cement, and fine and coarse aggregates) and exterior variables (R.H.). Through comparison with the previous test results, the applicability of GA technique for optimum mix proportions is verified. Assuming the intended service life and environmental conditions, the intended CO_2 diffusion coefficient is calculated. Finally, the mix proportions which satisfy the intended CO_2 diffusion coefficient are derived through GA technique. This technique can be utilized for performance-based concrete mix design. The techniques for carbonation prediction and optimization of mix proportions are dealt with in this paper.

BACKGROUND OF GA AND INFLUENCING PARAMETERS ON CARBONATION

Overview of GA
Unlike conventional search technique, GA technique constructs arbitrary solutions in initial group, and then the fittest solution is derived through modification of the solutions. GA technique is mainly utilized in the field of mechanical and electrical engineering and recently applied to civil engineering such as design optimization for structures, line network analysis, and concrete mix design for strength. This technique can provide more accurate results than other algorithms having many local solutions [20, 21]. GA technique starts with an initial set of random solutions called population. Each individual in the population is called a chromosome, which represents a solution to the problem at hand. The evolution operator simulates Darwinian evolution process to create population from generation to generation. The availability of genetic algorithm depends on its ability to keep existing parts of solution, which have a positive effect on

the outcome, and proceed with optimizing the no optimal part. The transition rules which combine and change those samples for improving the solutions are probabilistic and not deterministic. This enables genetic algorithm to reach a global optimum without being fixed in local optima [18]. In the selection stage, GA fundamentally starts from Darwinian natural selection and the initial individuals are selected in this process. Selection provides the driving force in genetic algorithm, and selection pressure is critical in it. The selection directs genetic algorithm search toward promising regions in the search space [18]. The second stage, crossover is the most important genetic operator in which the bit-strings of two (or more) parents are cut into two (or more) pieces and the parts of bit-string are crossed over. The point where the parents are cut is randomly determined. Through the crossover operator, a new child population has been created using inherited values from the parent population. Mutation operator is used to insert new information into the new population, preventing GA from getting stuck in certain regions of the parameter space [18]. Mutation consists of making slight changes in parameters of child population after they have been generated by crossover. More detailed information on GA can be found in many researches [17–21]. The process of GA is presented in Figure 1.

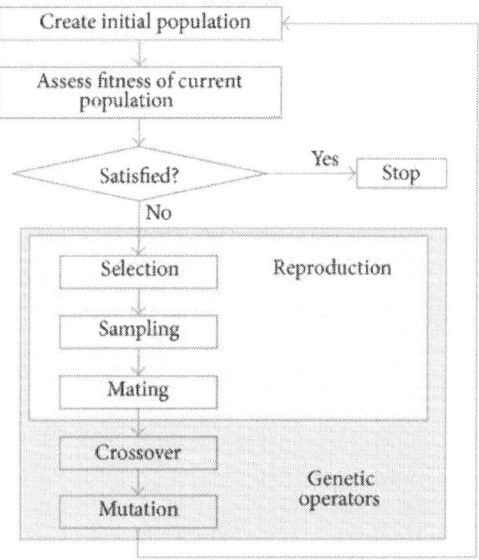

Figure 1: Genetic algorithm process [29].

Study of Carbonation Parameters and Prediction Techniques

Influencing parameters on carbonation can be classified into two groups. One is for external parameters regarding environmental conditions and the other is for internal parameters regarding diffusion coefficient and carbon table materials. Considering these parameters, many carbonation prediction techniques have been proposed in semi empirical form. These equations assume that carbonation depth is proportional to square root of exposed time. This assumption was verified through experiments, field investigations [1, 2, 9, 22], and analytical solution [6, 23]. In Table 1, carbonation parameters are summarized. Conventional techniques for carbonation prediction are listed in Table 2 [1, 2].

Table 1: Influencing parameter on carbonation behavior

Internal parameter (mixture)	Low w/c and large unit cement amount	(i) Holding pH in alkali through producing large amount of hydration of CSH and $Ca(OH)_2$
		(ii) Low CO_2 diffusion through dense pore structure
	Aggregate	CO_2 intrusion through artificial light weight aggregate
	Mineral admixture	(i) Small amount of $Ca(OH)_2$ due to pozzolanic reaction and latent hydraulic reaction
	(slag and fly ash)	(ii) Low diffusion coefficient of CO_2
	Mixed chloride content	Rapid carbonic reaction due to high pH from ion dissociation
	Alkali	(i) Rapid carbonic reaction due to high alkali cement
		(ii) Residual metallic oxide (K_2O, Na_2O)
External parameter	CO_2 concentration	Rapid carbonation through higher concentration of CO_2
	Temp.	Increasing activity energy due to high temperature (Arrhenius law)
	R.H.	(i) Decreasing carbonation in low R.H. due to insufficient H_2O
		(ii) Decreasing carbonation in high R.H. due to low CO_2 diffusion
	Induced chloride ion	Rapid carbonation due to dissociated chloride ion (cation)

Table 2: Semiempirical equations for carbonation process

Researcher	Equations		
Syrayama	$$t = \alpha\beta\gamma\delta\varepsilon \frac{5000C^2}{(x-38)^2}$$		
	C: carbonation depth, x: w/c ratio		
	$\alpha, \beta, \gamma, \delta, \varepsilon$: factors for admixtures, cement type, exposure condition, and so forth		
Kishitani	$$t = \frac{0.3(1.15 + 3w/c)C^2}{R^2(w/c - 0.25)^2}$$	(w/c ≥ 0.6)	
	$$t = \frac{7.2C^2}{R^2(4.6w/c - 1.76)^2}$$	(w/c < 0.6)	
	C: carbonation depth		
	R: factor for cement type, aggregate type, and surface treatment		
Hamada	$$t = \frac{kC^2}{R}, k = \frac{0.3(1.15 + 3x)}{(x - 0.25)^2}$$		
	C: carbonation depth, x: w/c ratio		
	R: factor for cement type, aggregate type, and surface treatment		
Ida	$$t = \alpha\beta\gamma \frac{KC^2}{(100x - 18)^2}$$		
	C: carbonation depth, x: w/c ratio		
	K: factor for exposure and cement type		
	α, β, γ: quality, retardation, and environmental condition		

The flowchart for this study is shown in Figure 2. Through this study, fitness function for diffusion coefficient, intended diffusion coefficient for service life, and mix proportions satisfying intended diffusion coefficient are derived using GA technique.

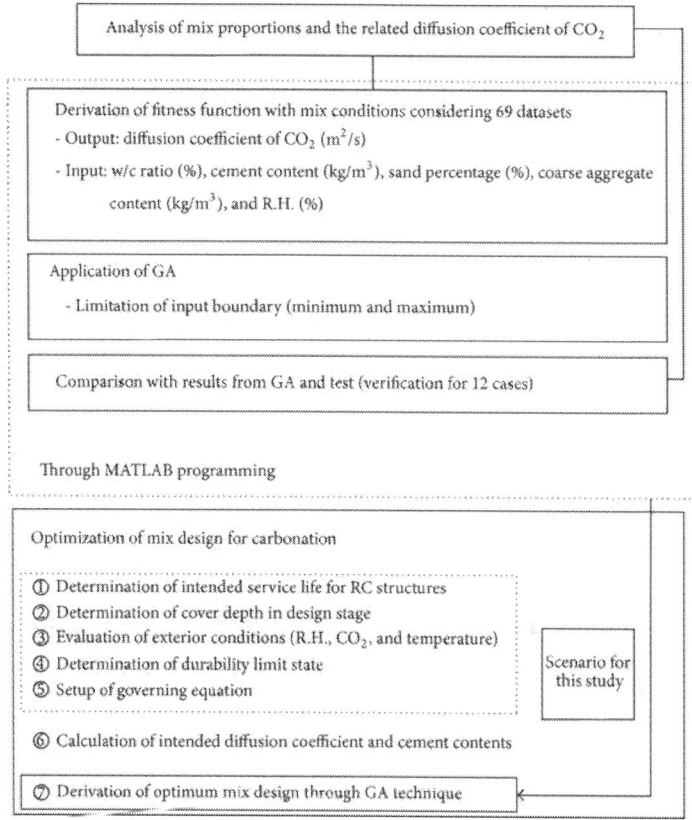

Figure 2: Flowchart for this concrete mix optimization.

CONCRETE MIX OPTIMIZATION USING GA

Fitness Function for Diffusion Coefficient
Previous Test for Diffusion of CO_2 [14, 16]
For the derivation of fitness function, the previous test results are adopted. In the test, CO_2 diffusion coefficients were measured through diffusion cell. Three different mix conditions and 4 different R.H. were considered [14]. So far, several researches have been reported for the test of CO_2 diffusion coefficients; however, they are not for concrete but for cement

mortar or cement paste [6, 13]. Very limited cases are reported for CO_2 diffusion coefficient in concrete. In Table 3, the procedures for the adopted test are summarized. Mix proportions and cement properties are listed in Table 4. The test adopted in this paper covers only OPC (Ordinary Portland Cement, type I) concrete since concrete with mineral admixtures has different carbonation behavior due to the decreased diffusion coefficient and pozzolan reaction [3, 7].

Table 3: Summary of test setup [14]

Steps
(a) Installation of test equipment in room (20°C)
(b) Measurement of concrete sample thickness and diameter
(c) Installation of sample (concrete disk) in cell
(d) Applying N_2 gas and CO_2 gas to different cells with same pressure
(e) Measurement of CO_2 concentration when CO_2 concentration in N_2 gas keeps constant (steady state)
$$D_{CO_2} = \frac{Q f_{CO_2} L}{(1 - f_{CO_2}) A}$$
D_{CO_2}: diffusion coefficient of CO_2; Q: flow rate of gas
f_{CO_2}: mol fraction in $N_2 + CO_2$; L: thickness of disk; A: area of disk

Table 4: Mix proportions for CO_2 diffusion measurement [14]

			(a)		
Case	w/c (%)	Cement (kg/m³)	Water (kg/m³)	Sand (kg/m³)	Coarse aggregate (kg/m³)
1	42	425	179	714	895
2	50	315	158	748	1,076
3	58	277	161	726	1,117

	(b)		
	Aggregate properties		
Type	Specific gravity	Absorption (%)	Fineness modulus
Fine	2.56	2.18	2.85
Coarse	2.6	0.94	6.51

Derivation of Fitness Function

The adopted test was performed considering 4 different R.H. as 10%, 45%, 75%, and 90%. In order to obtain more reasonable fitness function, several previous test results [15, 24, 25] are considered. Carbonation process is very sensitive to R.H. since concrete with high saturation allows active carbonation reaction but low diffusion of CO_2, and concrete with low saturation allows high diffusion of CO_2 but it has little H_2O for carbonic reaction. With higher R.H. and lower w/c ratio, CO_2 diffusion coefficients decrease as in Figure 3. In Figure 3, several results [25, 26] are obtained from reverse analysis based on measured carbonation depth with constant R.H.

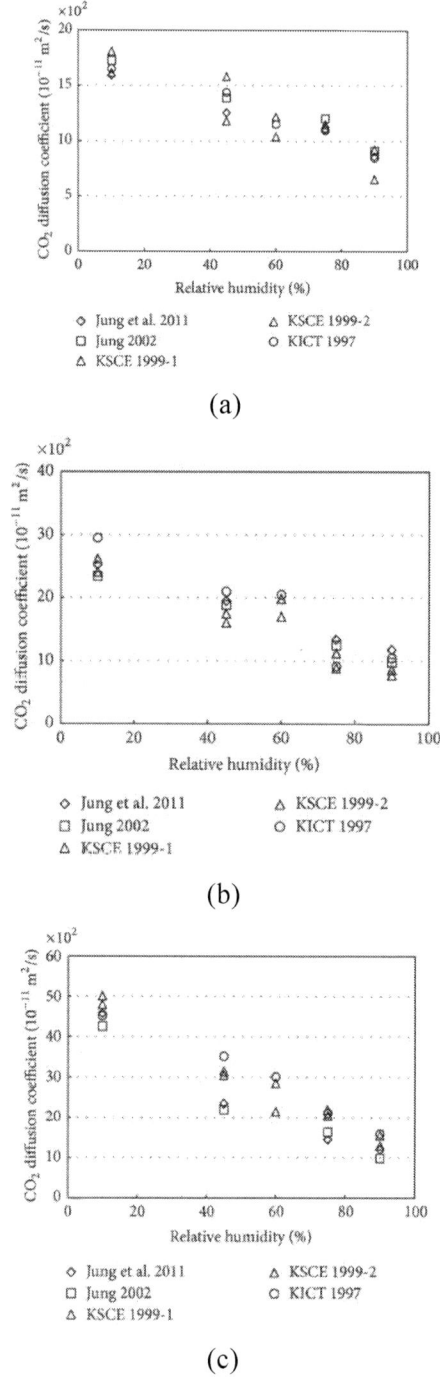

Figure 3: CO_2 diffusion coefficients with w/c ratios and R.H.

For the relation with mix proportions and CO_2 diffusion coefficient, fitness function with mix components should be obtained. In the previous researches, fitness function for strength was derived through linear multiregression curve, which contained the variables of mix components like w/c ratio and unit amount of cement [18]. Unlike the fitness function for strength, R.H. is very critical to CO_2 diffusion coefficient, so that both mix components (w/c ratio, content of cement, sand ratio, and content of coarse aggregate) and R.H. are considered as variables in the fitness function in this analysis.

In the optimization technique, many local solutions can be obtained. For avoiding convergence to local solution, initial variables (starting variables) and wide ranges for each solution are necessary. Even wide ranges of solutions are considered, local solutions may be obtained because of the initial variables in conventional optimization techniques, so that GA technique is preferred for searching solution in overall ranges. The variables and the related ranges are listed in Table 5.

Table 5: Variables for fitness function

Type	w/c (%)	C (kg/m³)	S/a (%)	G (kg/m³)	R.H. (kg/m³)	a and b	Constant
Max	100	0	10	10	30	1	200,000
Min	−100	−100	−10	−10	−30	−1	0

With larger unit content of cement, diffusion coefficient decreases, so that C in Table 5 is set to have below zero. w/c ratio is assumed to have a range of −100~100. S/a (sand to total aggregate) and G have relatively small effect on CO_2 diffusion, so that they are assumed to have small range of −10~10. CO_2 diffusion coefficient is much dependent on R.H., so that the range of R.H. is assumed as −30~30.

Typical multiregression analysis is shown in (1). In (2), additional term for the consideration of R.H. is added. Averages of relative error are evaluated to be 17.3% from (1) and 7.6% from (2), respectively. The results of multiregression curves are listed in Table 6. For the derivation of constant in (1) and (2), GA technique is utilized. Consider

$$D_{CO_2} = I + A\left(\frac{W}{C}\right) + B(C)$$
$$+ C\left(\frac{S}{a}\right) + D(\text{Agg}) + E(\text{RH}),\qquad(1)$$

$$D_{CO_2} = \left[I + A\left(\frac{W}{C}\right) + B(C) + C\left(\frac{S}{a}\right)\right.$$
$$\left. + D(\text{Agg}) + E(\text{RH})\right](a\text{RH} + b).\qquad(2)$$

Table 6: Variables and constants in regression analysis

	I	A	B	C	D	E
Equation (1)	1018.56	16.08	−1.42	2.75	0.6	−9.75
Equation (2)	15,427.82	49.562	−29.60	5.48	−4.42	9.79
	a: −0.005825; b: 0.7542					

w/c: w/c ratio (%); C: cement content (kg/m³); S/a: sand percentage (sand/total aggregate) (%); Agg: coarse aggregate content (kg/m³); R.H: relative humidity (%); I: Intersection (constant).

In Figure 4, the results of regression analysis ((1) and (2)) and test results (averages) are compared. As shown in Figure 4, when (1) is selected, it provides a big error for the case of high w/c ratio (w/c 58%), so that (2) is selected for fitness function for this study. For the case of w/c 58%, (1) shows the relative error range of 10.6~24.3% but (2) shows 3.3~14.2%. If a better fitness curve can be defined through nonlinear regression analysis, it would provide the more reasonable mix proportions based on the test dataset.

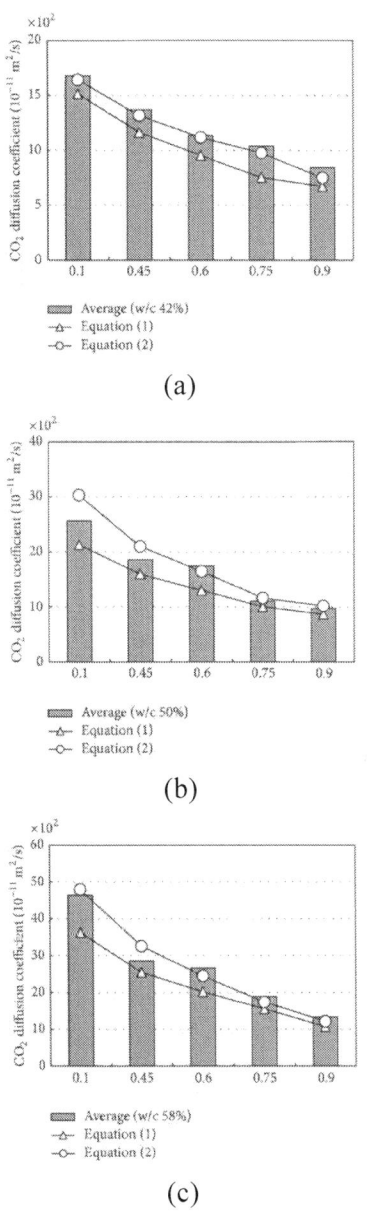

Figure 4: Test and simulated results for diffusion coefficient.

CO_2 diffusion coefficients are strongly dependent on mix proportions. The contours are shown in Figure 5. Case 1 represents w/c 42% in Table 4, where unit cement content is changed from 277 kg/m³ to 425 kg/m³ and R.H. is changed from 10% to 90%. Cases 2 and 3 show the simulations of

CO_2 diffusion coefficient in w/c 50% and 58%, respectively With larger cement content and higher R.H., CO_2 diffusion coefficients decrease in every case.

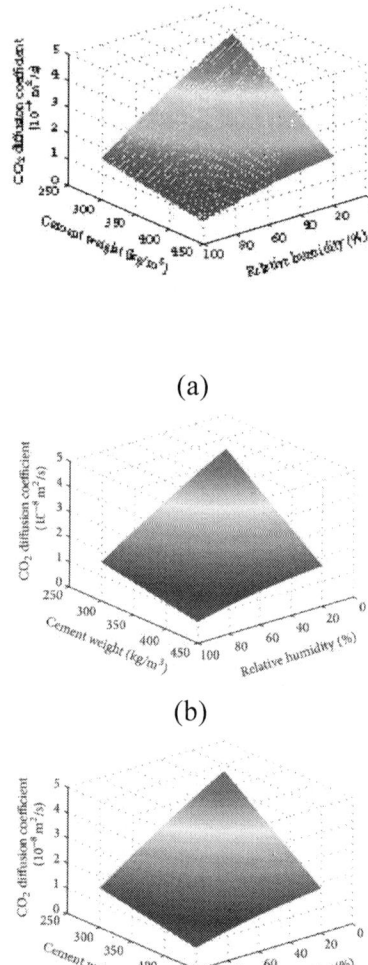

Figure 5: Diffusion coefficient contour with cement content and relative humidity.

Evaluation of GA Applicabilit y to Generating Mix Proportion

In order to evaluate the applicability of GA, verification is performed for 3 different cases (w/c 42% with R.H. 10%, w/c 50% with R.H. 75%, and w/c 58% with R.H. 90%). Population size is set as 20 and the number of generation is set as 10,000 for avoiding early convergence. For formation of 1st generation, uniform function is adopted. For parent selection for next generation, stochastic uniform function is utilized and two superior chromosomes are transferred to next generation. Crossover function of two-point is adopted and normal distribution is considered for mutation operator with mutation ratio of 0.8.

Determination of up/down boundary conditions is important to obtain each mix component and this needs user's experience. The range of boundary conditions, obtained mix components through GA, and the range of relative errors are listed in Table 7. Output results are mix component and CO_2 diffusion coefficient. R.H. is set to fix since it can be known from exterior condition. The fixed R.H. is made through letting up/down boundaries have the same R.H. value.

Table 7: Comparison with results from test and simulation from GA

w/c (%)	Diffusion coefficient (10^{-11} m²/sec)	w/c (%)	R.H. (%)	Cement	S/a (%)	Coarse aggregate (kg/m³)
42	1574	42	10	425	44.4	895
Input range	—	40–45	10-Oct	400–450	39–45	800–895
Result from GA	1580.5	41.5	—	419.2	41.8	874.5
Relative error (%)	0.4	−1.2		−1.4	−5.9	−2.3
42	1257	42	45	425	44.4	895
Input range	—	40–45	45-45	400–450	39–45	800–895
Result from GA	1194.2	42.1	—	420.8	42.2	869.7
Relative error (%)	−5.0	0.24		0	−5.0	−2.9
42	1105	42	75	425	44.4	895
Input range	—	40–45	75-75	400–450	39–45	800–895
Result from GA	1088.4	42.2	—	421.2	42	882.1

Relative error (%)	−1.5	0.5		−0.9	−5.4	−1.4
42	862	42	90	425	44.4	895
Input range	—	40–45	90–90	400–450	39–45	800–895
Result from GA	855.2	41.8	—	426.1	43.2	887.2
Relative error (%)	−0.8	−0.5		0.3	−2.7	−0.9
50	2520	50	10	315	41	1076
Input range	—	47.5–52.5	10-Oct	290–330	39–45	950–1100
Result from GA	2775.2	49.8	—	311.8	42.6	1085.3
Relative error (%)	10.1	−0.4		−1.0	3.9	0.9
50	1950	50	45	315	41	1076
Input range	—	47.5–52.5	45-45	290–330	39–45	950–1100
Result from GA	2124.2	51.3	—	318.5	41.2	1092.5
Relative error (%)	8.9	2.6		1.1	0.5	1.5
50	1503	50	75	315	41	1076
Input range	—	47.5–52.5	75-75	290–330	39–45	950–1100
Result from GA	1452.2	49.3	—	322.2	41.8	1044.7
Relative error (%)	−3.4	−1.4		2.3	2	−2.9
50	1105	50	90	315	41	1076
Input range	—	47.5–52.5	90-90	290–330	39–45	950–1100
Result from GA	1127.3	48.2	—	318.9	39.5	1068.5
Relative error (%)	4	−3.6		1.2	−3.7	−0.7
58	4480	58	10	277	39.4	1117
Input range	—	55–60	10-Oct	260–330	39–45	950–1300
Result from GA	4922.3	57.2	—	266.2	39.7	1204.2
Relative error (%)	9.7	−1.4		−3.9	0.8	7.8
58	2350	58	45	277	39.4	1117

Input range	—	55–60	45-45	260–330	39–45	950–1300
Result from GA	2472.3	58.9	—	270.5	40.1	1200.8
Relative error (%)	5.2	1.6		−2.4	1.8	7.5
58	1450	58	75	277	39.4	1117
Input range	—	55–60	75-75	260–330	39–45	950–1300
Result from GA	1377.2	57.8	—	277.7	42.2	1208.7
Relative error (%)	−5.0	−0.3		0.3	7.1	8.2
58	1172	58	90	277	39.4	1117
Input range	—	55–60	90-90	260–330	39–45	950–1300
Result from GA	1150.2	59.4	—	278.5	41.9	1187.5
Relative error (%)	−0.9	2.4		0.5	6.4	6.3

The bold numbers are results from GA.

For 12 data [14], the comparison with test and simulated results from GA are shown in Figure 6(a) with regard to CO_2 diffusion coefficient, which shows a reasonable agreement. In Figure 6(b), the comparison of relative errors is shown for 12 cases and the average relative error of each component is shown in Figure 6(c). The processes of searching the optimum solutions are plotted in the case of w/c 42% with R.H. 10% from Figure7 (a) to Figure 7(e).

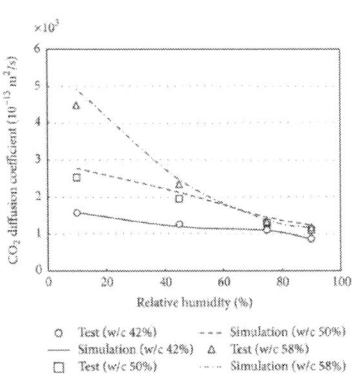

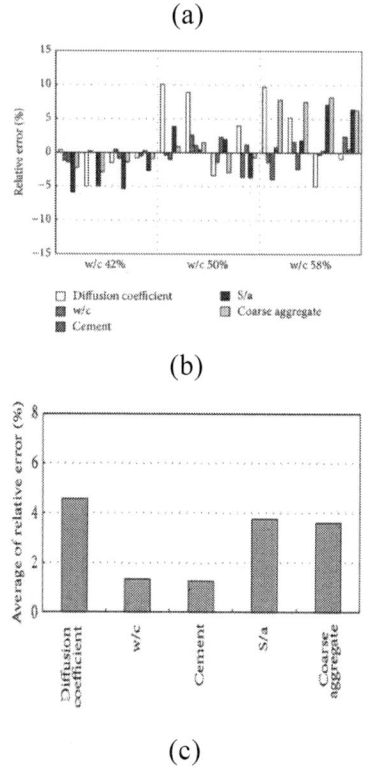

Figure 6: Comparison with results of CO_2 diffusion coefficient and relative errors.

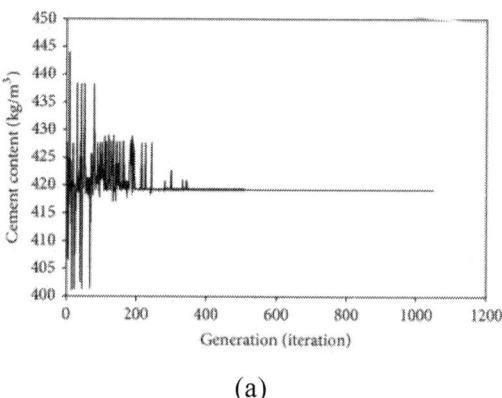

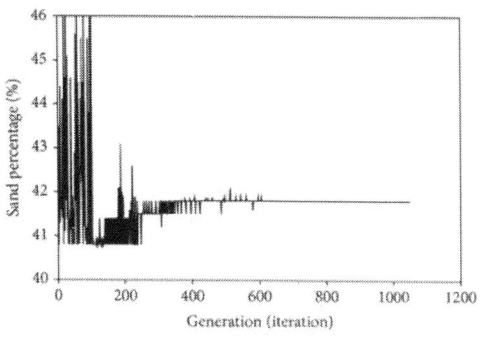

(b)

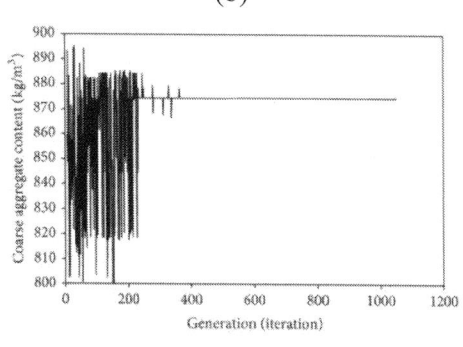

(c)

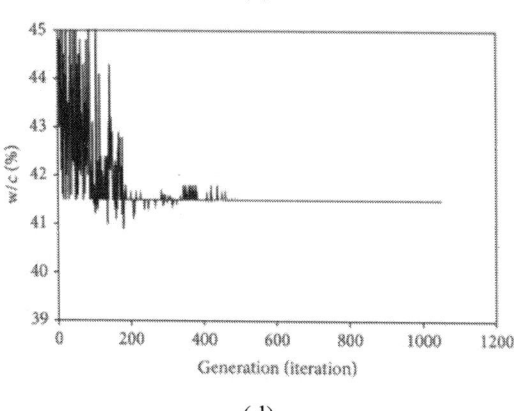

(d)

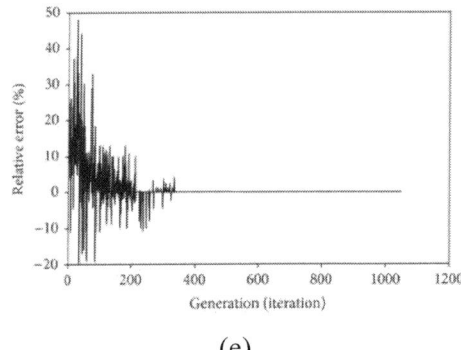

(e)

Figure 7: Simulated process through GA (w/c 0.42 and R.H. 10%).

As listed in Table 7, this technique reasonably estimates the CO_2 diffusion coefficients and mix proportions with −5.0~10.1% of relative errors.

DESIGN OF CONCRETE MIX PROPORTIONS FOR CARBONATION

Scenario for Mix Design Considering Carbonation

In this section, concrete mix design is performed considering exterior condition-carbonation. The design flow is as follows:

a) determination of intended service life,
b) determination of design cover depth,
c) evaluation of exterior condition,
d) determination of durability criteria,
e) determination of governing equation,
f) Mix optimization through GA.

If reduction factors or safety factors are considered [26, 27], conservative design can be induced. However, intended diffusion coefficient is derived assuming 1.0 of reduction and safety factor in this paper. Generally, underground site and urban area are reported to be the environments where durability design for carbonation is necessary since they have relatively high CO_2 concentration and normal R.H. (50%~70%). In the previous research [25], durability design for carbonation is strongly recommended over 300 ppm of CO_2 concentration. In urban cities, CO_2 concentration over 350 ppm is reported; furthermore, CO_2 concentration over 650 ppm is reported in underground sites like subway structures [25]. Several specifications [26–28] guide durability design for carbonation in urban cities and underground structures.

Mix Design Considering Exterior Conditions and Design Parameters
Scenario for Concrete Mix Design
Based on the design flow in Section 4.1, concrete mix proportions are simulated. The target structures are assumed as underground structures and two types (A and B) are considered. A structure has 75 years and B structure has 100 years for intended service life. Design cover depths are assumed as 50 mm for A structure and 30 mm for B structure. A structure has 65% of R.H. and 12.7°C of temperature. B structure has 75% of R.H. and 22°C of temperature, which are normal exterior conditions in underground site. Extremely high CO_2 concentration of 2,700 ppm is assumed for A structure and 980 ppm which is normal condition in underground structure is assumed for B structure. Durability limit state is determined as the condition when carbonation proceeds to steel location [23, 27]. For governing equation, mesolevel equation from CEB [23] is adopted as follows:

$$d_C = \sqrt{2k_1 k_2 k_3 \Delta c} \cdot \sqrt{\frac{D_{CO_2} t}{a}} \left(\frac{t_0}{t}\right)^n, \qquad (3)$$

where d_C is carbonation depth (mm), k_1 is constant for local condition, k_2 is constant for curing condition, k_3 is constant for locally different w/c ratio, Δc is CO_2 concentration (kg/m³), D_{CO_2} is CO_2 diffusion coefficient (m²/sec), a is carbonation reaction function with hydrate amount, n is constant for cyclic drying and wetting, t_0 is reference time (1 year), and t is exposed period (year).

For considering the effect of temperature on carbonation, a parameter like (4) is considered [4]. Consider

$$f(T) = D_{ref} \exp\left[\frac{U}{R}\left(\frac{1}{T_{ref}} - \frac{1}{T}\right)\right], \qquad (4)$$

where D_{ref} is referential CO_2 diffusion coefficient, U is activation energy of CO_2(8500 Cal/mol·K), R is universal gas constant, T_{ref} is reference temperature (298 K), and T is exterior temperature (K).

In (3), the target structure is assumed to have normal construction level and to be sheltered from rain and k_1, k_2, k_3, and n can be assumed as 1.0 and 0.0, respectively [23]. Considering the temperature parameter, (3) can be written as follows:

$$d_C = \sqrt{2\Delta c} \cdot \sqrt{\frac{D_{CO_2} f(T)}{a}} t, \qquad (5)$$

where a can be expressed as follows (CEB 1997):

$$a = 0.75 \cdot C \cdot CaO \cdot \alpha_H \frac{M_{CO_2}}{M_{CaO}}, \qquad (6)$$

where C is unit content of cement (kg/m3), CaO is content of CaO(calciumoxide, 0.65), α_H is hydration rate (0.85), and M is molar weight (CO2: 44 g/mol, CaO: 56 g/mol).

In (5), Δc, a, t, and T are given by design parameter. Considering the durability limit state (carbonation depth = cover depth), intended diffusion coefficient can be calculated.

The design parameters above are summarized in Table 8.

Table 8: Design parameters for carbonation design

(a)		
Type	Structure A	Structure B
Intended service life (year)	75	100
Design cover depth (mm)	50	30
Exterior condition	R.H.: 65%	R.H: 75%

	Temp.: 12.7°C	Temp.: 22°C
CO_2 concentration (ppm)	2,700	980
Durability limit state	carbonation depth = cover depth	

(b)

Assumed cement weight (kg/m³)	300	330	335	370
Intended diffusion coefficient × 10^{-11} (m²/sec)	1,742	1,916	1,248	1,378

In (5), two unknown variables exist so that unit content of cement is assumed referring to conventional mix proportions in domestic condition [25]. Four different contents of cement are assumed and the related intended diffusion coefficients are derived through (5).

Derivation of Optimum Mix Proportions

In this section, optimum mix proportions are derived through GA technique. The fitness function of (2) is utilized for obtaining mix components with fixed R.H. and cement content.

The results of mix proportions are listed in Table 9.

Table 9: Mixture design through proposed GA technique

Case		Intended diffusion coefficient (10^{-11} m²/sec)	w/c (%)	Cement (kg/m³)	S/a (%)	Coarse aggregate (kg/m³)	R.H. (%)
A	Structure A	1,742	42.3	300	38.4	1191	65
	Input range		42–58	300-300	37–43	800–1,200	65-65
B	Structure A	1,916	52.7	330	38.1	960.2	65
	Input range		42–58	330-330	37–43	800–1,200	65-65

C	Structure B	1,248	49.6	335	41.2	1172.8	75
	Input range		42–58	335-335	37–43	900–1,200	75-75
D	Structure B	1,378	51.4	370	40.1	875.3	75
	Input range		42–58	370-370	37–43	800–1,200	75-75

As shown in Table 9, intended diffusion coefficient and unit cement content are given and mix proportions for concrete can be obtained through GA technique. When this technique is applied, convergence of relative error to 0.0 should be checked.

In this paper, fitness function for CO_2 diffusion coefficient is derived based on the previous test results, and then concrete mix design is proposed through GA technique. However, this technique is only for OPC concrete mix design and has limitation of range for mix proportion. The applicable ranges of unit content of cement and w/c ratio are 277 kg/m^3~425 kg/m^3 and 0.42~0.58 since both the fitness function and the process for generating each mix proportion are governed by test dataset which is previously adopted.

With more data-set containing CO_2 diffusion coefficient and an accurate fitness function, the proposed technique would be much improved. This technique is applied for mix proportion of concrete under carbonation. With similar procedures, this can be applied to generation of mix proportions which can guarantee the service life of RC structures exposed to different deteriorations like chloride attack, freezing and thawing action, and sulfate attack.

CONCLUDING REMARK

The conclusions on concrete mix optimization technique for service life of RC structures under carbonation using genetic algorithm are as follows.

1. Based on the previous experimental results, fitness function for CO_2 diffusion coefficient containing the variables like mix proportions (w/c ratio, unit content of cement, sand/aggregate ratio, and unit content of coarse aggregate) and R.H. (relative humidity) is derived. Through

consideration of the parameters of R.H., variation of relative errors decreases.
2. Through GA technique, three concrete mix proportions are simulated for verification. The simulated results provide below 10.1% of relative errors for each mix component such as w/c ratio, unit content of cement, sand ratio to total aggregate, and unit content of coarse aggregate.
3. Assuming the exposure conditions of carbonation and design parameters, intended diffusion coefficients are determined and optimum concrete mix proportions which satisfy intended service life are obtained through GA technique. The results from this study are only applicable to OPC concrete. If data-set with mineral and chemical admixtures is prepared, this technique can be applied more widely to durability design for RC structures under carbonation.

ACKNOWLEDGMENTS

This research was supported by a grant (Code 11-Technology Innovation-F04) from Construction Technology Research Program funded by Ministry of Land, Infrastructure and Transport.

REFERENCES

1. K. Kobayashi and Y. Uno, "Mechanism of carbonation of concrete," Japan Society of Civil Engineers, vol. 1, no. 1, pp. 139–151, 1990.
2. I. Izumi, D. Kita, and H. Maeda, Carbonation, Kibodang Publication, 1986.
3. T. Ishida and K. Maekawa, "Modeling of PH profile in pore water based on mass transport and chemical equilibrium theory," Japan Society of Civil Engineers, vol. 1, no. 37, pp. 151–166, 2001.
4. T. Saeki, H. Ohga, and S. Nagataki, "Change in micro-structure of concrete due to carbonation," Japan Society of Civil Engineers, vol. 1, no. 18, pp. 1–11, 1991.
5. CEB, Durable Concrete Structures-Design Guide, Thomas Telford, London, UK, 2nd edition, 1992.
6. V. G. Papadakis, C. G. Vayenas, and M. N. Fardis, "Physical and chemical characteristics affecting the durability of concrete," ACI Materials Journal, vol. 88, no. 2, pp. 186–196, 1991.
7. T. Ishida and K. Maekawa, "Modeling of durability performance of cementitious materials and structures based on thermo-hygro physics," in

Proceedings of the 2nd International RILEM Workshop on Life Prediction and Aging Management of Concrete Structures, 2003.
8. F. P. Glasser, J. Marchand, and E. Samson, "Durability of concrete—degradation phenomena involving detrimental chemical reactions," Cement and Concrete Research, vol. 38, no. 2, pp. 226–246, 2008.
9. S. J. Kwon, S. S. Park, S. H. Nam, and H. J. Cho, "A study on survey of carbonation for sound, cracked, and joint concrete in RC column in metropolitan city," Journal of the Korea Institute For Structural Maintenance and Inspection, vol. 11, no. 3, pp. 116–122, 2007.
10. H. W. Song, S. J. Kwon, K. J. Byun, and C. K. Park, "Predicting carbonation in early-aged cracked concrete," Cement and Concrete Research, vol. 36, no. 5, pp. 979–989, 2006.
11. O. B. Isgor and A. G. Razaqpur, "Finite element modeling of coupled heat transfer, moisture transport and carbonation processes in concrete structures," Cement and Concrete Composites, vol. 26, no. 1, pp. 57–73, 2004.
12. S. J. Kwon and U. J. Na, "Prediction of durability for RC column with crack and joint under carbonation based on probabilistic approach," International Journal of Concrete Structures and Materials, vol. 5, no. 1, pp. 11–18, 2011.
13. Y. F. Houst and F. H. Wittmann, "Influence of porosity and water content on the diffusivity of CO_2 and O_2 through hydrated cement paste," Cement and Concrete Research, vol. 24, no. 6, pp. 1165–1176, 1994.
14. S. H. Jung, M. K. Lee, and B. H. Oh, "Measurement device and characteristics of diffusion coefficient of carbon dioxide in concrete," ACI Materials Journal, vol. 108, no. 6, pp. 589–595, 2011.
15. S. H. Jung, Diffusivity of carbon dioxide and carbonation in concrete through development of gas diffusion measuring system [Ph.D. thesis], Deptartment of Civil Engineering, Seoul National University, Seoul, Republic of Korea, 2002.
16. S. J. Kwon and H. W. Song, "Analysis of carbonation behavior in concrete using neural network algorithm and carbonation modeling," Cement and Concrete Research, vol. 40, no. 1, pp. 119–127, 2010.
17. I. C. Yeh, "Computer-aided design for optimum concrete mixtures," Cement and Concrete Composites, vol. 29, no. 3, pp. 193–202, 2007.
18. C. H. Lim, Y. S. Yoon, and J. H. Kim, "Genetic algorithm in mix proportioning of high-performance concrete," Cement and Concrete Research, vol. 34, no. 3, pp. 409–420, 2004.
19. S. J. Kwon and S. C. Kim, "Concrete mix design for service life of RC structures exposed to chloride attack," Computers and Concrete, vol. 10, no. 1, pp. 687–607, 2012.

20. E. Cantú-Paz and D. E. Goldberg, "Efficient parallel genetic algorithms: theory and practice," Computer Methods in Applied Mechanics and Engineering, vol. 186, no. 2–4, pp. 221–238, 2000.
21. D. E. Goldberg, Genetic Algorithms in Search, Optimization and Machine Learning, Addison-Welsley, Reading, Mass, USA, 1989.
22. Y. Abe, "Result of reference review on crack width effect to carbonation of concrete," in Proceedings of the Symposium on Rehabilitation of Concrete Structures, 1999.
23. CEB Task Group 5.1, 5.2, New Approach to Durability Design, CEB, Sprint-Druck, Stuttgart, Germany, pp. 53–62, May 1997.
24. KICT-Korea Institute of Construction Technology, "Durability improvement of concrete with sea/normal sand," Tech. Rep. R&D/97-0001, section 1–3, Bon Press, Seoul, Republic of Korea, 1997.
25. KSCE-Korea Society of Civil Engineering, "Study on durability improvement for underground structure," Tech. Rep., Seoul Metropolitan Government Office of Subway Construction, KSCE Press, Seoul, Republic of Korea, 1999.
26. KCI-Korea Concrete Institute, Concrete Specification-Durability, 2004.
27. RILEM, "Durability design of concrete structures," Report of RILEM Technical Committee 130-CSL, E&FN, 1994, pp. 75–78.
28. JSCE-Concrete Committee, Standard Specification for Concrete Structures, 2002.
29. Mathworks, Genetic Algorithm and Direct Search Toolbox 2, User's Guide, 2007.

CITATION

Seung-Jun Kwon, Byung Jae Lee, and Yun Yong Kim, "Concrete Mix Design for Service Life of RC Structures under Carbonation Using Genetic Algorithm," Advances in Materials Science and Engineering, vol. 2014, Article ID 653753, 13 pages, 2014. doi:10.1155/2014/653753.

CHAPTER 12

Achievements of Truss Models for Reinforced Concrete Structures

Panagis G. Papadopoulos, Hariton Xenidis, Panos Lazaridis, Andreas Diamantopoulos, Periklis Lambrou, and Yannis Arethas

Department of Civil Engineering, Aristotle University of Thessaloniki, Thessaloniki, Greece

ABSTRACT

Achievements are presented for truss models of RC structures developed in previous years: 1) Two constitutive models, biaxial and triaxial, are based on regular trusses, with bars obeying nonlinear uniaxial σ-ε laws of material under simulation; both models have been compared with test results and show a dependence of Poisson ratio on curvature of σ-ε law; 2) A truss finite element has been used in the nonlinear static and dynamic analysis of plane RC frames; it has been compared with test results and describes, in a simple way, the formation of plastic hinges; 3) Thanks to the very simple geometry of a truss, the equilibrium equations can be easily written and the stiffness matrix can be easily updated, both with respect to the deformed truss, within each step of a static incremental loading or within each time step of a dynamic analysis, so that to take into account geometric nonlinearities. So the confinement of a RC column is interpreted as a structural stability effect of concrete. And a significant role of the transverse reinforcement is revealed, that of preventing, by its close spacing and sufficient amount, the buckling of inner longitudinal concrete struts, which would lead to a global instability of the RC column; 4) The proposed truss model is statically indeterminate, so it exhibits some features, which are not met by the "strut-and-tie" model.

INTRODUCTION

In 1967, in a pioneering work [1], D. Ngo and A. C. Scordelis presented a detailed finite element model for a RC beam, in which separate finite elements are used for concrete and steel reinforcement. The material nonlinearities of the reinforcement can be easily described by the nonlinear uniaxial σ-ε law of a bar element. However, it is difficult to represent the nonlinear biaxial or triaxial stress-strain behavior of concrete or any other material. The relevant problems are discussed in two state-of-theart reports on nonlinear finite element analysis of RC structures, one by P. G. Bergan and I. Holand in 1979 [2] and another in a special publication of ASCE in 1982 [3], written by specialists on this field, under the co-ordination of A. C. Scordelis. Also, the difficulties appearing in the application of finite elements to nonlinear problems have been discussed in the series of three Conferences F. E. No. Mech. (Finite Elements in Nonlinear Mechanics), organized by J. H. Argyris in the Institute of Statics and Dynamics, University of Stuttgart, Germany in the years 1978, 1981, 1984 [4].

In order to describe the nonlinear biaxial or triaxial stress-strain behavior of a structure by the Finite Element Method, constitutive models for the structural materials have to be developed in order to be embodied in the individual finite elements. Efforts to develop such constitutive models have been made by many researchers, e.g. plasticity models by W. F. Chen [5] and Z. Mroz [6], the plastic-fracturing model of Z. P. Bazant [7], as well as the more practical contributions by D. Darwin [8] and K. I. Willam [9], for nonlinear, biaxial and triaxial, respectively, stress-strain behavior of concrete.

In 1977 [10], N. J. Burt and J. W. Dougill presented a random network constitutive model, in order to describe the nonlinear biaxial stress-strain law of a material, and noticed that equivalent results can be obtained by use of simple regular networks. By applying this idea, P. G. Papadopoulos developed in 1984 and 1986 [11,12] a biaxial and a triaxial network constitutive model, based on a regular plane octagon and a regular space rhombic dodecahedron, respectively, in which sides and diagonals are bars obeying the nonlinear uniaxial σ-ε laws of the material under simulation. Results from the above network constitutive models have been found in satisfactory approximation with corresponding published test results [13-15].

Trusses have been used not only in constitutive models, but also in finite elements of structures. In 1978 [16], E. Absi, in his "theorie des equivalences" stated that simple truss finite elements give equivalent results with the usual more complicated continuum finite elements. This idea was extended to problems with material nonlinearities and to the nonlinear static and dynamic analysis of plane RC frames by P. G. Papadopoulos [17,18]. A simple truss finite element was proposed, based on a plane rectangle in which all sides and diagonals are bars obeying nonlinear uniaxial σ-ε laws of concrete or steel. So, the nonlinear biaxial stress-strain behavior of the element is, in a simple way, described, thus the embodying of a constitutive law in the individual finite elements is no more needed. Results from nonlinear static analysis for cyclic loading, as well as nonlinear seismic dynamic analysis of simple plane RC frames, by the proposed truss RC element, were compared with corresponding published test results and found in a satisfactory approximation with them [19,20]. As the bars of the proposed finite element include the main material nonlinearities of concrete and steel, that is concrete tensile cracking and ultimate compressive strength, as well as tensile yield of reinforcement, the proposed truss model can, in a simple way, describes the formation of plastic hinges in a RC frame.

Afterwards, some other versions of E. Absi ideas for truss finite elements were developed for plane structures, under various names but all similar to each other, e.g. "truss analogy" in 1997 [21] for steel structures, "lattice model" in 1997 [22] and "lumped stress model" in 2002 [23], the latter two for RC structures.

In 1987 [24] J. Schlaich invented the so called "strutand-tie" model, which is a statically determinate truss model, consisting of concrete and steel bars. These bars include the main material nonlinearities of a RC structure. So, the "strut-and-tie" model can effectively describe the main stress-strain states of a RC structure, that is bending, shear and even torsion in 3D, thus it has been proved as a very useful practical tool in analysis and design of RC structures.

The "strut-and-tie" model has been further developed by other researchers, as by T. T. Hsu in 1993 [25], by F. J. Vecchio and M. P. Collins in 1993 [26], as well as by ASCE-ACI Committee on shear and torsion in 1998 [27].

The proposed here truss model is a statically indeterminate structure, so it exhibits some features that are not met by the statically determinate "strut-and-tie" model:

1) It can describe lateral expansion (Poisson ratio) effect.

2) It takes into account geometric nonlinearities, by writing the equilibrium equations and updating the stiffness matrix, both with respect to the deformed truss, within each step of a static incremental loading or within each time step of a dynamic analysis. This is easily achieved thanks to the very simple geometry of a truss..

By this proposed truss model which includes geometric nonlinearities, the confinement of a RC column is interpreted as a structural stability effect of concrete [28-30].

And, beyond the already known roles of the transverse reinforcement [31-33] (that is, shear transfer, reduction of concrete spalling, preventing of buckling of longitudinal reinforcement, increase of compressive stiffness, strength and ductility of the confined concrete core), another significant role of the transverse reinforcement is revealed by the proposed truss model with structural instability, that of retarding and even preventing, by its close spacing and sufficient amount, the buckling of inner longitudinal concrete struts, which would lead to a global instability of the RC column. Results from the application of this proposed model with structural instability on RC column confinement have been found in a satisfactory approximation to Codes requirements [34-36], regarding the spacing and amount of transverse reinforcement, which, in turn, are based on test results, too.

In the following, some of the achievements of the above proposed truss models for nonlinear analysis of structures, mainly RC structures, proposed in previous years, will be described in more detail.

TRUSS CONSTITUTIVE MODELS

A biaxial and a triaxial constitutive model for the nonlinear stress-strain law of a material have been developed [11,12], based on a regular plane octagon and a regular space rhombic dodecahedron, respectively, in which sides and diagonals are bars obeying the nonlinear uniaxial σ-ε law of the material under simulation. So, in a simple way, by the nonlinear uniaxial σ-ε laws of the bars, the nonlinear biaxial or triaxial stress-strain behavior of the whole truss is described. Results from the above truss constitutive models, for various loading histories, have been found in satisfactory approximation to corresponding published test results [13-15]. Both above truss constitutive models show a dependence of the Poisson ratio value ν on the curvature κ of the nonlinear uniaxial σ-ε law of the material under consideration, as shown in Figure 1.

TRUSS FINITE ELEMENT FOR PLANE RC FRAME

A truss finite element is proposed for beams of a plane RC frame, based on a rectangle, in which all sides and diagonals are bars, obeying nonlinear uniaxial σ-ε laws of concrete or steel, as shown in Figure 2. The σ-ε law of concrete bars includes tensile cracking, ultimate compressive strength, as well as loading-unloading rules after compressive yield. Whereas, the σ-ε law of a steel bar includes ultimate tensile and compressive strengths, as well as loading-unloading rules after tensile or compressive yielding.

DETERMINATION OF BAR SECTIONS

In the above proposed truss finite element for beams of plane RC frames, the cross-section areas of steel bars are reasonably and easily determined as sums of sections of the corresponding steel reinforcing bars. Whereas, in order to determine the cross-sections areas A_1, A_2, A_3 of the concrete bars of the truss element as shown in Figure 3(b), we have to compare it to the corresponding continuum concrete beam element of Figure 3(a), as regards three representative stress-strain states in the linear elastic region. And we chose, as such characteristic states, the pure bending, the confined axial deformation as well as the confined transverse deformation.

For the pure bending shown in Figure 3(c), the curvature angle of the beam element is $\Delta\varphi = Ml/EJ$ where $J = wh^3/12$, whereas for the truss element $\Delta\varphi = 2\Delta l/h$

Where $\Delta l = \dfrac{M}{h} l / A_1$. By combining the above equations we obtain $A_1 = wh/6$.

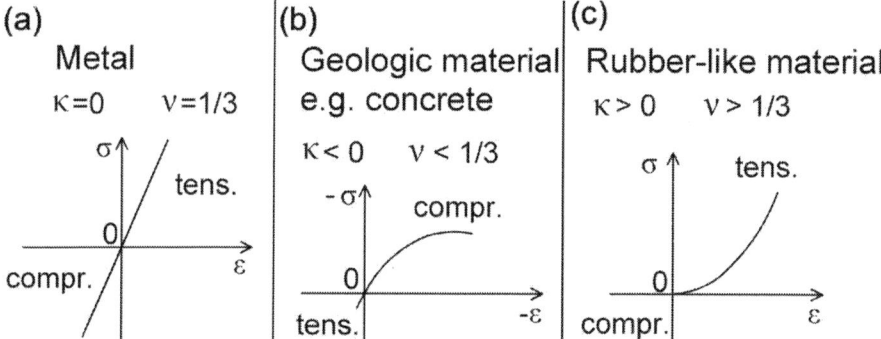

Figure 1: Dependence of Poisson ratio ν on the curvature κ of nonlinear uniaxial σ-ε law of the material. (a) Metal κ = 0 g ν = 1/3; (b) Geologic material e.g. concrete κ < 0 g ν < 1/3; (c) Rubber-like material κ > 0 g ν > 1/3.

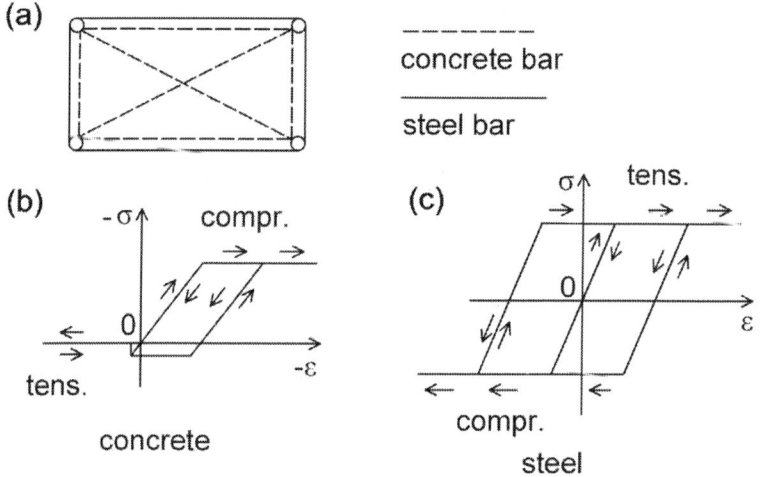

Figure 2: (a) Truss finite element for beam of a plane RC frame, with concrete and steel bars; (b) Nonlinear uniaxial σ-ε law of concrete bars; (c) Nonlinear uniaxial σ-ε law of steel bars.

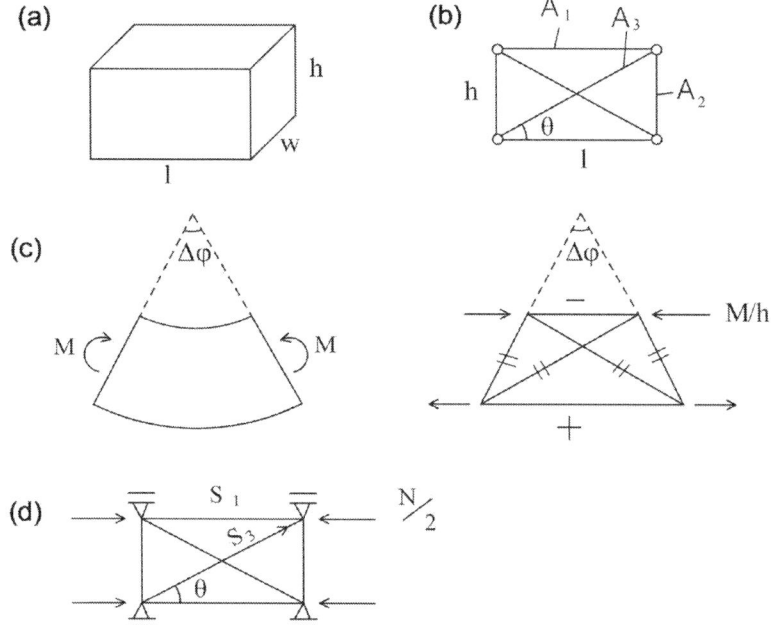

Figure 3: Comparison between characteristic stress-strain states of the concrete beam element and the corresponding truss element in order to determine the concrete bar sections. (a) Concrete beam element; (b) Corresponding truss element; (c) Pure bending; (d) Confined axial compression.

For the confined axial deformation, the elasticity theory gives $\sigma_x = \dfrac{E}{1-v^2}\varepsilon_x$.

For $v \approx 0.2, 1-v^2 \approx 1$, thus $\sigma_x = N/wh$ and $\varepsilon_x = \Delta l / l$., in the corresponding state of the truss element shown in Figure 3(d), we have $S_1 + S_3 \cos\theta = N/2$

Where $S_1 = \dfrac{EA_1}{1}\Delta 1$ and $S_3 = \dfrac{EA_3}{1/\cos\theta}\Delta 1\cos\theta$.

From combination of above equations, we obtain
$$A_3 = wh/3\cos^3\theta$$
From similar considerations for confined transverse deformation, we obtain
$$A_2 = wl/2 - (wh/3)3tg^3\theta.$$

Obviously, when the angle θ tends to zero, $\theta \to 0$, the sections tend to $A_3 \to wh/3$ and $A_2 \to wl/2$.

NONLINEAR STATIC ANALYSIS

The incremental loading of the structure is preferably performed by strain control, which is a more stable procedure than stress control. The material nonlinearities are taken into account by the variations of the elasticity moduli E of the bars during the loading. Whereas, in order to take into account geometric nonlinearities, the equilibrium equations are written and the global stiffness matrix updated, both with respect to the deformed truss, within each step of incremental loading. The local stiffness matrix of a bar in 2D, consisting of elastic and geometric part, is:

$$\mathbf{k} = \mathbf{k}_E + \mathbf{k}_G = \frac{EA}{l_o}\begin{pmatrix} c_x^2 & c_x c_y \\ c_x c_y & c_y^2 \end{pmatrix} + \frac{N}{l}\begin{pmatrix} c_y^2 & -c_x c_y \\ -c_x c_y & c_x^2 \end{pmatrix}$$

where A section, l_o undeformed length, l present length, N axial force and c_x, c_y direction cosines of the bar.

Whereas, the global stiffness matrix of the truss is:

$$\mathbf{K} = \mathbf{B}\operatorname{diag}(\mathbf{k}_i)\mathbf{B}' \quad i = 1\cdots n_b$$

where B Boolean linkage matrix and n_b number of bars of the truss.

Based on the proposed algorithm, a very short computer program, with only about 200 FORTRAN instructions, has been developed, for the nonlinear static analysis of a truss model of a plane RC frame.

NONLINEAR DYNAMIC ANALYSIS

A lumped mass is assigned to every free node of the truss. Zero damping and zero initial velocities are assumed. The resulting initial value problem:

$$\dot{\mathbf{y}} = \mathbf{q}(t,\mathbf{y}), \quad \mathbf{y}(0) = \mathbf{y}_0,$$

Where the state vector is y={r,v,c} with r, v positions and velocities of nodes and c constitutive variables of the bars, is solved by the step-by-step algorithm of trapezoidal rule, which coincides with the Newmark's algorithm of constant average acceleration:

$$\mathbf{y}_{n+1} = \mathbf{y}_n + \frac{1}{2}\left[\mathbf{q}(t_n,\mathbf{y}_n) + \mathbf{q}(t_{n+1},\mathbf{y}_{n+1})\right]\Delta t,$$

Combined with a predictor-corrector technique with two corrections per step, PE(CE)² [37]. So, there is no need to solve an algebraic system within each step of the algorithm.

The stability criterion of the algorithm is $\omega_{max}\Delta t < 2.0$ rad and the accuracy criterion is $\omega_{max}\Delta t < 0.5$ rad, that is $\Delta t < T_{min}/4\pi$, which dictates the choosing of the time step-length Δt of the algorithm.

An upper bound for the normal frequencies can be found from the norm of the matrix $M^{-1}K$, where M mass matrix and K stiffness matrix of the structure:

$$\omega_{max} < \|\mathbf{M}^{-1}\mathbf{K}\|$$

Based on the proposed algorithm, a very short computer program has been developed, with only about 150 FORTRAN instructions, for the nonlinear dynamic analysis of a truss model of a RC frame.

APPLICATIONS TO ANALYSIS OF SIMPLE PLANE RC FRAMES

The above proposed truss finite element for plane RC frames, as well as the proposed algorithms for nonlinear static and dynamic analysis, have been applied to the nonlinear static analysis of a simple plane RC frame for cyclic loading [17], as well as to the nonlinear dynamic seismic analysis of a simple plane RC frame [18]. The results of these analyses have been found in satisfactory approximation with corresponding published test results [19, 20].

As the nonlinear uniaxial σ-ε laws, of the bars of the proposed truss model, include all the main material nonlinearities of a RC structure, that is tensile cracking and ultimate compressive strength of concrete, as well as tensile yielding of steel reinforcement, the formation of plastic hinges in a RC frame is, in a simple way, described, as shown in Figure 4.

APPLICATION TO CONFINEMENT OF A RC COLUMN

In order to take into account geometric nonlinearities, by the proposed truss model, the equilibrium equations are written and the stiffness matrix is updated, both with respect to the deformed truss, within each step of a static incremental loading or within each time step of a dynamic analysis. This is easily achieved thanks to the very simple geometry of a truss.

As the proposed truss model includes geometric nonlinearities, it interprets the confinement of a RC column as a structural stability effect of concrete [28-30].

In Figure 5(a), the compressive axial σ-ε diagram of a confined RC column is shown. An early small drop of the stress $-\Delta\sigma$ is observed, which is due to a local instability because of spalling (buckling) of outer concrete. As the loading further increases, preferably by strain control, for a significant value of the compressive axial deformation, the stress σ suddenly drops to zero, which is an obvious mark of global structural instability, observed in experiments and verified by the proposed truss model, too.

In Figure 5(b), a part of a confined RC column, between two successive sets of transverse reinforcement, is shown. The longitudinal reinforcement is omitted, for simplicity. As the compressive axial loading N gradually increases, a lateral expansion of concrete occurs. For a significant compressive axial deformation, because of the large lateral expansion of concrete, a tensile yielding of the transverse reinforcement occurs, which implies a further lateral expansion of concrete. So, wide longitudinal-vertical concrete cracks are formed, and between such successive concrete cracks, inner longitudinal-vertical concrete struts are formed, which tend to buckle, leading to a global instability of the RC column.

Beyond the already known roles of transverse reinforcement in a RC column [31-33] (that is shear transfer, reduction of concrete spalling, preventing of buckling of longitudinal reinforcement, increase of compressive axial stiffness, strength and ductility of the confined concrete core), another significant role of the transverse reinforcement is revealed by the proposed truss model with structural instability, that of retarding and even preventing, by its close spacing s and sufficient amount ρ (mechanical ratio), the buckling of the above inner longitudinal-vertical concrete struts, which would lead to a global instability of the RC column.

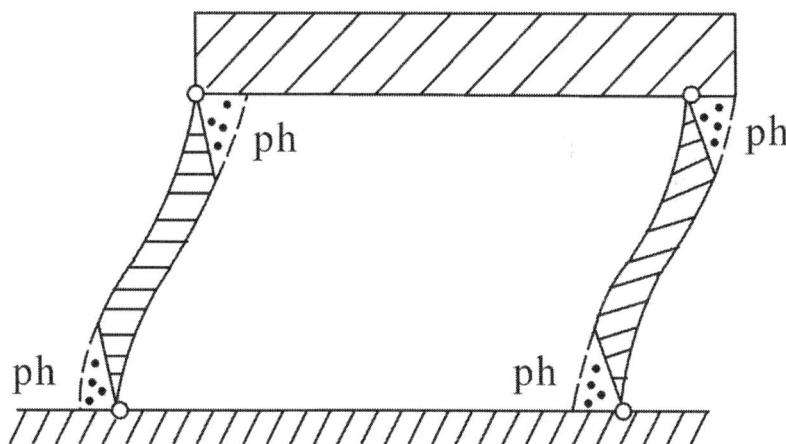

Figure 4: Description of formation of plastic hinges, in a RC frame, by the proposed truss model. "……." cracked concrete. "------" reinforcement in tensile yielding. "//////" rigid parts. "ph" plastic hinges.

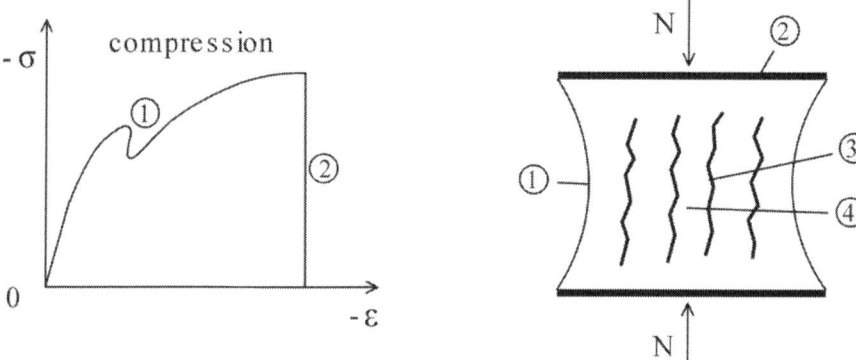

Figure 5: (a) Compressive axial σ-ε diagram of a confined RC column. 1. Early small drop of stress due to spalling of outer concrete. 2. For a significant value of compressive axial deformation, the stress σ suddenly drops to zero, which is a mark of global structural instability; (b) A part of a confined RC column between two successive sets of transverse reinforcement. 1. Spalling of outer concrete. 2. Transverse reinforcement in tensile yielding. 3. Longitudinal concrete cracks. 4. Longitudinal concrete struts.

Results, from application of the proposed truss model with structural instability to the confinement of RC columns, have been found in satisfactory approximation with corresponding requirements of codes [34-36], regarding the spacing s and the mechanical ratio ρ of transverse reinforcement; these requirements are, in turn, based on test results, too.

CONCLUSIONS

Some achievements have been presented for truss models of structures, mainly RC structures, which have been developed in previous years and found in satisfactory approximation with test results and Codes requirements:

1) N. J. Burt and J. W. Dougill developed in 1977 [10] random network constitutive models and stated that equivalent results can be, in a simple way, obtained by regular networks. This idea was realized in 1984 [11] and 1986 [12] by two network constitutive models, a biaxial and triaxial one, based on a regular plane octagon and on a regular space rhombic dodecahedron, respectively, in which sides and diagonals are bars obeying the nonlinear uniaxial σ-ε law of the

material under simulation. Both models show a dependence of Poisson ratio on the curvature of the nonlinear uniaxial σ-ε law of the material.

2) E. Absi in 1978 [16], in his "theorie des equivalences", stated that simple truss finite elements give equivalent results with the usual more complicated continuum finite elements. This idea was extended in 1988 [17, 18] to structures with material nonlinearities, and applied particularly to the nonlinear static and dynamic analysis of plane RC frames. As the individual bars of the proposed truss finite element include, in their uniaxial σ-ε laws, the main material nonlinearities of a RC structure, that is the concrete tensile cracking, the reinforcement tensile yield, as well as the ultimate compressive strength of concrete, the formation of plastic hinges in a RC frame can be, in a simple way, described.

3) Compared to the "strut-and-tie" model for RC structures, invented by J.Schlaich in 1987 [24] and further developed by other researchers, which proved as a very effective tool in the analysis of RC structures, the proposed here truss model exhibits the difference that it is a statically indeterminate, whereas the "strut-and-tie" model is statically determinate. So, the proposed truss model has some features that are not met by the "strut-and-tie" model: a) It can describe lateral expansion (Poisson ratio) effect. b) It takes into account geometric nonlinearities, by writing equilibrium equations and by updating stiffness matrix, both with respect to the deformed truss, within each step of a nonlinear static or dynamic analysis. So, it interpreted in 1999 [31] the confinement of a RC column as a structural stability effect of concrete. And revealed a significant role of transverse reinforcement, that of retarding and even preventing, by its close spacing and sufficient amount, the buckling of inner longitudinal concrete struts, which would lead to a global instability of the RC column.

REFERENCES

1. D. Ngo and A. C. Scordelis, "Finite Element Analysis of Reinforced Concrete Beams," ACI Journal, Vol. 64, 1967, pp. 152-163.
2. P. G. Bergan and I. Holand, "Nonlinear Finite Element Analysis of Concrete Structures," Computer Methods in Applied Mechanics and Engineering, Vol. 17-18, 1979, pp. 443-467.doi:10.1016/0045-7825(79)90027-6

3. A. C. Scordelis, Editor, ASCE Task Committee on Concrete and Masonry Structures, "State-of-the-Art Report on Finite Element Analysis of Reinforced Concrete," ASCE Special Publication, 1982.
4. J. H. Argyris, Organizer, International Conferences F.E.No.Mech. (Finite Elements in Nonlinear Mechanics). Institute for Statics and Dynamics, University of Stutgart, Germany, I.30 August-1 September 1978, II. 25-28 August 1981. III. 10-13 September 1984.
5. W. F. Chen and E. C. Ting. "Constitutive Models for Concrete Structures," Journal of Engineering Mechanics Division ASCE, Vol. 106, No. 1, 1980, pp. 1-19.
6. Z. Mroz, V. A. Norris and O. C. Zienkiewicz, "Application of an Anisotropic Hardening Model in the Analysis of Elastic-Plastic Deformation of Soils," Geotechnique, Vol. 29, 1979, pp. 1-34. doi:10.1680/geot.1979.29.1.1
7. Z. P. Bazant and S. S. Kim, "Plastic-Fracturing Theory for Concrete," Journal of Engineering Mechanics Division ASCE, Vol. 105, No. 3, 1979, pp. 407-428.
8. D. Darwin and D. A. Pecknold, "Analysis of Cyclic Loading of RC Structures," Computers and Structures, Vol. 7, No. 1, 1977, pp. 137-147. doi:10.1016/0045-7949(77)90068-2
9. K. J. Willam and E. P. Warnke, "Constitutive Model for the Triaxial Behavior of Concrete," Proceedings of IABSE, Structural Engineering Report 19, Section III, 1975, pp. 1-30.
10. N. J. Burt and J. W. Dougill, "Progressive Failure in a Model Heterogeneous Medium," Journal of Engineering Mechanics Division ASCE, Vol. 103, 1977, pp. 365-376.
11. P. G. Papadopoulos, "Biaxial Network Constitutive Model," Journal of Engineering Mechanics ASCE, Vol. 110, No. 3, 1984, pp. 449-464. doi:10.1061/(ASCE)0733-9399(1984)110:3(449)
12. P. G. Papadopoulos, "A Triaxial Network Constitutive Model," Computers and Structures, Vol. 23, 1986, pp. 497-501. doi:10.1016/0045-7949(86)90093-3
13. H. B. Kupfer, H. D. Hilsdorf and H. Rusch, "Behavior of Concrete under Biaxial Stresses," ACI Journal, Vol. 66, No. 8, 1969, pp. 656-666.
14. R. Palaniswamy and S. P. Shah, "Fracture and StressStrain Relationships of Concrete under Triaxial Compression," Journal of Structural Division ASCE, Vol. 100, 1974, pp. 901-916.
15. R. Scavuzzo, T. Stankowski, K. Gerstle and H.-Y. Ko, "Stress-Strain Curves for Concrete under Multiaxial Load Histories," University of Colorado, Boulder, 1983.
16. E. Absi, "Méthodes des Calcus Numerique en Elasticité," Eyrolles, Paris, 1978.

17. P. G. Papadopoulos, "Nonlinear Static Analysis of Reinforced Concrete Frames by Network Models," Advances in Engineering Software, Vol. 110, No. 3, 1988, pp. 114-122.doi:10.1016/0141-1195(88)90010-1
18. P. G. Papadopoulos and C. G. Karayannis, "Seismic Analysis of R/C Frames by Network Models," Computers and Structures, Vol. 28, No. 4, 1988, pp. 481-494.doi:10.1016/0045-7949(88)90022-3
19. K. Stylianidis and G. Penelis, "Experimental Study of, bare and Infilled by Wall, One Story Frames under Cyclic shear Loading," 7th Greek Conference on Concrete, Vol. 2, Patra, 1985, pp. 47-55.
20. P. Hidalgo and R. W. Clough, "Earthquake Simulator Study of a Reinforced Concrete Frame," EERC Report 74-13, University of California, Berkeley, 1974.
21. S. C. Goel, B. Stojadinovicz and K. H. Lee, "Truss Analogy for Steel Moment Connections," Engineering Journal, Second Quarter 1997, pp. 43-53.
22. E. Schlangen and E. J. Garboczi, "Fracture Simulations of Concrete Using Lattice Models: Computational Aspects," Engineering Fracture Mechanics, Vol. 57, No. 2-3, 1997, pp. 319-332. doi:10.1016/S0013-7944(97)00010-6
23. F. Fraternali, M. Angelilo and A. Fortunato, "A Lumped Stress Method for Plane Elastic Problems and the Discrete Continuum Approximation," International Journal of Solids and Structures, Vol. 39, 2002, pp. 6211-6240. doi:10.1016/S0020-7683(02)00472-9
24. J. Schlaich, K. Schäfer and M. Jennewein, "Towards a Consistent Design of Structural Concrete," PCI Journal Special Report, Vol. 32, No. 3, 1987, pp. 75-150.
25. T. T. C. Hsu, "Unified Theory of Reinforced Concrete," CRC Press, 1993.
26. F. J. Vecchio and M. P. Collins, "Compression Response of Cracked Reinforced Concrete," Journal of Structural Engineering ASCE, Vol. 113, 1993, pp. 3590-3610.doi:10.1061/(ASCE)0733-9445(1993)119:12(3590)
27. ASCE-ACI Committee 445 on Shear and Torsion, "Recent Approaches to Shear Design of Structural Concrete. State-of-the-Art Report," Journal of Structural Engineering ASCE, Vol. 119, No. 12, 1998, pp. 1375-1417.
28. P. G. Papadopoulos and H. C. Xenidis, "A Truss Model with Structural Instability for the Confinement of Concrete Columns," Journal of EEE (European Earthquake Engineering), Part 2, 1999, pp. 57-79.
29. P. G. Papadopoulos, H. Xenidis, C. Karayannis, A. Diamantopoulos and P. Lambrou, "Confinement of Concrete Column Interpreted as a Structural Stability Effect," 6th GRACM (Greek Association of Computational Mechanics) Conference, Thessaloniki, 19-21 June 2008.

30. P. G. Papadopoulos, H. Xenidis, D. Plasatis, P. Kiousis and C. Karayannis, "Concrete Stability Achieved by Confinement in a RC Column," 12th International Conference on Civil, Structural and Environmental Engineering Computing, Coordinator B.H.V. Topping, Madeira, Portugal, 1-4 September 2009.
31. K. Park, M. J. N. Priestley and W. D. Gill, "Ductility of Square Confined Concrete Columns," Journal of Structural Division ASCE, Vol. 108, No. 4, 1982, pp. 929-950.
32. S. Watson, F. A. Zahn and R. Park, "Confining Reinforcement for Concrete Columns," Journal of Structural Engineering ASCE, Vol. 120, No. 6, 1984, pp. 1798-1849.
33. J. B. Mander, M. J. N. Priestley and R. Park, "Theoretical Stress-Strain Model for Confined Concrete," Journal of Structural Engineering ASCE, Vol. 114, No. 8, 1988, pp. 1804-1826.doi:10.1061/(ASCE)0733-9445(1988)114:8(1804)
34. Uniform Building Code 2, "Structural Engineering Design Provisions," Chapter 19. Concrete, 19.2.1. Reinforced Concrete Structures Resisting Forces Induced by Earthquake Motions 19.2.14. Frame Members Subjected to Bending and Axial Load, 1994, pp. 237-239.
35. New Zealand Standards 3101, "Code of Practice for the Design of Concrete Structures," Chapter 17, Members Subjected to Flexure and Axial Loads, Additional Seismic Requirements, 1989.
36. Eurocode 8, "Earthquake Resistant Design of Structures," Part 1-3. General Rules and Rules for Buildings. 2, Specific Rules for Concrete Buildings. 2.8. Provisions for Columns, Brussels, 1993, pp. 35-46.
37. P. G. Papadopoulos, "A Simple Algorithm for the Nonlinear Dynamic Analysis of Networks," Computers and Structures, Vol. 18, No. 1, 1984, pp. 1-8. doi:10.1016/0045-7949(84)90074-9

CITATION

P. Papadopoulos, H. Xenidis, P. Lazaridis, A. Diamantopoulos, P. Lambrou and Y. Arethas, "Achievements of Truss Models for Reinforced Concrete Structures," *Open Journal of Civil Engineering*, Vol. 2 No. 3, 2012, pp. 125-131. doi: 10.4236/ojce.2012.23018.

Index

A
Accurately determine, 196
American Concrete Institute (ACI), 222
Aramid fiber reinforced plastic (AFRP), 127
Automatically stops, 218

B
Behaviour, 212

C
Calcium carboaluminate, 255
Carbon fiber reinforced plastic (CFRP), 127
Cementitious matrix, 180, 181
Compressive Packing Model, 255
Compressive yield, 149
concrete impact, 1, 2, 13, 33, 35
Concrete-filled steel tube (CFST), 71, 72
Confined columns, 197
Confinement, 145, 148, 154, 155, 156

Confinement effect, 74, 83, 84, 85

D
Different behaviour, 197

E
Economical impacts, 218
Empirical equations, 1, 2, 13, 25, 29, 30
Experimental data, 1, 6, 7, 8, 9, 10, 11, 12, 17, 19, 20, 30
Experimental investigation, 221
Experimental variable, 221, 222

F
Fibre reinforced polymer (FRP), 125

G
Genetic algorithm (GA), 289
Ground granulated blast-furnace, 253, 254
Ground granulated blast-furnace slag, 253

H

High Strength Concrete, 254
High-modulus steel fiber, 223
High-performance concrete (HPC), 162
Homogenous microstructure, 254
HPFRCC material, 223, 224

I

industrial wastes, 38

L

Laboratory mixer, 227
Linear Packing Density Model, 255

M

Maximum paste thickness (MPT), 162
Mechanical homogeneity, 162
Mercury intrusion porosimetry (MIP), 181
Moment-curvature relationships, 205

N

Normal strength concrete, 255, 268, 271
Nuclear power industry, 2, 7, 30

P

Paulay and Priestley, 208
Polycarboxylic, 256
Pozzolanic cementitious materials, 173
Pozzolanic reactions, 162

Q

Quartz sand, 165

R

Rankine method, 73
Rapid chloride ion penetrability (RCPT, 183
Reactive Powder Concrete, 254
Reinforced concrete (RC), 195, 289
reinforcement, 125, 126, 130, 131, 132, 133, 134, 135, 136, 137, 138, 139, 140, 141

S

Silica fume, 162, 164, 165, 254, 281, 283
Solid Suspension Model, 255
Sorptivity coefficients, 183
steel reinforcement, 125, 126, 137
Stress–strain, 228, 229, 233
Stress-strain behavior, 146, 147, 149
Superplasticizer (SP), 162, 165

T

Targeted experimental program, 225
Test data, 1, 2, 3, 6, 8, 9, 10, 11, 13, 18, 19, 20, 21, 22, 23, 24, 25, 29
Three dimensional modelling, 198
Transverse reinforcement, 146, 148, 154, 155, 156
Triaxial constitutive model, 148
Truss models, 145, 148, 156
Typical apartment buildings, 200

U

Ultra-High Performance Concrete, 253, 254, 256, 279, 286, 287

Ultra-high performance concrete (UHPC), 162, 163

V
Various technique, 227

Volume of permeable voids (VPV), 181

W
waste foundry sand, 39